Mining Engineering

Mining Engineering

Edited by **Beth Thorpe**

SYRAWOOD
PUBLISHING HOUSE
New York

Published by Syrawood Publishing House,
750 Third Avenue, 9th Floor,
New York, NY 10017, USA
www.syrawoodpublishinghouse.com

Mining Engineering
Edited by Beth Thorpe

International Standard Book Number: 978-1-68286-257-5 (Hardback)

Printed in China.

Contents

Preface

This book aims to highlight the current researches and provides a platform to further the scope of innovations in this area. This book is a product of the combined efforts of many researchers and scientists from different parts of the world. The objective of this book is to provide the readers with the latest information in the field.

Mining Engineering is an emerging branch of engineering focusing upon the extraction and processing of minerals from their natural state. It is intertwined with other engineering disciplines like geotechnical engineering, metallurgy and mineral science. It concerns not only the production of minerals but the sustainability of resources as well. The book compiles the diverse practices of mining simulation and mining equipment. It takes into account the exploration, discovery and determination of minerals through different techniques. This book covers various methodologies and models involved in the study of mining engineering collated by eminent industry experts and academicians from across the globe.

I would like to express my sincere thanks to the authors for their dedicated efforts in the completion of this book. I acknowledge the efforts of the publisher for providing constant support. Lastly, I would like to thank my family for their support in all academic endeavors.

Editor

HEAVY METAL CONTAMINATION OF LANDSCAPE COMPONETS IN SURROUNDINGS OF CLOSED MERCURY DEPOSIT IN MALACHOV (SLOVAKIA)

Jana DADOVÁ[1], *Jiří KUPKA* [2], *Peter BARTO* [3], *Erika REMEŠICOVÁ*[4], *Tomáš ŠTRBA*[5]

[1] *RNDr., PhD., Department of Environmental Sciences, Faculty of Natural Sciences, Matej Bel University, Tajovského 52, Banská Bystrica*
e-mail jana.ruskova72@gmail.com

[2] *Ing. Ph.D., Institute of Environmental Engineering , Faculty of Mining and Geology, VŠB –Technical University of Ostrava, 17. listopadu 15, Ostrava*
e-mail jiri.kupka@vsb.cz

[3] *Mgr., Department of Geography, Faculty of Natural Sciences, Matej Bel University Tajovského 52, Banská Bystrica*
e-mail peter.barto@umb.sk

[4] *Ing. Institute of Environmental Engineering , Faculty of Mining and Geology, VŠB –Technical University of Ostrava, 17. listopadu 15, Ostrava, e-mail: remesicova.e@gmail.com*

[5] *Mgr., Department of Environmental Sciences, Faculty of Natural Sciences, Matej Bel University Tajovského 52, Banská Bystrica*
e-mail strbi@sbb.sk

Abstract

The present study investigates mercury contamination of soil, groundwater, surface water, vegetation and food commodities (fruit, vegetable, mushroom, and fish) close to the abandoned Hg deposit in Malachov, Central Slovakia. The soil was classified as cambi-soil (rendzina). Maximum concentrations of Hg (44.24 ppm) were found in the soil from the area nearby old mining activities in the Veľká Studňa locality. In the groundwater, 0.84 $\mu g.L^{-1}$, and in the surface water, even 394 $\mu g.L^{-1}$ of Hg content was found. Also plant tissues and mushrooms are Hg-contaminated (in the root of *Salix fragilis*, the Hg content was 22 $mg.kg^{-1}$). The Hg concentration decreases in general in the following order: root, branch/stem, and leaf/needle. In the food commodities, the Hg content does not exceed the Slovak law limits. The Hg content in trout muscle (from the local Malachovský brook) is 252 $\mu g.kg^{-1}$ and in liver 402 $\mu g.kg^{-1}$.

The calculation of an AMD formation potential (neutralisation potential, total acidity production, net neutralisation potential) is discussed as well. The value of the total acidity production potential (sensu Sobek et al., 1978) is low (1.562 – 3.125; with exception of a single sample, in which it increased to 9.375). On the other hand, the abundant presence of dolomite rocks causes a high neutralization potential value (up to 812.84). This result suggests that the assumption of the AMD production is excluded and the environmental risk from this viewpoint might be neglected.

Abstrakt

Kontaminace půdy, podzemních a povrchových vod, vegetace a potravin (ovoce, zelenina, houby, ryby) rtutí v okolí uzavřeného dolu Malachov na středním Slovensku byla zkoumána v této studii. Půdní typ byl klasifikován jako kambizem (rendzina). Maximální koncentrace rtuti (44,24 $mg.kg^{-1}$) byly zjištěny v půdě poblíž starých hornických aktivit na lokalitě Veľká Studňa. Koncentrace rtuti v podzemních vodách je 0,84 $\mu g.L^{-1}$ a v povrchových vodách byla popsána koncentrace až 394 $\mu g.L^{-1}$. Rovněž tkáně rostlin a houby jsou rtutí kontaminovány (v kořenech *Salix fragilis* byl stanoven obsah Hg 22 $mg.kg^{-1}$). Koncentrace Hg obecně klesá v pořadí: kořen - větve/stonek - listy/jehlice. Obsah rtuti v potravinách nepřekročil limity dané slovenskou legislativou. Obsah rtuti ve svalovině pstruhů (z místního Malachovského potoka) činil 252 $\mu g.kg^{-1}$ a v játrech 402 $\mu g.kg^{-1}$.

Také výpočet potenciálu tvorby AMD (neutralizační potenciál, celková tvorba acidity, čistý neutralizační potenciál) je diskutován. Hodnota celkového potenciálu tvorby acidity (sensu Sobek et al., 1978) je nízká (1,562 - 3,125; s výjimkou jednoho vzorku, kde byl zaznamenán nárůst na 9,375). Na druhé straně bohaté zastoupení dolomitických hornin vykazuje vysoký neutralizační potenciál (až do 812,84). Tento výsledek naznačuje, že předpoklad produkce AMD je vyloučen a tudíž riziko pro životní prostředí je z tohoto hlediska zanedbatelné.

Key words: mercury, soil, groundwater, surface water, plant, food, speciation

1 INTRODUCTION

The eastern slopes of the Kremnica Mts. are built by effusive and extrusive rocks of Neogene volcanism. In consequence of denudation, in surroundings of Malachov, older Mesozoic carbonates (limestones and dolomites) and sand-silt series of Paleogene period are uncovered. Hg mineralization is connected with young structures in the N-S to NE-SW directions and is genetically related to the post-volcanic stage of a Neogene stratovolcano. Cinnabar and metacinnabar accumulations, accompanied by pyrite, marcasite and arsenopyrite were formed in tectonically crushed carbonates of Mesozoic age and in Paleogene sandstones, below impermeable layers of argillite tuffs (Ivančík & Mitáček, 1986; Bancík & Jeleň, 1999).

The Hg-As mineralization at the Malachov deposit is situated in the Eastern rim of the Kremnica Mts., west from Banská Bystrica. Numerous occurrences and anomalies are concentrated on an area of 80 – 100 km^2: Veľká Studňa, Nemecký Hill, Dolná Cliff, Trávny Ždiar, Mútne, Ortuťový Hill, Cipkove Jamy, Pri Jazere etc. (Koděra et al., 1990; Jeleň et al., 2010). The main mineral assemblages of Hg ores in the Malachov ore field are cinnabar – HgS and metacinnabar. Arsenopyrite – FeAsS, pyrite – FeS_2, marcasite – FeS_2, auripigment – As_2S_3, realgar – As_4S_4 and arsenolite – As_2O_3 are less frequent minerals. In the gangue mineral assemblage, various modifications of SiO_2: quartz, chalcedony, opal and carbonates: calcite – $CaCO_3$ and dolomite – $CaMg(CO_3)_2$ and gypsum – $CaSO_4.2H_2O$ prevail(Koděra et al., 1990).

In the past, cinnabar veins were exploited in galleries and voluminous accumulations of ore close to the surface by shallow shafts (Bergfest, 1955; Jeleň et al., 2010). The Hg content in ore was about 0.5-1.0 %. Mineralization is found in Paleogene conglomerates, sandstones and Mesozoic carbonates, rarely also in Miocene volcanites. The Hg-As mineralization is connected to faults in the N-S direction (Ivančík & Mitáček, 1986).

Montaneous forms of the country relief were created by long-term mining activities, mainly from the 14th to the 17th centuries and have been being noticeably visible until present times in the localities of Pri Jazere, Ortúty and Veľká Studňa. The last exploitation was realised from 1980 to 1990 in the locality of Veľká Studňa (Ivančík & Mitáček, 1986). The deposit has been closed since 1990 (Bancík & Jeleň, 1999).

2 EXPERIMENTAL PART

In the surroundings of Malachov, 11 soil and sediment samples (OP-1 to OP-8B) were collected during 2012 (Fig. 1). In several cases, the sampling was realised both from A and B horizons to have a possibility to compare the heavy metal contents in separate horizons and receive the data about the preferential bonding of metals to selective soil components.

The soil samples for the Mg, Ti, Co, Ni, As, Sb, Mn, Rb, Sr, V, Ba, Cr, Zr, S determination were melted with a Li-metaborate and Li-tetraborate mixture, and consecutively dissolved in diluted nitric acid. In order to determine Fe, Pb, Zn and Cu contents, 0.5 g of a sample was lixiviated by aqua regia. The determination of individual elements was realised by an ICP–MS analysis at the ACME Analytical Laboratories Vancouver Ltd. in Canada. The presented data represent an average value calculated from four parallel determinations.

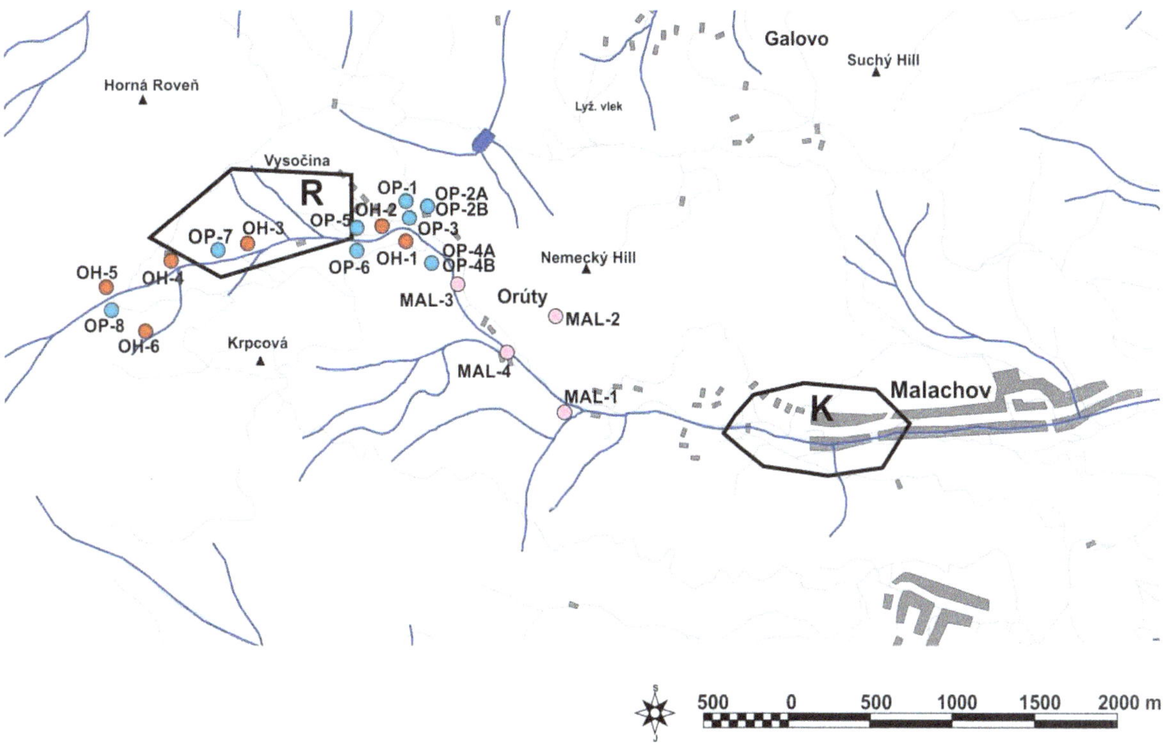

Fig. 1 Map of soil, water and plant samples localization
Explanations: OP – soil and sediment samples, MAL – groundwater, 0H – surface water;
R – area of plant sampling, K – area of food commodities sampling

The groundwater sample set was taken from the Žltý spring on the Malachov brook bank (MAL-1), from the Hnedý spring on the Nemecký Hill slope (MAL-2), from mining water at Cipkové Jamy (Krvavá spring, sample MAL-3) and from the Turistický spring (MAL-4). In addition, a set of 8 surface water samples from the Malachov brook were collected. Both the groundwater and surface water samples of a 1000mL volume were stabilized by 10 mL of concentrated HCl. The water samples were analysed using a flame atom absorption analysis with a hydride system in the National Water Reference Laboratory in Slovakia at the Water Research Institute in Bratislava.

The plant samples were divided into roots, branches/stems, and leaves/needles and dried at a laboratory temperature. 0.5 g of a vegetation sample was after split digestion in concentrated HNO_3 and then in aqua regia analysed by ICP-MS for ultralow detection limits in the same laboratory as soils.

Also two fishes from the Malachov brook, vegetable and fruit samples were subject the flame atom absorption analysis in a laboratory of the Regional Institute of Public Health in Banská Bystrica. The localization of sampling sites is presented on the attached map (Fig. 1).

The Hg content in soil, plant and fish samples were determined by a thermomechanical analysis using the device AMA 254, KEI, at the Faculty of Environmental Engineering of the Technical University in Zvolen (Ing. Emília Hroncová, PhD).

Paste and rinse pH (as well as Eh) was determined in soils in a suspension of a 20g sample and 50 mL of distilled water or 1 M KCl, respectively, according to the method described by VanReeuwijka (1995) using the pH meter WTW Multi 3420 with the combined redox-electrode SenTix ORP and with a reference system Ag/AgCl containing 3 M KCl electrolyte. Also the pH and Eh values in groundwater and surface water were measured by the same pH meter.

To calculate the acidity production potential (AP) and the neutralization potential (NP), the Eh and pH values of soil and sediments, as well as the sulphur and carbon contents, must be determined. For the acidification risk calculation, the method was used which was published in the Acid Mine Drainage Prediction EPA530-R-94-036 (1994) and adapted by Lintnerová and Majerčík (2005).

The total carbon ($C_{tot.}$), organic carbon ($C_{org.}$) and inorganic carbon ($C_{inorg.}$) contents in the sediments were IR analysed using the furnace Ströhlein C-MAT 4000 in laboratories of the Geological Institute of the Slovak

Academy of Sciences (Alžbeta Svitáčová). A static test of the total acid potential was realized according to Sobek et al. (1978).

The acidification risk was calculated according to the method published by Sobek et al. (1978).

The neutralisation potential in soil samples was determined by titration to pH 3.5 using hydrochloric acid, and the obtained value was recalculated to $CaCO_3$ $kg.t^{-1}$ in laboratories of the Geological Institute of the Slovak Academy of Sciences in Banská Bystrica.

2 RESULTS

2.1 Heavy metals distribution in soil

Tab. 1 Paste and rinse pH/Eh and soil analyses

Sample	pH	Eh (mV)	pH	Eh (mV)	Fe	Mg	Zn	Hg	Cu	Pb
	H_2O		1M KCl		%		$mg.kg^{-1}$			
OP-1	6.93	-43	6.80	-36	3.05	4.17	50	6.31	12	21
OP-2A	7.04	-52	6.81	-35	0.89	6.85	64	19.32	11	117
OP-2B	7.34	-60	7.31	-65	0.53	10.09	30	6.00	6	29
OP-3	7.42	-69	6.60	-21	3.80	4.38	73	2.49	8	12
OP-4A	6.18	2	5.54	46	2.37	2.36	67	15.95	11	48
OP-4B	7.02	-49	6.57	-22	2.27	3.06	54	15.51	11	35
OP-5	7.26	-62	6.35	-7	4.10	3.73	75	2.36	10	12
OP-6A	7.43	-70	6.95	-45	3.03	3.76	102	32.13	22	44
OP-7	7.75	-85	7.70	-84	0.70	11.37	17	44.24	15	7
OP-8A	6.11	9	5.71	36	5.73	2.48	108	0.56	17	27
OP-8B	6.93	-43	6.80	-36	3.92	1.30	92	0.43	10	33

Tab. 1 (Continuation) paste and rinse pH/Eh and soil analyses

Sample	Zn	Ni	Co	Cr	As	Sb	Mn	V	Sr	Zr	Ba	Ti
	$mg.kg^{-1}$											
OP-1	50	15	14	38	37	4	937	89	111	53.6	170	2770
OP-2A	64	6	5	17	33	16	300	23	80	18.8	82	810
OP-2B	30	3	3	12	23	13	222	11	82	10.4	48	480
OP-3	73	8	15	20	8	2	894	87	161	75.7	312	3150
OP-4A	67	15	11	32	42	6	686	60	86	45.8	173	2370
OP-4B	54	13	10	27	38	6	601	61	98	43.7	171	2380
OP-5	75	8	15	21	7	2	921	88	167	78.4	316	3360
OP-6A	102	14	14	47	16	7	756	70	100	57.1	226	3070
OP-7	17	7	2	11	10	3	156	14	78	9.7	51	530
OP-8A	108	6	23	19	4	3	1332	142	220	125.4	412	5720
OP-8B	92	8	15	20	6	2	2741	73	227	104.6	426	3800

The Hg contents in soil (Tab. 1) vary from 0.56 to 44.24 ppm. The highest content was found from a dump material (sample OP-7), and the lowest in the stream sediment sample OP-8. The Hg contents in stream sediments are lower as those from soils. An amazingly low Hg (6.314 $mg.kg^{-1}$) content is in stream sediments, mainly in the sample OP-5 from the Malachov brook. The A soil horizon seems to be Hg enriched.

In the soil and stream sediment samples from Malachov surroundings, high Fe contents (0.53 – 5.73 %; Tab. 1) were determined; in average – 2.81 %. Next abundant metals are: Mg (1.30 – 11.37 %), Mn (156 – 2741 $mg.kg^{-1}$) and Pb (6.9 – 117.0 $mg.kg^{-1}$). Also Sr (78 – 227 $mg.kg^{-1}$), Ba (51 – 426 $mg.kg^{-1}$), and Zr (9.7 – 125.4 $mg.kg^{-1}$) contents are relatively high. Cu, Zn, Co, Cr, V, As, Sb and Ti are present only in several tens of $mg.kg^{-1}$ (Tab. 1).

2.2 Heavy metal contamination of water

The groundwater analyses (Tab. 2) showed that the samples MAL-1 and MAL-2, of so called „*rough water*", respect the law standards for drinking water (sensu SR Government Decree No. 354/2006 Coll. on drinking water) from a viewpoint of selected analysed elements with exception of Fe. The Eh value 140 in the sample MAL-2 indicates toxic conditions. The Fe and Mn contents are higher than the recommended values for „*raugh water*" (according to Decree of the Ministry of Environmental Protection of the Slovak Republic No. 496/2010 Coll.). On the other hand, the Hg contents in the water samples MAL-3 and MAL-4 respect the regulations.

Tab. 2 Heavy metal concentrations in groundwater

Sample	pH_{H2O}	Eh_{H2O}	°C	Fe	Hg	As	Sb	Mn	Pb	Cd	Cu	Co	Ni	Zn
				$mg.L^{-1}$	$\mu g.L^{-1}$									
MAL-1	6,15	19	16.6	9.95	0.84	1.16	1.07	282	7	0.2	3.6	1.0	1.5	15
MAL-2	3.90	140	10.8	8.45	0.32	<1	2.04	185	5	0.1	2.6	4.2	<1	32
MAL-3	6.80	24	9.4	8.12	0.53	1.02	2.03	150	6	0.1	2.5	3.1	<1	15
MAL-4	5.12	40	12.0	7.33	0.24	0.98	0.99	144	3	0.2	2.2	2.0	<1	16

The heavy metal contents in surface water are presented in Tab. 3. The water pH is close to a value of 7 and is slightly alkaline (pH up to 7.87). The As concentrations were below the detection limit (0.01 $\mu g.L^{-1}$).

Tab. 3 Heavy metal concentrations in surface water (Malachov brook)

Sample	pH_{H2O}	Eh(mV) H2O	pH_{KCl}	Eh(mV) KCl	°C	Hg	As
						$\mu g.L^{-1}$	
OH – 1	7.82	-86	7.26	-76	14.5	n.d	n.d.
OH – 2	7.68	-80	7.15	-70	11.0	<2	n.d.
OH – 3	6.99	-46	6.45	-32	7.9	394	<0.01
OH – 4	7.02	-50	6.89	-41	19.5	265	<0.01
OH – 5	6.71	-29	6.60	-16	11.2	<2	n.d
OH – 6	7.87	-88	7.54	-80	18,3	40	n.d

The Hg concentrations in water of the Malachov brook exceed the standard for Hg contents in surface water (1 $\mu g.L^{-1}$) recommended by Decree of the Government of the Slovak Republic No. 269/2010 Coll. in the samples from surroundings of old mining activities in the Veľká Studňa locality.

2.3 Heavy metal contamination of plants and selected food commodities

In plant tissues from the area of the Veľká Studňa locality, high Hg contents (in roots of *Salix fragilis* up to 22 $mg.kg^{-1}$) were determined. The results of thermomechanical Hg analyses of individual plant tissues for *Pinus sylvestris*, *Salix fragilis* and *Juncus articulatus* are presented in Tab. 4. The contamination generally decreases in the following order: root – branch/stem – leaf/needle.

Tab. 4 Mercury contents in plant samples from Veľká Studňa area

Plant	Part of plant	Hg
		$mg.kg^{-1}$
Pinus sylvestris	root	9.85
	branch	6.02
	needles	2.30
Salix fragilis	root	22.00
	branch	0.21
	leafs	1.04
Juncus articulatus	root	5.94
	steem	0.64
	leafs	0.51

The Hg contents in analysed samples of selected food commodities from Malachov (Tab.5) did not exceed the standards (limits) given by Decree of the Ministry of Health of the Slovak Republic 608/3/2004-100 Coll. and Decree No. 14300/2007-OL.

Tab. 5 Hg content in selected food commodities and animals

Food commodity/animal	Part of plant/animal	Hg	Limit*
			$\mu g.kg^{-1}$
apple	fruit	<1.6	30
bean	hull	1.7	50
onion	leafs	8.2	50
nut	fruit	5.1	50
carrot	root	3.0	50
parsley	root	4.7	50
		6.8	50
potato	tuber	2.9	20
salmon trout (*Salmo trutta*)	muscles	252	50
	vitals	402	-
bullhead Siberian (*Cottus poecilopus*)	muscles	136	50
	vitals	473	-

Note: * Decree of the Ministry of Health of the Slovak Republic No. 608/3/2004-100 Coll. and Decree No. 14300/2007-OL.

The analysis of salmon trout (*Salmo trutta*) captured in the brook in the Malachov village proved a relatively high Hg contents both in muscles (252 $\mu g.kg^{-1}$) and in vitals (in liver 402 $\mu g.kg^{-1}$, Tab. 5). We obtained similar results also from the analysis of the second fish - bullhead Siberian (*Cottus poecilopus*). Although these contents don't exceed the standards given in Decree of the Ministry of Health of the Slovak Republic 608/3/2004-100 Coll. and Decree No. 14300/2007-OL for Hg (0.5 $mg.kg^{-1}$ for overall product and 0.1 $mg.kg^{-1}$ for vitals), they indicate that the excess of the limit is colourable and probably depends on the living-space of the fishes (animals).

2.4 Acidification of landscape components

In order to assess a risk of acidity production, a set of 10 samples of soil, sediments and stream sediments representing the landscape were used. The $pH_{(H2O)}$ in the sediments and soil from these samples, determined in the distilled water leachate in the heap field, moves in a range of 6.11 – 7.75 (Tab. 1). The $Eh_{(H2O)}$ values are relatively low (up to -70). Only in the samples OP-2 and OP-4, we can see higher Eh values in the A horizon vs. B horizon (-52 vs. -60 and 2 vs. -49) which indicate more oxic conditions in surface soil layers (and sediment) in comparison with the B horizon. The Eh values in 1M KCl lixivium show similar results (Tab. 1).

The content of sulphur is another important piece of data for the acidification risk calculation. This ranges from 0.05 to 0.30 % (Tab. 6). The result indicates that the content of the primary sulphide minerals, the oxidation of which can form the acidity, is relatively low.

The acidity production potential (AP) is mainly influenced by the inorganic carbon content, which reflects the content of carbonates able to cause the neutralization of acidity. The total carbon ($C_{tot.}$) content is fortunately high (0.69 to 13.62 %); the average $C_{tot.}$ content is 7.31 % (Tab. 6). The unfavourable assignment is caused by that most of the carbon is represented by organic carbon ($C_{org.}$). In spite of this negative effect, it is possible to state that the inorganic carbon content is satisfactory (in average 4.15 %).

The average $pH_{(H2O)}$ value in surface water is 6.99 (Tab. 3) and the average $pH_{(KCl)}$ value is nearly the same 6.82. Such pH rates close to the neutral value indicate that the formation of acid mine drainage water is not probable. The negative Eh values (up to -96) are characteristic for anoxic and anaerobic conditions.

Indeed the Eh values are more unstable (they range in groundwater from -11 to -108 and in surface water from -29 to -88; Tabs. 2 and 3), they generally reflect reducing conditions.

The pH and Eh values in water indicate that the acidification risk is not acute.

Tab. 6 Values of acidity production (AP), neutralization potential (NP) and net neutralization potential (NNP)

Sample	$S_{tot.}$	$C_{tot,}$	$C_{org,}$	$C_{inorg,}$	CO_2	$CaCO_3$	AP	NP	NP:AP
	%								
OP-1	0.05	6.95	3.67	3.28	12.02	27.33	1.562	226.73	145.20
OP-2A	0.05	9.86	6.02	3.84	14.07	31.99	1.562	265.39	169.87
OP-2B	0.05	12.40	0.64	11.76	43.09	97.98	1.562	812.84	519.38
OP-3	0.05	3.48	0.73	2.75	10.08	22.91	1.562	190.06	121.68
OP-4A	0.10	13.62	11.95	1.67	6.12	13.91	3.125	115.40	36.93
OP-4B	0.05	5.24	4.46	0.78	2.86	6.50	1.562	53.92	34.52
OP-5	0.05	2.96	0.62	2.34	8.57	19.50	1.562	161.77	103.57
OP-6A	0.05	6.98	2.58	4.40	16.12	36.66	1.562	304.13	194.71
OP-7	0.30	10.96	0.26	10.70	39.20	89.15	9.375	739.59	78.89
OP-8	0.05	0.69	0.68	tr.	tr.	tr.	1.562	n.d.	n.d

In Malachov surroundings there are the main minerals which can potentially cause acidification –cinnabar – HgS and pyrite – FeS_2. The present condition of acidification seems to be not acute, but it is necessary to realize that in many mining regions, the acidification process has started only after several decades since the time when the mine had been closed.

The recalculation of S % to H_2SO_4 gives the maximum amount of the acid per ton of soil or sediment which can be produced. The total acidity production (AP) is calculated using the following formula: AP = 31.25 x % S and corresponds to the amount of the acid that can be produced potentially by the soil and sediment material. In Malachov, the AP varies in a range of 1.562 – 9.375, on average 2.5, respectively (Tab. 6).

The neutralization potential (NP) which gives the content of neutralizing substances in the soil and sediments that are able to neutralize the acidity produced by the material varies in surroundings of Malachov from 0 to 812.84; x = 286.98 (Tab. 6). The highest NP (35.80) is in the sample OP-2B (Tab. 6) with relatively high contents of $C_{tot.}$ (12.40 %) and C_{inor} (11.76 %), and after the recalculation it corresponds to 97.98 $kg.t^{-1}$ of $CaCO_3$.

The value of NP:AP informs about the risk of production of acid in mine drainage water the best. In the Malachov region, the range of the NP:AP ratio is 0 – 519.38 (in average 140.48). If this ratio is close to the value of 1, the probability of the acidity formation is high; if the value is equal to or higher than 3, the risk of AMD production is negligible (Sobek et al., 1978). All the values of NP:AP in Malachov with exception of that in the sample OP-8 are so high (34.52 – 519.38) that according to this calculation, the risk of the acidity production is excluded.

4 DISCUSSION

The source of Fe in soils and sediments is predominantly from pyrite (FeS_2) and marcasite (FeS_2). As a result of their degradation, the formation of hydrogoethite („limonite" – $Fe_2O_3.nH_2O$) occurs. The Pb and Ba origin is less apparent. Probably both of them come from abundant dolomite rocks. The increased Zr contents is possibly derived from zirconium.

According to Beneš and Pabianová (1987), the Hg contents in soil is in a range from 0.02 to 0.2 $mg.kg^{-1}$ and the average Hg content is 0.098 $mg.kg^{-1}$ (Kabata-Pendias & Pendias, 2000). The soil in the Malachov brook drainage is cambi-soil (rendzina). The Hg content in soils and in sediments is not very high despite of the vicinity of the mercury deposit (0.558 – 44.238 $mg.kg^{-1}$), although exceeds the limits (Hg 0.5 $mg.kg^{-1}$) sensu Decree No. 220/2004 Coll. of the Slovak Republic, on agricultural soil utilization (for sandy-clay and clay soil). The results can be compared with those of Rusková (2009) which for a mining locality exploited in 17th century, described 0.411 ppm Hg content in soil mixed with technogenous sediments from the dump.

According to Greenwood and Earnshaw (1990), Kabata-Pendias and Pendias (2000), Hg is in soil present in several forms: a) elementary - Hg^0 (volatile and only slightly soluble in water), b) Hg^{2+}, which forms CH_3Hg^+ compounds in a form of cations. In soil, it is present mainly in a form of slightly mobile organocomplexes. Its mobility is influenced by the type of soil/sediment, mainly by the value of physical and chemical parameters as pH and Eh values. The oxidation of the metallic Hg starts at Eh >390 mV (Greenwood & Earnshaw, 1990), while the highest Eh value in Malachov (sample OP-4A, Tab.1) was only 52 mV.

Sensu pedological classification (Čurlík & Šefčík, 1999; McNeill, 1992, VanReeuwijk, 1995), the soils in Malachov (classification based on pH_{H2O}) are slightly acid (samples OP-4A, OP- 8), neutral (samples OP-1, OP-2A, OP-4B, OP-5) to slightly alkaline (OP-3, OP-6A, OP-7; Tab. 1).

The increased Hg contents in the A horizon are probably caused by cinnabar grains or by Hg bonding on mould (eventually on carbon and sulphur) and the data are conformable with the information that Hg ions are not well sorbed on clay minerals (Kabata-Pendias & Pendias, 1992), but Hg^{2+} show a good sorption property on mould (Kolář, 1988). A higher sorption capacity was described at pH 4-5 (Babčan & Ševc, 1994). As the mould horizon is resistant against the degradation, only a negligible part of Hg is released and permeates to the B horizon rich in clay minerals.

The main Hg mineral – cinnabar (HgS) is not well soluble during the oxidation process. Mechanically disintegrated grains are usually accumulated in the stream sediments (Polański & Smulikowski, 1978). This assumption was confirmed by the analyses of stream sediments, published by Matúš (2004, in Rusková, 2009). A high As (±Sb) content was described as well.

Stream sediments with high sludge and organic components have usually higher Hg contents in comparison with sandy or gravel stream sediments. The brooks and creeks in the surrounding of Malachov have predominantly a gravel bed with boulders bottom so there is a great chance that the water takes away the released Hg to the Hron River where it is dispersed in a great volume of water and causes no environmental risk. In spite of this situation, the stream sediments in the Malachov brook contain a great portion of cinnabar grains because of its weight. The Hg content in the stream sediments varies from 0.37 to 45.4 ppm (Grofčík, 2009, Rusková, 2009). This Hg can migrate from the sediments to the fytobenthos and from them to the zoobenthos.

The sorption of Hg on organic matter is very good (Kabata-Pendias & Pendias, 1992; Melicherčík & Melicherčíková, 2010), so Hg is often transformed to one of the most soluble and most toxic compounds – methylmercury (CH_3Hg^+) which contaminates the water, poisons water organisms, and through them the whole surrounding ecosystem (Virčíková & Pálfy, 1997; Kafka & Punčochářová, 2002).

In fresh water there is the main methyl form of Hg the CH_3HgOH and in sea-water there are CH_3HgCl compounds (Moore & Ramamoorthy, 1984). Methylmercury (CH_3Hg^+) is very mobile and thus contaminates the plants. Bacteria can cause the reduction of Hg^{2+} to Hg^0 and the following Hg volatilization from soil (Kabata-Pendias & Pendias, 1992). The dominant inorganic Hg form in surface water is $Hg(OH)_2$ (Moore & Ramamoorthy, 1984). The Hg speciation in aqueous systems influences mainly chlorides and sulphides. Under acid and oxidic conditions, chlorides prevail, under alkaline oxidic conditions mainly oxides and hydroxides and under reductive conditions, sulphides are dominant. Under alkaline conditions, the solubility of HgS increases in consequence of complex anion $[HgS_2]^{2-}$ formation (Fergusson, 1990). Hg forms stable complexes with various organic ligands, mainly with those which contain ligands as cysteine, amino-acids and hydroxyl-carboxyl acids.

High concentrations of methylmercury were described in humid countries (Gray, 2001). Methylmercury degradation in water surrounding is realized due to a microbiological activity and oxidation operations. The degradation of methylmercury by germs produce Hg^{2+} and CH_4, while following the oxidation process CO_2, CH_4 and Hg^{2+}, final products are formed (Bailey et al., 2001).

The mercury content in groundwater from the surrounding of the mine galleries does not exceed in most cases the allowed limit sensu Decree of the Government of Slovak Republic No. 496/2010 Coll. on drinking water or sensu Decree of the Ministry of Environmental Protection of the Slovak Republic No. 636/2004 Coll. on „rough water“ used in water-supplies. In some analyses (Rusková, 2009), an excess of limits for Fe, Mn and As concentrations was proved. Also in the surface water, the Hg contents are higher as the allowed limits imposed by Decree of the Government of the Slovak Republic No. 269/2010 Coll.

According to the previous investigations (Matúš 2004 in Rusková, 2009) in Malachov, Hg contamined mainly stream sediments, benthos, fishes, plants and mushrooms. It is interesting that the present study proved a relatively low Hg content in the stream sediment of the sample OP-5 (Tab. 1). The Hg, As and Sb contents in plant tissues decrease in the following order: root → branch/stem → leaf/needle.

Fishes as a top of the food chain under aqueous conditions contain the highest Hg contents. In animals, the highest Hg accumulations are in kidneys and in livers. The lowest Hg contents are in muscles (Kafka & Punčochářová, 2002; Kaneko & Ralston, 2007). This tendency was confirmed also in Malachov. The Hg content in muscles (255 and 136 $\mu g.kg^{-1}$) was two times lower than in livers (402 and 473 $\mu g.kg^{-1}$). Vitals of fishes from fresh waters are not used for consumption so it is necessary to follow only the Hg (and other heavy metal) contents in muscles. The Hg accumulations in fish-tissues depend on Hg concentrations in aqueous medium and environs. The water temperature, concentration of water-dissolved oxygen and less also the pH values are important as well. The ability of Hg accumulation increases at lower temperatures (Kaneko & Ralston, 2007). In surroundings of Malachov, the situation is favourable for Hg accumulations because of relatively cold climate (average temperature is 7.04 – 8.30 °C, but even lower temperatures are also common: 1.9, 3.6 or 4.3 °C).

According to Bukový (1990 in Rusková, 2009), the average Hg concentration increases in benthos from 56 to 118 $\mu g.kg^{-1}$ (the Hg content in benthos above the Veľká Studňa locality is 699 $\mu g.kg^{-1}$ and beneath the village Malachov decreases to 362 $\mu g.kg^{-1}$). The Hg contents in muscles of *bullhead Siberian* and *salmon trout* vary in a range from 136 to 252 $\mu g.kg^{-1}$. In vitals, the Hg contents were higher (473 and 402 $\mu g.kg^{-1}$) and we can assume that the concentrations increase during the life of fish. Although Hg contents in muscles and livers don't exceed the standards given in Decree of the Ministry of Health of the Slovak Republic 608/3/2004-100, Hg (0.5 $mg.kg^{-1}$, i.e. 500 $\mu g.kg^{-1}$), they indicate that the Hg accumulations in fish bodies are near the limit values and probably such localities exist where they can exceed the limit. In the present time, the mentioned Decree is not valid any longer, and is gradually replaced by several more benevolent regulations. The excess of the allowed limit concentrations in vitals (liver) doesn't represent any health risk because the parts of fishes are not consumed.

The Hg contents (tab. 5) in food commodities (fruits and vegetables) respect the limits for Hg given in Decree of the Ministry of Health of the Slovak Republic 608/3/2004-100 Coll. and Decree No. 14300/2007-OL. Only the mushrooms exceed the allowed limits for Hg (0.25 $mg.kg^{-1}$), As (0.5 $mg.kg^{-1}$) and Sb (0.3 $mg.kg^{-1}$).

The acidification process in mining landscapes cause increasing acidity, which has range of direct undesirable consequences such as soil degradation (e.g. mould, Ca and Mg removal), an increase in heavy metal mobility and consequently increasing contamination of the landscape, depressing metabolic rates, depressing the immune responses of organisms etc. The acidification is a multi-stage process which starts with the acid-producing material decomposition (usually sulphide minerals, mainly pyrite). Its kinetics depends on numerous factors as a bacterial activity impact, abundance of O_2, presence of water etc. During the initial phase, the acidification is attenuated by neutralisation efficiency of surrounding rocks (carbonates, clay minerals, organic matter...). After exceeding the pufration efficiency of neutralisation agents (mainly carbonates), usually (if there is sufficient acid producing material) the massive acidification starts.

The pH of soil and water are the first values which are important for the acidification process study. The pH in surface water is close to the neutral value 7, and suggests very limited or no risk of acidification. A relatively low pH value in groundwater from the spring from Nemecký Hill (sample MAL-3, pH 3.90; Tab. 2) and from the Turistický spring (MAL-4, pH 5.12; Tab. 2) indicates only a local process of acidification, connected with the oxidation of Hg-mineralization. The limited volume of water from the springs cannot substantially influence the acidity of the supergene processes in a wider region.

Measuring pH in soil is realised by several methods. All of them have certain advantages, as well as disadvantages. The pH is measured usually in a soil solution (suspension). The pH value changes also seasonally. During the dry season, the acid components (SO_4^{2-}, NO_3^-, Cl^-) aren't't removed by percolating waters. Such an increased concentration has no influence on the pH value measured in water. If the salts are removed during the wet period, the pH increases. This dilution effect causes an increase in pH in the soil suspension. It is the reason why some pedologists prefer measuring pH in solutions of 0.01M $CaCO_3$ or 0.1M KCl.

The pH_{KCl} and Eh_{KCl} is so-called pufration pH/Eh. This value indicates the cation exchange capacity and the cation sorption complex composition. In acid soils, the values are lower than the pH/Eh value in water. The D value, calculated using the formula pH: $DpH = pH_{H2O} - pH_{KCl}$, enables to estimate the presence of a prevailing charge on the colloids surface. The positive D value indicates the presence of negative charges, and vice versa, the negative D value indicates the presence of positively charged colloids. For soil in Malachov, the D is calculated from the average of the pH_{H2O} and pH_{KCl} values 0.38. It suggests that in soil colloids with a negative surface charge prevail. This determination suggests also a relatively limited ability of soil to bind anions which represent usually the neutralization agents under natural conditions.

The comparison of the $C_{tot.}$ content with the neutralization potential (NP) indicates that it depends mainly on the $C_{inorg.}$ content. The highest $C_{tot.}$ content is in the sample OP-4A (13.62 %), but the $C_{inorg.}$ content here is lower (10.70 %) than in the sample OP–2B (11.76 %) so the NP for the sample OP-4A is lower (739.59) than the NP for the sample OP-2B (812.84).

The acidity production potential (AP) is low, on average, 2.5. On the other hand, the neutralization potential (NP) values are high, on average, 286.98. The value of AP:NP, which is in the Malachov range 140.48 on average, informs about the acidification risk the best. According to Sobek et al. (1978), if the value is equal to or higher than 3, the risk might be neglected.

5 CONCLUSIONS

In soils/sediments, relatively high Hg contents (varying from 0.56 to 44.24 $mg.kg^{-1}$), as well as Fe (0.53 – 5.73 %), Mg (1.30 – 11.37 %), Mn (156 – 2741 $mg.kg^{-1}$) and Pb contents (6.9 – 117.0 $mg.kg^{-1}$) were described. Also increased Sr (78 – 227 $mg.kg^{-1}$), Ba (51 – 426 $mg.kg^{-1}$) and Zr (9.7 – 125.4 $mg.kg^{-1}$) contents were proved. Cu, Zn, Co, Cr, V, As, Sb and Ti contents are low. Mercury is bounded preferentially on mould in the A soil horizon.

The Hg content in groundwater, in most cases, does not exceed the limit sensu Decree of the Government of the Slovak Republic No. 496/2010 Coll. on drinking water, eventually of Decree of the Ministry of Environmental Protection of the Slovak Republic No. 636/2004 Coll. The Fe, Mn contents and in some older samples (Rusková et al., 2009) also As contents exceeds the limits.

In surface water, the recommended Hg contents are exceeded in several cases (sensu Decree of the Government of the Slovak Republic No. 269/2010 Coll.).

Mercury contaminates mainly stream sediments, benthos, fishes, plants and mushrooms. The Hg, As and Sb concentrations in plants decrease in the following order: root – branch/stem – leaf/needle.

Food commodities (fruits, vegetable and fishes) respect from the viewpoint the standards stated in Decree of the Ministry of Health of the Slovak Republic No. 608/3/2004–100 on food commodities and in Decree of the Ministry of Health of the Slovak Republic No.14300/2007-OL. On the other hand, the Hg contents in a great part of the mushrooms substantially exceed the allowed limits.

The low average value of the acidity production (AP) 2.5 and the extraordinary high values of neutralization potential (NP), on average, 286.98, indicate that the acidity production and the formation of acid mining waters and the subsequent acidification of landscape components (soil, water, plants...) seem to be out of the question. This assumption is confirmed also by the AP:NP ratio value (140.48).

ACKNOWLEDGEMENTS

This work was supported by the grant contracts APVV-0663-10. The authors thank to Dr. Jozef Krnáč for technical works.

REFERENCES

[1] BABČAN, J. & ŠEVC, J. Mercury (Hg) in systems with natural organic matter. *Ekológia*, Bratislava. 1994, 13, pp. 199-205.

[2] BAILEY, E. A., GRAY, J. E. & HINES, M. E. Mercury transformations in soils near mercury mines in Alaska, USA. *RMZ-Materials and geoenvironment.* Ljubljana. 2001, 48, 1, pp. 212-218.

[3] BANCÍK, T. & JELEŇ, S. *Mineralogické lokality širšieho okolia Banskej Bystrice.* Banská Bystrica : Brummer & Brummer, 1999. 118 pp.

[4] BENEŠ, S. & PABIANOVÁ, J. *Přirozené obsahy distribuce prvků v půdach.* Praha : VŠZ, 1987. pp. 123-149.

[5] BERGFEST, A. *Malachov Hg. (msc).* Banská Štiavnica : ŠÚBA, 1955. 128 pp.

[6] ČURLÍK J. & ŠEFČÍK P. *Geochemický atlas Slovenskej republiky, Pôdy.* Bratislava : MŽPSR, VÚPVR, 1999. 99 pp. ISBN 80-88833-14-0.

[7] FERGUSSON, J. E. *The heavy metals, chemistry. Environmental impact and health effect.* New Zealand : Pergamon Press, 1990. 614 pp. ISBN 3-936028-05-2.

[8] GRAY, J. E. An overview of mercury transport, cycling, and environmental effects of mercury mining. *RMZ-Materials and geoenvironment.* Ljubljana. 2001, 48, 1, pp. 2-7.

[9] GREENWOOD, N. N. & EARNSHAW, A. *Chemie der Elemente.* Würzburg, 1990. 1707 pp. ISBN 3-527-26169-9.

[10] IVANČÍK, Ľ. & MITÁČEK, J. *Nové poznatky o ložiskovo-geologických pomeroch Hg ložiska Zlatá studňa* (msc.). Banská Bystrica : archív GÚ SAV, 1986. 15 pp.

[11] JELEŇ, S., GALVÁNEK, J., ANDRÁŠ, P., BENDÍK, A., BELÁČEK, B., BOZALKOVÁ, I., GAÁL, Ľ., GAJDOŠ, A., HÁBER, M., KONEČNÝ, V., KRIŽÁNI, I., LUPTÁKOVÁ, J., MAZÚREK, J., MICHAL, P., SOTÁK, J., STAŇOVÁ, S., ŠIMO, V., ŠURKA, J. & WETTER, R. *Náučno-poznávací sprievodca po*

geologických a geografických lokalitách stredného Slovenska. Martin : Quick Print, 2010. 320 pp. ISBN 978-80-970413-4-2.

[12] KABATA-PENDIAS A & PENDIAS H. *Trace elements in Soils and Plants.* Boca Raton, Florida : CRC Press, 1992. 365 pp. ISBN 0849366437.

[13] KABATA-PENDIAS A & PENDIAS H. *Trace Elements in Soils and Plants.* Boca Raton, Florida : CRC Press, 2000. 432 pp. ISBN 978-0-8493-1575-6.

[14] KAFKA, Z. & PUNČOCHÁŘOVÁ, J. Těžké kovy v přírodě a jejich toxicita. *Chemické listy,* 2002. 96, pp. 611-617. ISSN 1213-7103.

[15] KANEKO, J. J. & RALSTON, N. V. C. Selenium and mercury in pelagic fish in the central north pacific near Hawaii. *Biological Trace Elements Research*, 2007. 119, pp. 242-254. ISSN 1559-0720.

[16] KODĚRA, M., ANDRUSOVOVÁ-VLČEKOVÁ, G., BELEŠOVÁ, O., BRIATKOVÁ, D., DÁVIDOVÁ, Š., FEJDIOVÁ, V., HURAI, V., CHOVAN,, M., NELIŠERIVÁ, V., ŽENIŠ, P., FEJDI, P., GREGOROVÁ, Z., GREGUŠ, J., HATÁR, J., HVOŽĎARA, P., CHOVANOVÁ, M., JUDINOVÁ, V., KAROLUSOVÁ, E., ONDRUŠOVÁ, S., ŠAMAJOVÁ, E. & VARČEKOVÁ, A. *Topographis mineralogy 2.* Bratislava : Veda, vydavateľstvo SAV, 1990. 518 pp.

[17] KOLÁŘ, L. Cesty k omezení vstupu těžkých kovů do zemědelské produkce. *Seminař CSVTS Cizorodé látky v půde a v rostlině*, Tábor, 1998. pp. 44-49.

[18] LINTNEROVÁ, O. & MAJERČÍK, R. Neutralisation potential of tailing of sulphide deposit at Lintich near Banská Štiavnica (in Slovak). *Mineralia Slovaca*, 2005. 37, 4, pp. 517-528.

[19] MCNEILL D. Rapid accurate mapping of soil salinity using electromagnetic ground conductivity meters. In: Topp G.C, Reynolds W.D. (Eds) *Advances in measureent of soil physical properties: Bringing theory into practice.* SSSA Special Publicatin. Madison, Wisconsin, USA, 1992. Number 30, 48 pp.

[20] MOORE, J. W. & RAMAMOORTHY, S. *Heavy metals in natural waters – Applied monitoring and impact assessment.* New York : Springer, 1984. 268 pp. ISBN 10: 1461297397.

[21] POLAŃSKI, A. & SMULIKOWSKI, K. *Geochémia.* Bratislava : SPN, 1978. 607 pp.

[22] RUSKOVÁ, J. *Uplatnenie environmentálnej politiky pri remediácii banských záťaží vo vybraných lokalitách okolia Banskej Bystrice (Špania Dolina, Malachov, Ľubietová).* Dizertačná práca, Univerzita Mateja Bela v Banskej Bystrici, 2009. 158 pp.

[23] SOBEK, A. A., SCHULLER, W. A., FREEMAN, J. R. & SMITH, R. M. *Field and laboratory methods applicable to overburden and minesoils.* Cincineti, Ohio : U. S. Environmental Protection Agency, 1978. EPA 600/2-78-054. 203 pp.

[24] VANREEUWIJK L.P. *Procedures for soil analysis. Internationam soil reference and information centre (ISRIC) a FAO OSN.* Technical report, 1995. 9 pp.

[25] VIRČÍKOVÁ, E. & PÁLFY, P. Prehľad technológií zneškodňovania nebezpečných odpadov s obsahom arzénu, kadmia a ortuti. *Aktuálne ekologické otázky E 97, Nebezpečné odpady.* Košice : BIJO Slovensko, 1997. Košice 25. 3. 1997. pp. 87-103.

RESUMÉ

Východní svahy Kremnických vrchů jsou budovány výlevnými a explozívními produkty neogénního vulkanizmu. Hg-mineralizace je prostorově spjatá s mladými poklesovými strukturami S-J až SV-JZ směrů a geneticky s postvulkanickou etapou vývoje neogénního stratovulkánu. Ložiskové akumulace cinabaritu a metacinabaritu, doprovázené pyritem, markazitem a arzenopyritem se tvořily v tektonicky rozdrcených karbonátech mezozoika a v pískovcích paleogénu, pod nepropustnými vrstvami argilitizovaných sopečných tufů (Ivančík & Mitáček, 1986; Bancík & Jeleň, 1999). Žilky cinabaritu se dříve dobývaly pomocí štol a úpadnic, větší shluky blíže k povrchu pomocí plytkých šachtic (Bergfest, 1955; Jeleň et al., 2010). Výskyty Hg-As mineralizace jsou vázány na zlomy, většinou S-J směru, které sloužily jako kolektor rudonosných roztoků (Ivančík & Mitáček, 1986). Ložisko je v současnosti uzavřené (Bancík & Jeleň, 1999).

V půdách v okolí ložiska se zjistily vysoké obsahy Hg (0,56 po 44,24 mg.kg^{-1}), Fe (0,53 – 5,73 %), Mg (1,30 – 11,37 %), Mn (156 – 2741 mg.kg^{-1}) a Pb (6,9 – 117,0 mg.kg^{-1}). Poměrně vysoké jsou také obsahy Sr (78 – 227 mg.kg^{-1}), Ba (51 – 426 mg.kg^{-1}) a Zr (9,7 – 125,4 mg.kg^{-1}). Obsahy Cu, Zn, Co, Cr, V, As, Sb a Ti jsou nízké. Rtuť se váže především na humusový A horizont půdy.

Obsahy Hg v podzemních vodách nepřesáhly ve většině případů přípustný limit (ve smyslu Nařízení vlády SR č. 496/2010 Sb., kterým se ustanovují požadavky na vodu určenou na lidskou spotřebu a kontrolu kvality vody určené pro lidskou spotřebu, případně Vyhlášky MŽP SR č. 636/2004 Sb., kterou se ustanovují požadavky na kvalitu surové vody a na sledování vody vo veřejných vodovodech. Zjistilo se překročení limitů pro Fe, Mn a v některých starších analýzách lokálně také As.

V povrchových vodách jsou doporučené hodnoty pro Hg v některých případech překročené (ve smyslu Nařízení vlády SR č. 269/2010 Sb).

Rtutí jsou kontaminovány především říční sedimenty, bentos, ryby, rostlinstvo a houby. Koncentrace Hg, As a Sb v rostlinných pletivech klesají v pořadí: kořen - větev – jehlice resp. kořen – stonek - list.

Potravinové komodity (ovoce, zelenina a ryby) z hlediska obsahu Hg splĺňují požadavky na hladinu přípustných koncentrací ve smyslu Vyhlášky MP SR a MZ SR č. 608/3/2004–100, kterým se vydává hlava Potravinového kodexu SR upravující kontaminanty v potravinách a Vyhláška MP SR a MZ SR č. 14300/2007-OL. Naopak, velká část hub vykazuje překročení přípustných limitů koncentrace Hg, přičemž klobouky hub bývají často více kontaminované než jejich třeň.

Nízká průměrná hodnota potenciálu tvorby kyselosti (AP) 2,5 a mimořádně vysoké hodnoty neutralizačního potenciálu (NP), v průměru 286,98, indikují, že tvorba kyselých baňských vod a následná acidifikace krajinných složek jsou prakticky vyloučené. Tento předpoklad jednoznačně potvrdzuje rovněž poměr AP : NP 140,48.

2

OVERVIEW OF RESEARCH AND USE OF INDICATOR GASES OF COAL SPONTANEOUS COMBUSTION IN CHINA

Xiyang FENG [1], Alois ADAMUS [2]

[1]*Institute of Safety Technology Management, Faculty of Prospecting and Mining Engineering, Yunnan Land and Resources Vocational College, 2. Jingniu Road, Yunnan, China,*

e-mail 68775351@QQ.com

[2] *Prof. Ing. PhD., Institute of Mining Engineering and Safety, Faculty of Mining and Geology, VŠB – Technical University of Ostrava, 17. listopadu 15, 708 33 Ostrava – Poruba, Czech Republic,*

e-mail alois.adamus@vsbcz

Abstract

The beginnings of research and use of results in the area of indicator gas application for early detection of spontaneous combustion of coal in China comes from the sixties of the last century. The significant development occurred in the nineties, when the program "Characteristics of the adsorption of coal oxygen and its application in the prevention of mine fires" had been completed. The paper briefly describes an overview of the historical development, achievements, experimental methods, laboratory equipment, legislation, and the outlook for the use of indicator gases of coal spontaneous combustion in China.

Abstrakt

Počátky výzkum a využití výsledků v oblasti aplikace indikačních plynů pro včasnou detekci samovznícení uhlí pochází v Čině z 60tých let minulého století. Ke značnému rozvoji došlo v devadesátých létech, kdy byl dokončen program „Charakteristika adsorpce kyslíku uhlí a její aplikace v prevenci důlních požárů". Předložený článek stručně popisuje přehled historického vývoje, dosažené výsledky, experimentální metody, laboratorní vybavení, legislativu a výhled vývoje využití indikačních plynů samovznícení uhlí v Číně.

Key words: coal, spontaneous combustion, indicator gas.

1 INTRODUCTION

The findings related to the application of indicator gases for early detection of coal spontaneous combustion globally come from the first half of the 20th century. It was primarily Graham [1,2], whose indicators have used in many of the world's coalfields up to now, and subsequently Kitagawa [3], the first author dealing with released gaseous hydrocarbons in spontaneous combustion. In Chinese coal mines, only CO was monitored till the end of the 50s in the given context. Mines, which mine coal with a content of pyrite and have considerable difficulties with the presence of endogenous fires, confirmed the existing practices of spontaneous combustion indications as insufficient at the time. For this reason, in the sixties, a number of Chinese mines started to investigate spontaneous combustion indicator gases with a focus on local mining conditions.

2 EXPERIENCE IN USING INDICATOR GASES IN CHINA

In 1966, a research team at the Yuanbaoshan Mine in the Liaoning Province was formed whose aim was to examine relations of development of spontaneous combustion temperatures and gas compositions in obtained gas samples, focusing mainly on the presence of CO and related factors (litre emission, decreased oxygen, increased CO_2), [4]. The experiments confirmed that the concentration of CO increases with temperature in the centre of spontaneous combustion with a simultaneous decrease in oxygen and an increase in CO_2. Two prediction methods of spontaneous combustion were set – the method of an increase in CO concentration and the method for tracking CO litre emission. One of the results of the research was the concentration limit values listed in Tab. 1. The threshold CO litre emission value of spontaneous combustion was set to 5.9 l/min.

Tab. 1 Limit values of the initial state of spontaneous combustion for the Yuanbaoshan Mine, [4]

	+CO	$-\Delta O_2$	$+CO_2$
limit for normal conditions	<0.0020%	<0.5%	<0.15%
characteristics	main indicator	subsidiary indicator	subsidiary indicator

The China Coal Research Institute (CCRI), made up a research team in cooperation with the Gusan Mine in 1974, which followed the findings learned at the Yuanbaoshan Mine, [5]. The results were the indicators – H1 (litre emission of CO) and H2 (Graham's interpretation of indicators), which were verified in 36 cases and evaluated as satisfactory, see Tab. 2.

$$H_1 = CO \times Q \tag{1}$$

$$H_2 = \frac{CO \times Q}{\Delta O_2 \times Q} \tag{2}$$

where Q – a volume flow rate of a return airway of an active face, [m^3/min],
CO – a concentration, [%],
ΔO_2 – a decrease in oxygen, [%].

Tab. 2 Indicators H1, H2 monitored at the Gushan Mine according [5]

	current conditions	prediction	fire
H1(m^3/min)	<0.0049	0.0049-0.0059	>0.0059
H2 (-)	<1.0	1.0-1.8	>1.8

In the eighties of the last century, Graham indicators had been tracked by the authors Zhong Weiren and Li Fengze at the Laohutai Mine in Fushung for many years [6]. Later, the authors of [7] tracked them as well. The partial results are shown in Tab. 3:

Tab.3 Results from monitoring the Graham indicators at the Laohutai Mine, [6]

state of spontaneous combustion	G ($CO/\Delta O_2$)
common condition	0
phase of low-temperature oxidation	0-0.45
phase of high-temperature oxidation	0.46-4
phase of starting burning	4.1-9
fire	>9

Gaseous hydrocarbons were first used in China as indicator gases of spontaneous combustion in 1983 at the Chaili Mine, in particular in more than one hundred successful cases. According to the authors of [8], their applications were associated with the following shortcomings – ethylene was detected at coal temperatures of 110-130 °C, which was not considered sufficiently early indication, other unsaturated hydrocarbons like propylene, butylene and other gases occurred in low concentration, easily susceptible by the dilution through the volume flow of ventilation. In order to extend the existing knowledge, the research team from the Chaili Mine ensured taking 18 coal samples from Guizhou, Henan, Hebei, Anhui and other provinces which underwent further testing [8]. The experiments confirmed the relationship between the dimensionless indicator C2H6/C2H4 and the temperature of coal. The indicator value decreased with an increase in the temperature of coal, see Tab. 4. The given values of the indicators were used successfully in May 1986.

Tab. 4 The indicator value of C_2H_6, C_2H_6/C_2H_4 depending on the temperature according to [8]

indicator	value /-/	temperature	Condition
C_2H_4	detected	110-130°C	early detection
C_2H_6/C_2H_4	2-3	140-190°C	preparation of repression
C_2H_6/C_2H_4	≤1	>200°C	emergency condition – fire repression

In 1988, the Mining Institute of Liuzhi investigated the dependence of gaseous hydrocarbons C_2H_6, C_3H_8, C_4H_{10} and their dimensionless indicators C_4H_{10}/CH_4, C_3H_8/C_2H_6 on an increase in temperature. The indicator C_3H_8/C_2H_6 was finally chosen as a suitable indicator for early detection of spontaneous combustion of coal for the actual mining conditions, [9]. This information was also published by the author of [7], Tab. 5.

Tab. 5 The indicator C_3H_8/C_2H_6 as an indicator of spontaneous combustion prevention for the Liuzhi Mine, [7]

indicator	common condition	hazardous condition	state of spontaneous combustion
C_3H_8/C_2H_6	0.02-0.06	0.1-0.12	0.15-0.18

The above research brought numerous data and objective results only for specific mines. The results were burdened by certain limits and bias. In order to achieve the standards applicable nationally, China elaborated the research of coal spontaneous combustion into the "Sixth", "Seventh" and "Eighth five-year plans for the development of national economy", through the "National Foundation of Research Programmes for Natural Science" in 1981-1995. The research yielded the following outputs.

In 1990, the research institute CCRI Qiying Min terminated the research project "Characteristics of the adsorption of coal oxygen and its application in the prevention of mine fires". During the three-year study, 900 experiments were conducted on over 61 coal samples taken from six representative mines, including lignite, bituminous coal and anthracite [10]. The research included the assessment of oxidation of coal mass (the dynamics of spontaneous combustion, tendency classification) and the characteristics of indicator gases (gas images, mine air samplings). The experiments related to the indicator gases of spontaneous combustion confirmed that the tracked released gaseous products of basic types of coal (brown, black and anthracite) contained basic major gases O_2, CO, CO_2, CH_4 and other minor gases from the gas hydrocarbons within a temperature range from low-temperature oxidation to open fire. Their concentrations corresponded to an increase in temperature. For example, the temperature interval of the first occurrence of ethylene was observed for brown coal, ranging from 110 °C to 120 °C, and for anthracite from 120 °C to 140 °C. The interval of the first occurrence of acetylene from 150 °C to 190 °C was observed for brown coal and for anthracite from 250 °C to 300 °C. The samples of brown coal showed the most intensive formation of indicator gases. Different types of coal showed different dynamics of gas releases. CO was determined as a basic indicator gas, then C_2H_4 as a gas indicating the ongoing process of spontaneous combustion, and C_2H_2 – warning gas – as an indicator gas of beginning open fire.

In 1992, within the research [11], 400 samples of all types of coal were tested according to coalification from brown coal to anthracite coal (brown coal, long flame coal, gas coal, fat coal, coking coal, lean coal, meager coal and anthracite). Characteristics of released gases of different lithotypes of coal were evaluated. Relationships of released amounts of CO, C_2H_4, C_3H_6, dimensionless indicators C_2H_4/C_2H_6, C_3H_8/C_2H_6, total amounts of gaseous hydrocarbons of alkanoic series (C_2H_6/C_2-C_4) were evaluated, in particular depending on changes in temperatures and types of coal with the following conclusions:

- Dimensionless indicators of alkane and alkene series were the main indicators of low metamorphic coal (e.g. brown coal, long flame coal, gas coal, fat coal), CO and CO- based indicators were the secondary indicators;

- CO and CO-based indicators were the main indicators of moderately metamorphic coal (coking coal, lean coal and meagre coal), dimensionless indicators based on C_2H_4 and ratios of alkanes/alkenes were the secondary indicators;

- Indicators of highly metamorphic coal (anthracite), including coal with high contents of sulphur were CO-based and CO-derived indicators only.

The example of graphical research outputs [11] of released gases depending on the temperature and coalification is given in Fig. 1– CO formation and Fig. 2 – C_2H_4 formation.

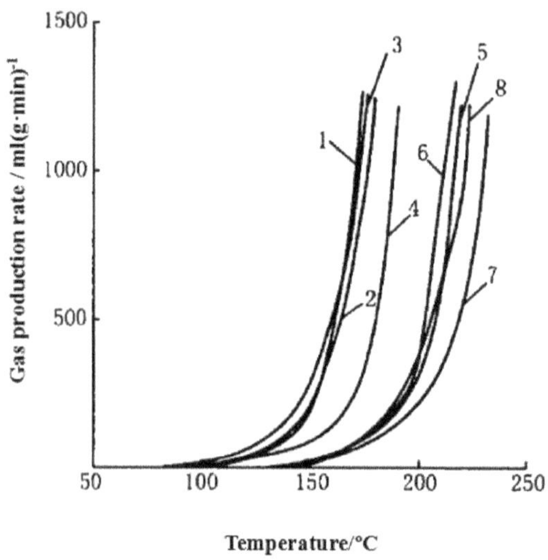

1 - lignite; 2 - long flame coal; 3 - gas coal; 4 - fat coal;

5 - coking coal; 6 - lean coal;7 - depleted coal; 8 - anthracite

Fig. 1 CO formation according to coalification, [11]

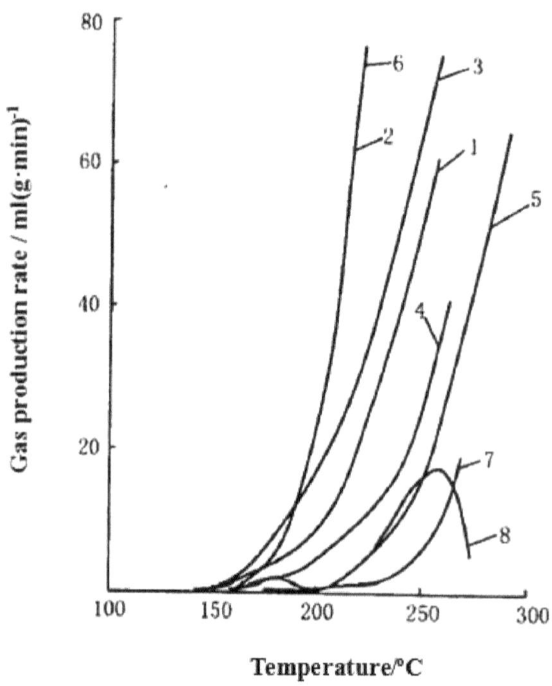

1 - lignite; 2 - long flame coal; 3 - gas coal; 4 - fat coal;

5 - coking coal; 6 - lean coal;7 - depleted coal; 8 - anthracite

Fig. 2 C_2H_4 formation according to coalification, [11]

The research also pointed out the necessity to consider a number of influencing factors and specific site conditions such as physical and chemical properties of coal, oxidation ability, point of spontaneous combustion, coal gas capacity, natural conditions, as well as conditions of the mine ventilation [11].

The research [12] investigated gas chromatographic analyses of indicator gases of different micro-petrographic components (virtrinit, durite, fusain) and three different types of coal samples taken at three mines. The research established the relationship of those groups of indicator gases to the mentioned factors, i.e. coal temperature, coal type and its petrography. The indicator gases were divided into three groups:

- Graham indicators, CO, CO/O_2,
- Saturated gaseous hydrocarbon C_2H_6, C_3H_8, C_4H_{10} and dimensionless indicators C_2H_6/CH_4, C_3H_8/CH_4, C_4H_{10}/CH_4, C_3H_8/C_2H_6, C_4H_{10}/C_2H_6,
- Unsaturated gaseous hydrocarbons C_2-C_4 (alkenes), C_2H_2 (alkyne) and dimensionless indicators C_2H_4/CH_4, C_3H_6/CH_4, C_4H_8/CH_4, C_2H_4/ C_2H_2.

The author of [12], He Ping, continued the research of coal samples taken from the Chaili Mine in 1995 [13]. Within the research, e.g. changes in the structure of oxygen functional groups during oxidation were analysed, aliphatic and aromatic hydrocarbons were studied. The experiments confirmed the relevance of the formation of the above first group of indicator gases (CO, CO/O_2) with considerable degradations of oxygen functional groups, and the connection of gas formation of the second and third groups with the loss of aliphatic hydrocarbons of coal mass.

In 2002, when simulating coal oxidation, the author Zhang Guangwen used the sorption concentrator technology for sampling released gases in order to determine the relationship of the contents of CO and C_2H_4 as a reliable indicator [14]. The conclusion of the research found that C_2H_4 can be captured through the concentrator at a temperature of 60 °C which is important at an early indication of early-stage spontaneous combustion, i.e. obtaining valuable time to control its centre. Similarly, in 2003, a research team led by the author Bao Zhonghong detected gases emitted from the beginning spontaneous combustion of coal left in a gob at the Nantun Mine [15]. The experiment confirmed that the use of concentrator technology had enabled to capture ethylene at 60 °C and propylene at 90 °C. Within the research, the dependency of selected dimensionless parameters on temperature was published, Fig. 3.

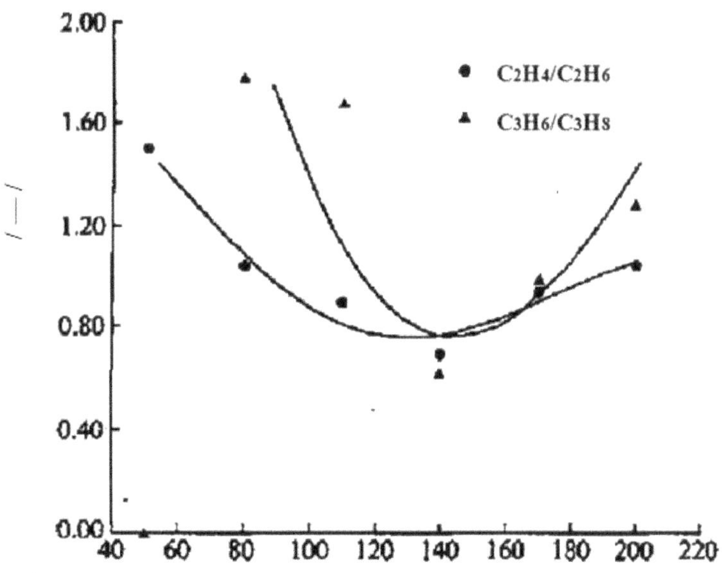

Fig. 3 Dimensionless indicators C_2H_4/C_2H_6 a C_3H_6/C_3H_8 according to [15]

The gas concentrator developed for practical applications was successfully tested in China and then registered as an authorized patent in 2001 under the number CN01238170.5.

In 2008, a research team led by the author Xiao Yang gathered coal samples from the mines Dongtan, Jiner and Nan Tun [16]. Oxidation tests in seven temperature ranges were supplemented by a thermal gravimetric analysis in order to express weight changes of coal mass during oxidation. The experiments confirmed the characteristic relationship between indicator gases and spontaneous combustion temperature. The critical temperature of the spontaneous combustion process of verified coal types ranged from 60 to 100 °C with an average of 85 °C. Carbon dioxides were recommended as indicator gases before reaching the critical temperature, while unsaturated hydrocarbons C_2H_4 a C_3H_6 were identified as characteristic gases from the temperatures above 150 °C.

In recent years, a large number of articles in the area of research and application of indicator gases of spontaneous combustion of coal have been published in China. Selected information from two of the research papers published in 2012 is presented here. The authors of [17] published the results from the laboratory monitoring of released indicator gases of heat-stressed coal, Fig. 4, with the following conclusions:

- A coal sample begins to release C_2H_6 at a coal temperature of 40 °C, a linear increase in the concentration of C_2H_6 is observed in the temperature range of 40-110 °C, the release intensity increases after exceeding 110 °C, the concentration of C_2H_6 equal to 85.32×10^{-6} was measured at a temperature of 180 °C;

- A coal sample begins to release C_3H_8 at a coal temperature of 90 °C and the concentration rises with increasing temperature, and the concentration of C_3H_8 equal to 68.29×10^{-6} was measured at a temperature above 180 °C.

- A coal sample begins to release C_2H_4 when the coal temperature exceeds 130 °C, the sample gets into a state of intense oxidation and the concentration slightly increases when the temperature of coal reaches 180 °C, the measured concentration of C_2H_4 was $3.735x10^{-6}$.

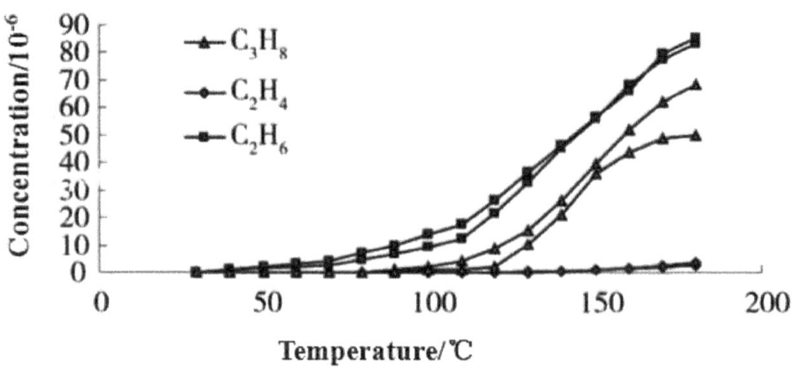

Fig. 4 Formation of C_2H_6 C_3H_8 and C_2H_4 depending on temperature according to [17]

The article [18] describes graphically dimensionless indicators C_2H_6/CH_4 and CO_2/CO, Fig. 5. The authors also state that the formation of CO, CO_2, CH_4, C_2H_6, C_2H_4, C_3H_8 increases with the temperature of coal. They also point out that CO is a characteristic indicator gas of coal spontaneous combustion, which evolves from 40 °C. The formation of CO_2 was captured in the laboratory from 24 °C. As soon as the temperature exceeded 70 °C, the concentration of CO and CO_2 increased, and the oxygen consumption and the released heat amount rose as well. The authors measured the concentrations of $CH_4 = 334\times10^{-6}$, $C_2H_6 = 4.76\times10^{-6}$ at a lab temperature of 24 °C . The concentrations of C_2H_6 rose significantly after the temperature had exceeded 110 °C. The ratio of gases C_2H_6/CH_4 had no regular behaviour; it did not characterize the trend of spontaneous combustion of a coal sample. The value of the gas ratio CO_2/CO decreased with an increase in temperature.

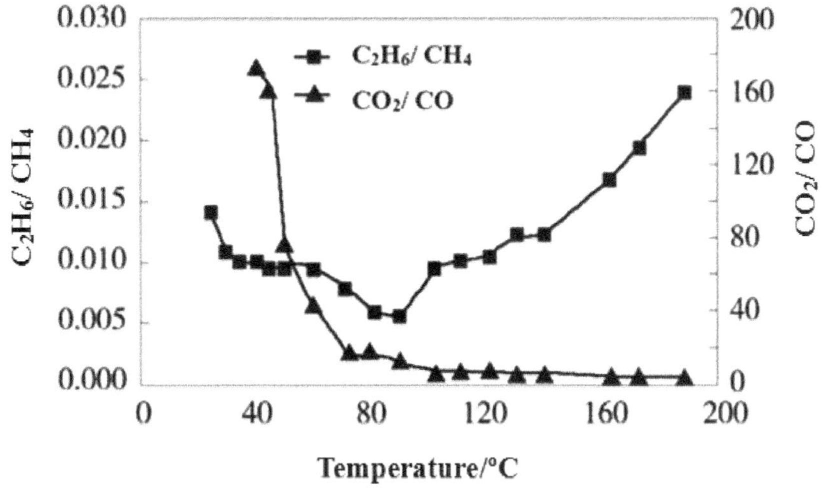

Fig. 5 Behaviours of gas ratios C_2H_6/CH_4 and CO_2/CO according to [18]

China has been dynamically developing the research of a system for preventing spontaneous combustion focused on key indicator gases like CO, C_2H_4, C_2H_2 and other auxiliary gases from the alkane and alkene series since the nineties of the last century. In recent years, hundreds of Chinese mines have been using mine air sampling for chromatographic analyses in accordance with the requirements of mining safety regulations and a number of technical documents in a form of mandatory and recommendatory technical standards.

3 EQUIPMENT AND PROCEDURES

In the fifties of the last century, detection tubes for detecting CO concentrations were widely used in China. In the sixties, some mines started to collect gas samples for analyses in surface laboratories with the intention to improve the accuracy of the analysis of low CO concentrations. Detection sensitivity increased to several tens of ppm from the previous sensitivity of the detection tubes. Portable CO detectors, COY-1 and COY-2 types, were developed. In the seventies, chromatographs started to be used for analysing mine air samples. Simultaneously, works were performed to improve the sensitivity of gas chromatography and to introduce gas monitoring through the use of transport tubes leading from the underground of the mine to the surface (a tube bundle system).

In recent years, the conditions for laboratory procedures for verification of indicator gases of spontaneous combustion gases have been standardized in China. The related equipment used must be equipped with a gas chromatograph, equipment for data logging, a programmable temperature control device for coal, a temperature controller-recorder, a system for dispensing gases, an automatic device for sampling gases, etc. The gas chromatograph must be equipped with a thermal conductivity detector and a flame ionization detector and must be able to analyse the major and minor gases including: O_2, N_2, CO, CO_2, CH_4, C_2H_6, C_3H_8, C_2H_4, C_3H_6, C_2H_2. A heat conductivity detector is used for analysing major gases; the signal should not be less than 0.5 mV. The hydrogen flame ionization detector is designed for the analysis of minor gases.

Currently, a system of program-controlled device intended for experimental observation of spontaneous combustion of coal mass, developed by the research institute CCRI, is widely used. The present apparatus for the oxidation of coal fines may be adjusted within the research by the user for the amount of a coal sample in a range from units of grams to hundreds of kilograms according to the dimensions of the used heat chamber. The device is able to simulate the growth of the spontaneous combustion process temperature to 300 °C. The system is designed to track the dynamics of coal oxidation, specific oxygen consumptions, heat release amounts, and the amount of released gases. One of the equipment used is shown in Fig. 6. [19].

Fig. 6 One of the program-controlled device for experimental coal oxidation, [19]

The verified coal sample of a set particle size is sucked through the oxidizing medium with a recommended oxygen content of 20.95 % (air) 12%, 10%, 7%, 5% and 3%. The temperature gradient of heating is determined at the following intervals: laboratory temperature-100°C: 0.5 °C/min, 100-200°C: 1.0 °C/min, 200-350°C: 2.0 °C/min, the temperature is registered 6 time per a minute, the time interval for the gas analysis of samples is determined by the formation of chromatographic analysis peaks, but the cycle should not exceed 20 min.

Today, in China, the chromatograph suitable for mining industry, type GC-4008B, is produced. The manufacturer is the Beijing East and West Analytical Instruments Co., (BEWA). The gas chromatograph is intended for coal mines and widely used in China. It is designed as laboratory equipment for the analysis of mine air gas components, it evaluates gas explosiveness, analyses gas components contained in outbursts, analyses spontaneous combustion indicator gases and gases of mine fires – H_2, N_2, O_2, CO, CO_2, alkanes, alkenes, alkynes, etc. It is equipped with a thermal conductivity detector, two flame ionization detectors and four special columns; the lowest analysed concentrations are as follows: CO, $C_2H_2 \leq 0.5$ppm, $C_2H_4 \leq 0.1$ppm, $H_2 \leq 5$ppm. An electron capture detector is designed for the tracer gas SF_6. The chromatograph is equipped with a computer program containing the explosion triangle, Fig. 7 [20].

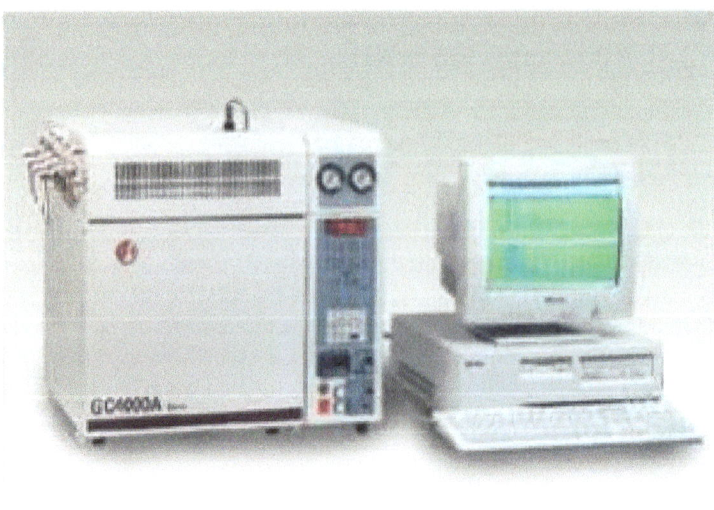

Fig. 7 Gas chromatograph GC-4008B designed for coal mines, [20].

Tube bundle systems connecting sampling points with analysers on the surface or in mines have been used in China since the eighties of the last century [21]. The mining tube bundle monitoring system based on the GC 8500 type chromatograph was developed in 1995. The analysis of mine gases was improved and extended to gases as follows: O_2, N_2 , CO, CH_4, CO_2, H_2, C_2H_4, C_2H_6, C_3H_8, C_3H_6, C_4H_{10}, C_2H_2, including SF_6 and fire gases [22]. At present, the tube bundle monitoring system JSG-9, developed by the research institute CCRI, is widely spread in China. The system monitors max 16 points at a distance up to 10 km at intervals of analysis of 2-20 min. It tracks caving areas, waste areas, and mine workings; it monitors CO, CO_2, O_2 and other gases of coal heating prevention. The system includes a computer for data backup, special software for the tube bundle monitoring system, a multi-gas analyser, and an assembly of suction vacuum pumps equipped with filters with water separators [23].

4 LEGISLATION AND TECHNICAL STANDARDS

The Chinese State Administration of Work Safety issued the "Coal Mine Safety Regulation" as of 1st November 2001[24]. Several provisions were amended on 14th December 2009. The last amendment, which came into effect on 1 March 2010, regulates indicator gases of spontaneous combustion as follows:

Article 228 – susceptibility of coal to spontaneous combustion is divided into three categories – low, medium and high. The coal susceptibility to spontaneous combustion must be verified before the commencement of mining. The degree of the susceptibility shall be determined by the state supervision in cooperation with a local authorized person. In case of occurrence of seams of middle and high susceptibility, measures of spontaneous heating prevention must be used.

Article 241 – in cases of exploiting seams of medium and high susceptibility, the points of frequent checks shall be carefully determined, a monitoring system shall be established, the indicator gases of spontaneous combustion shall be determined, and heating prevention shall be specified. All detection results must be recorded in the fire safety log, the recorded data must be regularly monitored and evaluated. If the indicators of spontaneous combustion exceed a set threshold limit, the current state will be immediately notified and measures will be taken to address the situation.

In February 2006, the Chinese State Administration of Work Safety issued the recommending technical standard AQ/T 1019-2006 "The method of the gas chromatography analysis and index optimization for mark gases of spontaneous combustion of coal stratum" as a tool for early detection of spontaneous combustion [25]. The standard contains recommendations for choosing indicator gases, their use and the method of analysis. It recommends to monitor CO, C_2H_6, C_2H_4, ratios of alkane/alkene (C_2H_4/C_2H_6), ratios of alkane/methane (C_2-C_4/CH_4), and ratios of alkane/ethane (C_3H_8/C_2H_6, C_4H_{10}/C_2H_6) as indicators of spontaneous combustion.

Coal sampling for laboratory verification of indicator gases is regulated by the Chinese recommended technical standard GB/T 482 "Sampling of Coal Seams" [26], requiring taking at least two coal samples from a characteristic seam area.

Taking gas samples of mine air shall be in accordance with the Chinese binding technical standard MT 142 "Coal Mine Underground Gas Collect Method", [27]. The standard imposes to complete the gas analysis within 12 hours after sampling; it allows to take air samples to bags, bottles or through the use of tube bundle systems.

Tube bundle monitoring systems in China are governed by the recommending technical standard MT/T 757 "General technical conditions of the tube bundle monitoring system for coal spontaneous combustion", [28]. The tube bundle monitoring systems are divided into surface and mine systems. The surface system uses a bundle of gas tubes led out to the surface; in the mine system, the sampling tubes lead to the nearest monitoring chamber and the evaluated data are then transmitted to the surface by the central monitoring system. The standard applies to both mentioned types of tube systems.

5 CONCLUSIONS

China has gained experience with the application of indicator gases in deep coal mining over the last forty years. The beginnings were focused only on basic gases, and then mainly gaseous hydrocarbons were investigated. In recent years, attention has been paid to setting standards for preventing coal spontaneous combustion, the choice of indicator gases of spontaneous combustion suitable for appropriate mines with respecting experimental analyses and practical applications. In order to improve the current knowledge in the detection, monitoring, and the choice of indicator gases of spontaneous combustion, neural network systems, GIS systems, the semiconductor technology, laser adsorption spectroscopy (DLAS), technology of adsorption enriching concentrators, and the technology of infrared spectroscopy were applied within the research, which provide partial results, but yet unapplied on a wider scale.

Several issues still exist in the area of indicator gases of spontaneous combustion and their use in practice as noted by the author of [29], which need to be addressed like:

- Identify different systems of spontaneous combustion indication for specific regional mining conditions, taking into account the genesis of deposits, layer conditions of coal seams, the nature and properties of coal, and the distinct accumulation of reserves in China;

- Establish a multi-level and interdisciplinary engineering dynamic model for risk prevention of endogenous fire occurrence;

- The applications of technologies should be based on a developed multi-cooperating system of knowledge with highly integrated application conditions of early detection of spontaneous combustion under conditions in a given mine.

The choice of appropriate indicators for specific conditions of a coal deposit is important for effective use of the technology of indicator gases of spontaneous combustion. That may reduce the risk of spontaneous combustion and increase safety of workers and operation in coal mines.

ACKNOWLEDGEMENTS

The article was written with support from the Technology Agency of the Czech Republic in the framework of the project No. TA01020351

REFERENCES

[1] GRAHAM, J.I. The normal production of carbon monoxide in coal-mines, *Transaction Institution of Mining Engineers*, vol.60, 1920-21, pp. 222-234.

[2] GRAHAM, J.I. The gaseous products resulting from fires and underground heating.*Transaction Institution of Mining Engineers*, vol.79, p. 92.

[3] KITTAGOWA, T. Detection of underground spontaneous combustion in its early stage with detectors. *In Proceedings of the 10th International Conference of Directors of Safetyin Mines Research*, Pittsburgh, PA, 1959.

[4] Prediction of coal spontaneous combustion. *(Chinese text),* Yuanbaoshan Mine of the Pingzhuang Bureau in Liaoning Province. *Coal Mine Safety*, June, 6/1966, pp. 6-10, ISSN 1003-496X.

[5] Prediction of coal spontaneous combustion in Gushan Mine. *(Chinese text),* Gushan Coal Mine of the Pingzhuang Bureau and Coal Research Institute of Liaoning Province., *Coal Mine Safety*, 2/1977, pp. 31-39. ISSN 1003-496X.

[6] ZHONG, W., LI, F. Graham coefficient (G) for the prediction of spontaneous combustion. *(Chinese text), Coal Mine Safety*, 10/1987, pp. 21-26. ISSN 1003-496X.

[7] WANG, Y. Indicator gases for prediction of spontaneous combustion of coal. *(Chinese text), Coal Mine Safety*, 6/2001, pp. 15, pp. 58. ISSN 1003-496X.

[8] LU, S., SONG, Y. Prediction of coal spontaneous combustion in Chaili Coal Mine. *(Chinese text), Coal Mine Safety*, 8/1995, pp. 5, pp. 13., ISSN 1003-496X.

[9] HUANG, Z. The ratio of alkane to alkene of spontaneous combustion in Liuzhi Mine. *(Chinese text), Coal Mine Safety*, 4/1988, pp. 52. ISSN 1003-496X.

[10] QI, Y., QIAN, G., LUO, H. The coal adsorption flow oxygen combustion characteristics and its application in mine fire prediction. *(Chinese text), Coal Mine Safety*, 9/1993, pp. 1-6, pp. 22, pp. 2-9, ISSN 1003-496X.

[11] LUO, H., QIAN, G. Research of indicator gases of spontaneous combustion associated to different types of coal. *(Chinese text), Coal Mine Safety*, S1/1992, pp. 86-89. ISSN 1003-496X.

[12] HE, P. et.al. Characteristics of gases formation in the process of coal oxidation and coal spontaneous combustion indicator gases choice. *(Chinese text), Journal of China Coal Society*, 6/1994, pp. 636-643, ISSN 0253-9993.

[13] HE, P. et.al. Relationship between the indicator gases produced from spontaneous combustion and coal chemical structure change in the oxidation process. *(Chinese text), Coal Analysis and Utilization*, 4/1996, pp. 12-16.

[14] ZHANG, G., WANG, X. Detect indicator gases of spontaneous combustion. (Chinese text), *Mining Safety and Environmental Protection*, 5/2002, pp. 7-8. ISSN 1008-4495.

[15] BAO, Z. et. al. Detection of concentrated indicator gases of coal spontaneous combustion. *(Chinese text), Jounal of Nanjing University of Technology*, 2/2003, pp. 57-60, ISSN 1671-7627.

[16] XIAO, Y. et. al. Correspondence between the research of indicator gases of coal spontaneous combustion and a characteristic temperature. *(Chinese text), Journal of Nanjing University of Technology*, 6/2008, pp. 47-51, ISSN 1671-7627.

[17] LI, M. et. al. Analysis of coal spontaneous combustion oxidation using multiple indicator gas system. *(Chinese text), Coal Mine Safety*, 1/2012, pp. 19-22, ISSN 1003-496X.

[18] TANG, J., ZHU, J. Experimental research for indicator gases of spontaneous combustion in coal seams at Zhujiadian Mine 4#. *(Chinese text), Coal Mine Safety*, 10/2012, pp. 21-24, ISSN 1003-496X.

[19] Mine Safety Technology Branch of China Coal Research Institute, [online].Updated December 2013 [viewed. 20 January 2013]. Available from: www.ccriaq.com/CCITPageManager.aspx?typeid=117

[20] Beijing East and West Analytical Instruments Co., [online].Updated December 2013, [viewed 21 January 2013]. Available from: www.instrument.com.cn/netshow/SH100293/C14420.htm

[21] Early prediction of mine fire bundle tube continuous monitoring system. *(Chinese text)*. China Coal Research Institute. *Coal Mine Safety*, 6/1983, pp. 9-14, pp. 40. ISSN 1003-496X.

[22] SU, J.: Development status and trends of the mine fire prevention technology. *(Chinese text), Journal of Ji Xi University*, 1/2005, pp. 52-53, pp.60. ISSN 1672-6758.

[23] Research Safety Equipment Branch of China Coal Research Institute, [online].Updated December 2013 [viewed 19 January 2013]. Available from: www.ccrise.cn/show.aspx?id=51&cid=49

[24] Coal Mine Safety Regulations no. 29/2010, State Administration of Work Safety,17-1-2011.

[25] The method of the gas chromatography analysis and index optimization for mark gases of spontaneous combustion of coal stratum. The technical standard AQ/T1019-2006, *State Administation of Work Safety*, 27-02-2006.

[26] Sampling of Coal Seams. The technical standard GB/T 482, *General Administration of Quality Supervision, Inspection and Quarantine of China*. 29-07-2008.

[27] Coal Mine Underground Gas Collect Method. The technical standard T 142, *State Ministry of Coal Industry*. 04-11-1986.

[28] General technical conditions of the tube bundle monitoring system for coal mine spontaneous combustion. The technical standard MT/T 757, *State Ministry of Coal Industry*. 30-12-1997.

[29] LUO, H., LI, Y. Present situation and prospect of coal spontaneous combustion prediction technology. *(Chinese text). China Safety Science Journal*, 3/2003, pp. 76-78, ISSN1003-3033.

RESUMÉ

Samovznícení uhlí je v hlubinných dolech Číny jedním z nejzávažnějších rizik. K jeho minimalizaci je v oblasti aplikace indikačních plynu věnována pozornost v čínském uhelném průmyslu již od šedesátých let minulého století. Prvopočátky výzkumu byly věnovány základním indikačním plynům, od let osmdesátých je výzkum a vývoj v daném směru intenzivně zaměřen na plynné uhlovodíky. Podstatnou roli sehrála čínská státní strategie podpory výzkumu v rámci šestého až osmého „Pětiletého plánu rozvoje národního hospodářství" v létech 1981-1995, kdy byl realizován výzkumný program „Charakteristika adsorpce kyslíku uhlí a její aplikace v prevenci důlních požárů", jehož výstupy přinesly poznatky z oblasti indikačních plynů samovznícení v různých směrech, např. ověření charakteristických indikačních plynů uhlí různého stupně prouhelnění, odlišných mikropetrografických složek, výzkum zaměřený na charakteristické podmínky těžených lokalit, ověření technologie plynových koncentrátorů aj. Souběžně byly v čině vyvinuty související technické prostředky, které jsou dnes reprezentovány v Číně vyráběným a široce používaným chromatografem GC-4008B určeným pro uhelné doly. Předmětné čínské poznatky a zkušenosti vyústily postupem doby v řadu závazných a doporučujících technických norem a znění závazného předpisu upravujícího prevenci samovznícení v uhelných dolech.

Článek byl napsán s podporou projektu Technologické agentury České republiky č. TA01020351.

3

EFFECT OF CALCINATION TEMPERATURE AND CALCINATION TIME ON THE KAOLINITE/TIO$_2$ COMPOSITE FOR PHOTOCATALYTIC REDUCTION OF CO$_2$

Martin RELI[1], Kamila KOČÍ[2], Vlastimil MATĚJKA[3], Pavel KOVÁŘ[4], Lucie OBALOVÁ[5]

[1] Ing, Department of Physical Chemistry and Theory of Technological Processes, Faculty of Metallurgy and Materials Engineering, VSB-Technical University of Ostrava
17. listopadu 15, Ostrava
e-mail martin.reli@vsb.cz

[2] doc., Ing., Ph.D, Department of Physical Chemistry and Theory of Technological Processes, Faculty of Metallurgy and Materials Engineering, VSB-Technical University of Ostrava
17. listopadu 15, Ostrava
e-mail: kamila.koci@vsb.cz

[3] Ing., Ph.D, Nanotechnology center, VSB-Technical University of Ostrava
17. listopadu 15, Ostrava
e-mail: vlastimil.matejka@vsb.cz

[4] Ing, The Czech Technological Centre for Inorganic Pigments, Nábřeží Dr. E. Beneše 24, Přerov
e-mail: pavel.kovar@precheza.cz

[5] prof., Ing., Ph.D, Department of Physical Chemistry and Theory of Technological Processes, Faculty of Metallurgy and Materials Engineering, VSB-Technical University of Ostrava
17. listopadu 15, Ostrava
e-mail: lucie.obalova@vsb.cz

Abstract

The kaolinite/TiO_2 composite (60 wt% of TiO_2) was prepared by thermal hydrolysis of a raw kaolin suspension in titanyl sulphate and calcined at different temperatures (600, 650 and 700°C) and for different times (1, 2 and 3 h). The obtained samples were characterized by XRPD, N_2 physical adsorption and SEM, and tested for photocatalytic reduction of CO_2. The different calcination conditions did not influence TiO_2 phase composition, only slightly changed the specific surface area, and significantly affected crystallite size of kaolinite/TiO_2 composite. A higher temperature and longer duration of calcination lead to higher crystallinity of the powder. The photocatalytic results showed that the crystallite size determined the efficiency of kaolinite/TiO_2 photocatalysts.

Abstrakt

Kompozit kaolinit/TiO_2 (60 hm% TiO_2) byl připraven termální hydrolýzou suspenze surového kaolinu v síranu titanylu a kalcinován při různých teplotách (600, 650 a 700°C) a po různou dobu (1, 2 a 3 h). Získaný vzorek byl charakterizován pomocí XRPD, N_2 fyzikální adsorpcí a SEM, a testován na fotokatalytickou redukci CO_2. Rozdíl kalcinačních podmínek neovlivnil složení fáze TiO_2, pouze se mírně pozměnila specifická povrchová plocha a výrazně byla ovlivněna velikost krystalitu kompozitu kaolinit/TiO_2. Vyšší teplota a delší doba kalcinace vedly k vyšší krystalitě prášku. Fotokatalytické výsledky ukázaly, že velikost krystality určuje účinnost fotokatalyzítoru kaolinit/TiO_2.

Key words kaolinite, TiO_2, photocatalysis, carbon dioxide, reduction

1 INTRODUCTION

The emission of carbon dioxide into the atmosphere is one of the most serious problems with regard to the greenhouse effect. Therefore, investigators make special effort to find effective methods to reduce CO_2 emissions, and moreover to convert CO_2 into more useful compounds (methane, methanol, ethanol, formic acid etc.). The efficient photocatalytic reduction of CO_2 with H_2O is one of the most desirable and challenging goals in research around the topic of the environment [1-8].

Nano-sized TiO_2 has received great attention as a photocatalyst due to a strong resistance to chemical and photocorrosion, high efficiency, lack of toxicity, and low cost. One of the possibilities regarding how to increase its photocatalytic activity is the introduction of TiO_2 nanoparticles into the clay mineral structure [9]. Preparation, characterization and photocatalytic efficiency for CO_2 photocatalytic reduction of kaolinite/TiO_2 composite was the subject of our recent work [10]. Significantly higher yields of CO_2 reduction products were observed in comparison to a commercial TiO_2 catalyst. Based on these promising results, we decided to provide the detailed study of an effect of kaolinite/TiO_2 composite preparation conditions on physicochemical and photocatalytic properties. Calcination temperature is one of the factors determining these properties.

Hamadanian et al. tested photocatalytic activity of the copper and sulfur codoped TiO_2 photocatalyst for degradation of methyl orange (MO) solutions [11]. The hotocatalytic activity of undoped TiO_2 calcinated at 500–700°C and 0.1% Cu, S-codoped TiO_2 calcinated at 500 and 700–850°C was tested under UV irradiation. Because of the sintering phenomenon, the size of nanoparticles increased while the calcination temperature was increased. Therefore, the surface area and photocatalytic activity decreased.

Mahdjoub et al. studied photoinduced degradation of a mono azo dye methylorange by titania nanopowders [12]. The samples were prepared with the sulphate process and were annealed in air at temperatures of up to 750°C, for 30 min. The X-ray diffraction indicates that the thermal annealing resulted in coarsening the average crystallite size from 13 to 72 nm. The highest photoactivity was observed in material treated at 500°C (average crystallite size was 20 nm).

Jia et al. probed photocatalytic activities of Ni, La-codoped $SrTiO_3$ photocatalysts [13]. The characterization results show that the calcination temperature has a strong influence on the physical–chemical properties of as-synthesized photocatalysts. The surface area and porosity, even the initial adsorption rate for malachite green (MG), decreased with increasing calcination temperature.

Kubacka et al. explored the simultaneous W–N doping of the anatase structure and, particularly, the influence of the calcination temperature in the photoactivity of the samples [14]. Gas-phase photodegradation tests were carried out with two organic pollutants (toluene and styrene). The maximum photoactivity was reached by calcination at 450°C and was concomitantly observed with near complete selectivity to partial oxidation products. Higher calcination temperatures yielded solids with significantly inferior photocatalytic performance.

Zheng et al. investigated photocatalytic degradation of methyl blue (MB) on a P-doped TiO_2 catalyst [15]. The rate constant of the P-doped TiO_2 increased gradually with the increasing calcination temperature of up to approximately 700°C, and decreased at higher calcination temperatures. After calcination at 900°C, the activity of P-doped TiO_2 was smaller, but its kinetic constant was still 2.5 times higher than that one during the calcination at 600°C. In contrast, the photocatalytic activity decreased greatly when the calcination temperature of pure TiO_2 was increased from 600 to 700°C.

The influence of calcination temperature of kaolinite/TiO_2 composite on its photocatalytic against carbon dioxide has not been studied yet. The aim of the presented paper is to investigate the influence of the temperature and duration of kaolinite/TiO_2 composite calcination on the yields of CO_2 photocatalytic reduction in both phases (liquid and gas).

2 EXPERIMENTAL PART

2.1 Kaolinite/TiO_2 composite preparation

The composite kaolinite/TiO_2 containing 60 wt.% of TiO_2 was prepared by a thermal hydrolysis of the suspension of raw kaolin (SAK47, LB minerals) in titanyl sulphate ($TiOSO_4$, Precheza a.s.), described in detail in a work published by Mamulová – Kutlaková et al [16]. The prepared composite was assigned KATI16, number *1* signals the sample in state after drying at 100°C, *6* depicts the content of TiO_2 (60 wt.%). For the purpose of the evaluation of the influence of the KATI16 heat treating conditions on its photodegradation ability against CO_2, KATI16 was calcined for 1h at 600°C to give sample KATI66; at 650°C to give sample KATI(65)6; at 700°C to give sample KATI76. With the aim to describe the influence of the duration of the KATI16 calcination at 600°C on its photodegradation ability against CO_2, the composite KATI16 was calcined at 600°C for 2 and 3 hours to give samples being assigned KATI66(2) and KATI66(3), respectively.

2.2 Characterization of kaolinite/TiO2 catalysts

The phase composition of the prepared samples was studied using the X-ray powder diffraction method (XRPD). The XRPD patterns were recorded under CoKα irradiation using a Bruker D8 Advance diffractometer (Bruker AXS, Germany) equipped with a fast position sensitive detector VÅNTEC 1. Measurements were carried out in a reflection mode, and powder samples were pressed in a rotational holder. The phase composition was evaluated using the database PDF 2 Release 2004 (International Centre for Diffraction Data). The crystallite size *Lc* of anatase was calculated using the Scherrer equation, lanthanum hexaboride (LaB6) was utilized as a standard [17].

The morphology of composite particles was observed on SEM Philips XL 30 (PHILIPS). Samples were coated with an Au/Pd film and the SEM images were obtained using a secondary electron detector. Elemental composition of samples was determined using the energy dispersive X-ray analysis (EDAX).

The specific surface area (SSA) of the powder samples was analyzed by nitrogen adsorption in a NOVA 4000e (QUANTACHROME INSTRUMENTS) nitrogen sorption apparatus. The samples were degassed for 3h at 105°C before the measurement. SSA was determined by the multipoint BET method using the adsorption data in the relative pressure range of 0.1 - 0.3.

The content of sulfur was determined using the CS244 carbon/sulfur determinator (LECO Corporation) equipped with the HT1000 induction furnace. The combustion process was carried out in ceramic crucibles filled with 1g of flux LECOCEL (tungsten particulates) and 1g of iron chip accelerator.

2.3 Photocatalytic reduction of CO2

The photocatalytic reduction of carbon dioxide was carried out in a homemade apparatus equipped with a stirred batch annular reactor with a suspended catalyst illuminated by a UV 8W Hg lamp (254 nm). The gas chromatograph (GC) equipped with flame ionization and thermal conductivity detectors (FID, TCD) were used for the analysis of gas and liquid reaction products. The details of the photocatalytic CO_2 reduction experiment and analytical methods were described in our previous publication [18]. It is important to minimize the influence of transport phenomena during kinetic measurements. The elimination of CO_2 diffusion from the bulk of gas through the gas–liquid interface in a laboratory batch slurry reactor was accomplished by saturating the liquid with pure CO_2 before the reaction started [2, 3]. Catalyst loading of 1g . dm^{-3} was chosen to avoid concentration gradients in the bulk of stirred liquid with kaolinite/TiO_2 suspension due to the scattering effect of light caused by the high catalyst concentration [3, 4, 19, 20]. The determination of a suitable volume of the liquid phase in our annular photoreactor to fulfill the requirement of perfect mixing has been published recently as 100 ml [21].

3 RESULTS AND DISCUSSION

3.1 Characterization of kaolinite/TiO_2 nanoparticles

The X-ray diffraction pattern registered for all the samples of kaolinite/TiO_2 reveals the presence of kaolinite, quartz, mica and anatase (Fig. 1). The kaolinite (001) basal diffraction line disappeared in all patterns of the calcined samples. This fact is attributed to the dehydroxylation of the kaolinite structure, which leads to the formation of an amorphous phase called metakaolinite. Another significant feature is the sharpening of anatase diffraction peaks related to the growth of the anatase crystallite size and increase of anatase content.

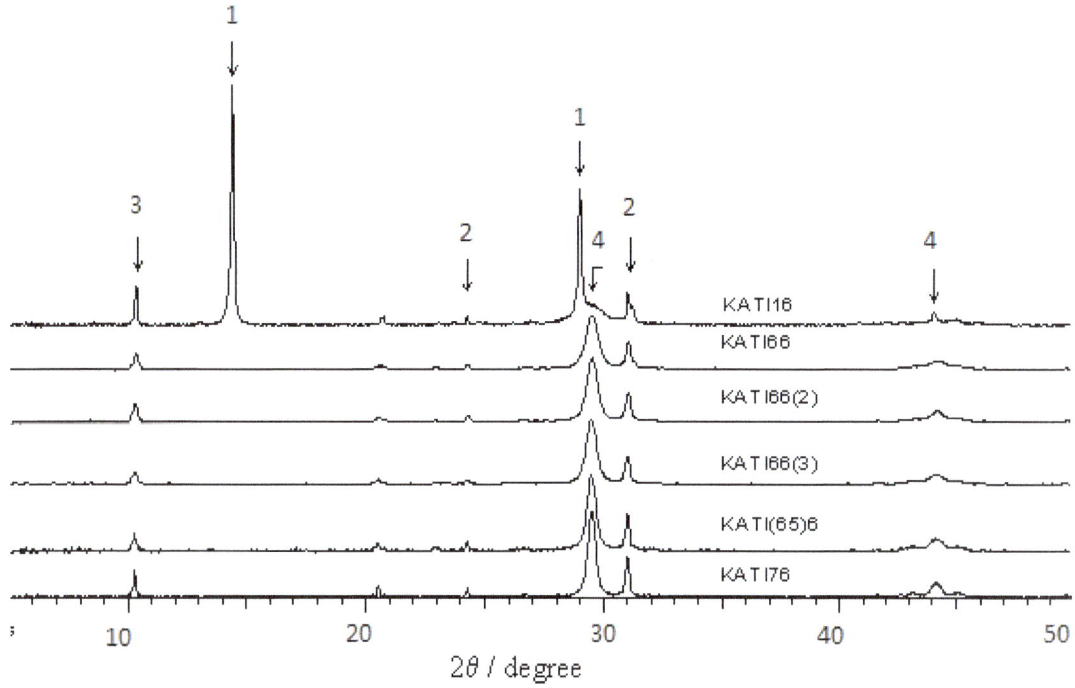

Fig. 1 XRPD of the original composite KATI16 and KATI16 calcined at the different temperatures (600°C, 650°C and 700°C) and for different calcination time (1, 2 and 3 hours). 1 … kaolinite, 2 … quartz, 3… mica, 4 anatase.

The crystallite size is closely related to the calcination temperature (Tab. 1) as well as to the duration of the calcination at a given temperature as shown in Tab. 2, and the same relationship is true for sulphur content.

Tab. 1 Selected characteristics of kaolinite/TiO2 catalysts calcinated at different temperatures

	Calcination temperature (°C)	Crystallite size (nm)	S_{BET} ($m^2 . g^{-1}$)	SO_3 (wt. %)
KATI16	dried	6*	65	3.50
KATI66	600°C	18	39	2.48
KATI(65)6	650°C	27	29	0.73
KATI76	700°C	34	24	0.35

*The calculation of crystallite size was not feasible due to the overlapping of the 101 anatase diffraction line and 002 kaolinite diffraction line. The value was extrapolated from a specific surface area–crystallite size relation.

The increase of the calcination temperature caused the decrease of the specific surface area and the increase of kaolinite/TiO_2 crystallite size. The dried catalyst KATI16 had the highest specific surface, and the catalyst calcinated at 600°C (KATI66) had the smallest crystallite size.

Tab 2. Selected characteristics of kaolinite/TiO2 catalysts calcinated at 600°C with different duration of calcination

	Duration of calcination (hour)	Crystallite size (nm)	S_{BET} ($m^2 . g^{-1}$)	SO_3 (wt. %)
KATI66	1	18	39	2.48
KATI66(2)	2	22	34	1.21
KATI66(3)	3	23	33	1.13

The increasing duration of calcination increased the crystallite size and decreased the sulphur impurity content of kaolinite/TiO_2 catalysts. The specific surface area decreased with the duration of calcination only slightly. The highest changes of the crystallite size, specific surface area and impurity content were registered for the catalyst with a calcination duration of 1 and 2 hours (KATI66 a KATI66(2)) while the changes, when calcination was longer than 2 hours, were minimal (KATI66(2) and KATI66(3)).

The morphology of the kaolinite sample and KATI66 sample that was studied using the scanning electron microscopy is illustrated in Fig. 2. It is evident from this image that TiO_2 particles preferably grow on the kaolinite particles' edges. This observation implies the idea that the rigid, smooth surface of kaolinite particles without cracks are not a suitable matrix for a TiO_2 nanoparticles growth. This observation is in contrast with the observation proposed by Chong et al. who prepared a 7 nm thick compact layer using titanium (IV) tetraisopropoxide as a TiO_2 precursor [22].

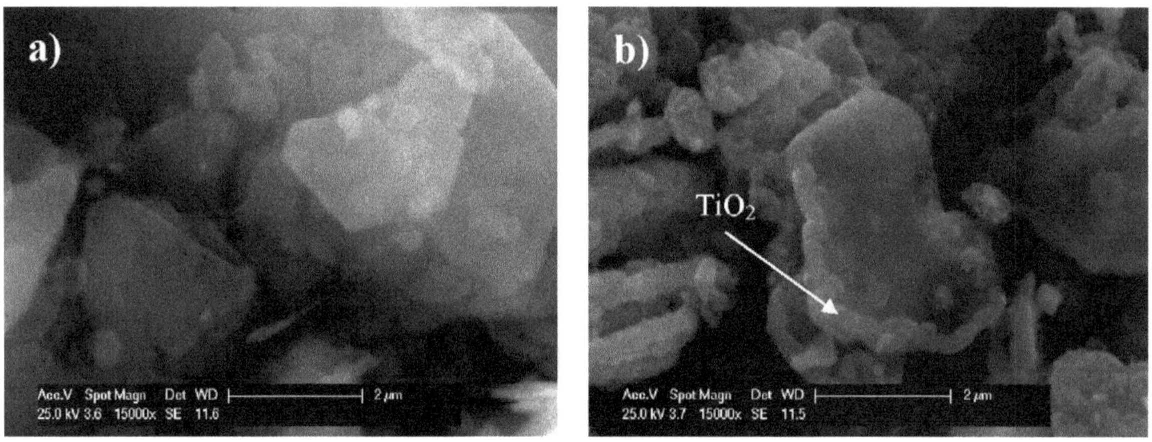

Fig. 2 SEM micrograph of kaolinite (a) and sample KATI66 (b).

3.2 The photocatalytic reduction of CO_2

The effect of irradiation time on the formation of CO_2 photocatalytic reduction products was investigated over a period of 0 – 24h on kaolinite/TiO_2 catalysts calcined at different temperatures and for different calcination time. Fig. 3 shows the evolution of all reaction products as the functions of the irradiation time for the KATI66 catalyst. Two main products were determined: methane in the gas phase and methanol in the liquid phase. Hydrogen and low amounts of carbon oxide were also detected. Other products such as formic acid, formaldehyde, ethane and ethylene could also be formed [1, 4], but they were undetectable. The observed order of yields (μmol . g^{-1} . cat) was: $H_2 > CH_4 > CH_3OH \geq CO$.

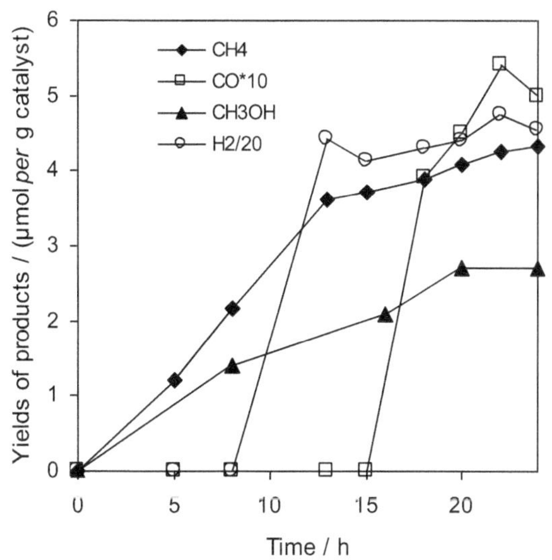

Fig. 3 Time dependence of the product yields of CO_2 photocatalytic reduction over the KATI66 catalyst.

3.3 Effect of calcination temperature

Comparisons of CH_4, CH_3OH and H_2, CO yields over kaolinite/TiO_2 calcined at different temperatures are shown in Fig. 4A, B and 5A, B, respectively. The yields of methane were negligible during the first 8h of irradiation in all cases with the exception of the most active KATI66 catalyst (Fig. 4A). A similar trend of CH_4 yield-time dependency was observed over all catalysts.

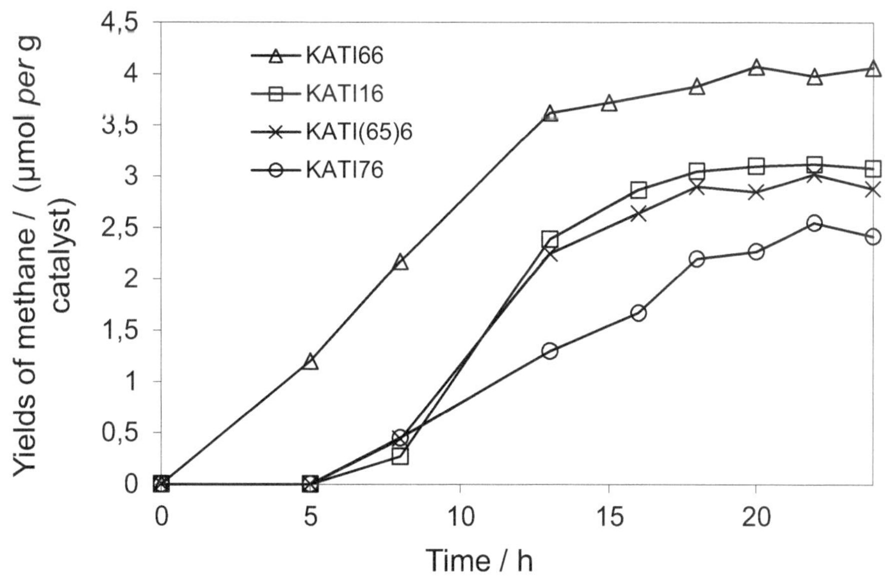

A

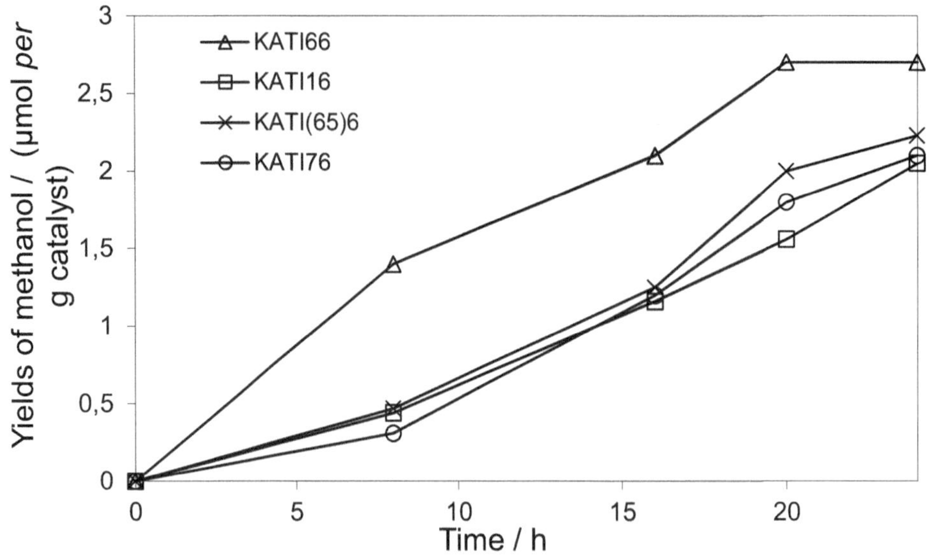

B

Fig. 4 Time dependence of methane (A) and methanol (B) yields over the kaolinite/ TiO_2 catalysts calcined at different temperatures.

The yields of methanol were lower than the yields of methane (Fig. 4B). Incipient methanol can act as a sacrificial agent and in part can be completely oxidized to carbon dioxide and H_2 [23, 24]. The data at 5 hours were measured but were under the limit of determinableness which was determined as 12 μg CH_3OH *per* L (i.e. 0.38 μmol CH_3OH *per* g catalyst). The steepest curve of methanol yield was measured for the KATI66 catalyst.

The hydrogen yields were not detected during the first 5 h of irradiation in all cases (Fig. 5A). After 5 h the hydrogen yields started to increase with the exception of the KATI76 catalyst. The yields of carbon monoxide increased after the first 5 h of the reaction over all catalysts with the exception of the KATI66 catalyst, formation of CO over this catalyst did not start until 16 h of the reaction. However, all CO yields were near the limit of detection (Fig. 5B). The most effective catalyst regarding main products was KATI66, which had a calcination temperature of 600°C.

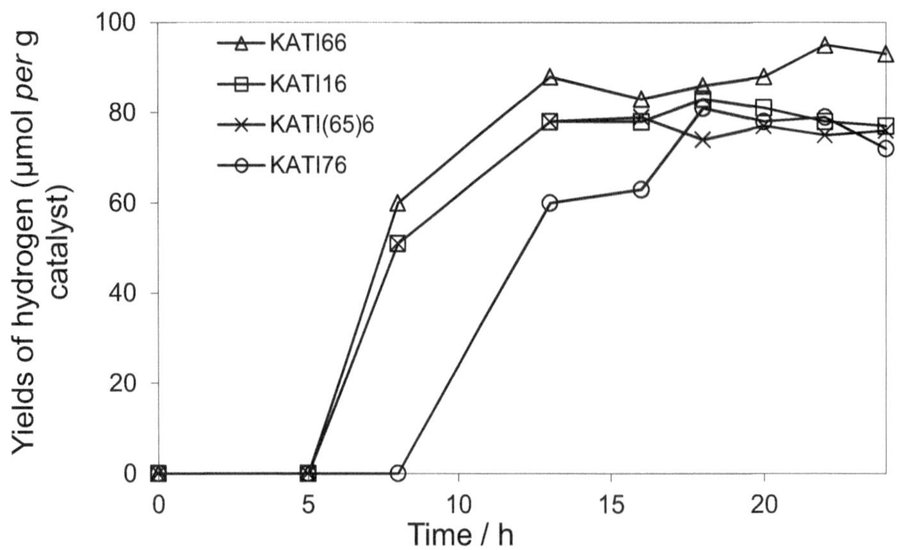

A

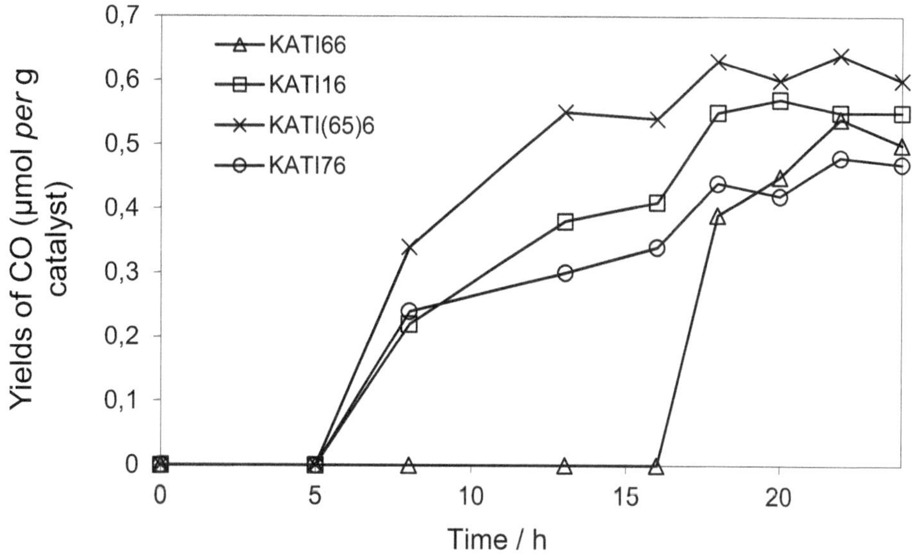

B

Fig. 5 Time dependence of the hydrogen (A) and the carbon monoxide (B) yields over the kaolinite/ TiO_2 catalysts calcined at the different temperatures.

3.4 Effect of calcination duration

The catalyst with the best photocatalytic activity – KATI66 – was calcinated for a different time (1, 2 and 3 h) at 600°C. Comparisons of CH_4, CH_3OH and H_2, CO yields are shown in Fig. 6A, B and 7A, B, respectively. The yields of methane were negligible during the first 8 h of irradiation for the catalysts KATI66(2) and KATI66(3) (Fig. 6A). The highest yield of methane after 24 h of irradiation was observed over the KATI66(2) catalyst. The yields of methanol were almost the same (Fig. 6B).

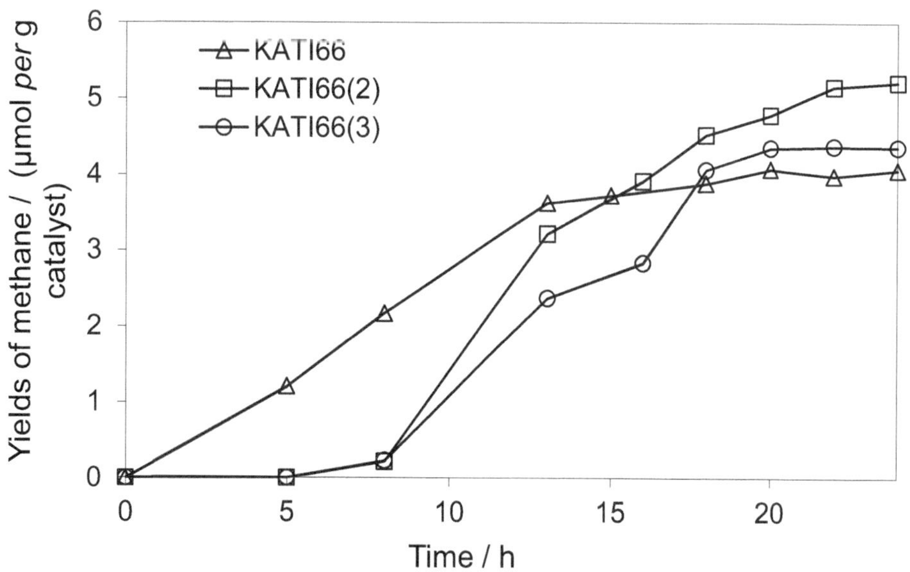

A

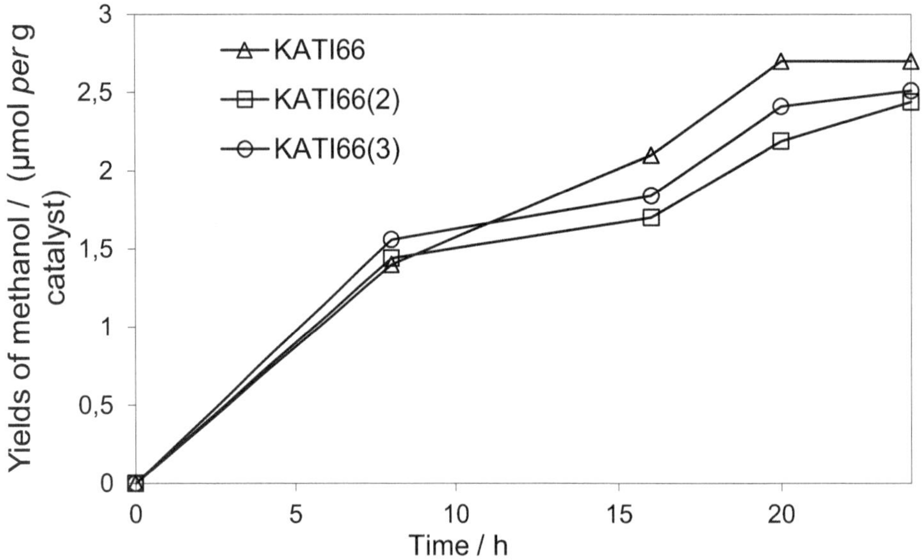

B

Fig. 6 Time dependence of methane (A) and methanol (B) yields over the kaolinite/TiO_2 catalysts calcined for different calcination time.

The hydrogen yields and yields of carbon monoxide are showed in Fig. 7A, B. A similar trend of yield-time dependencies of both products was observed for both KATI66(2) and KATI66(3). These yields on KATI66(2) and KATI66(3) are higher than over KATI66 in the case of CO and lower in the case of H_2.

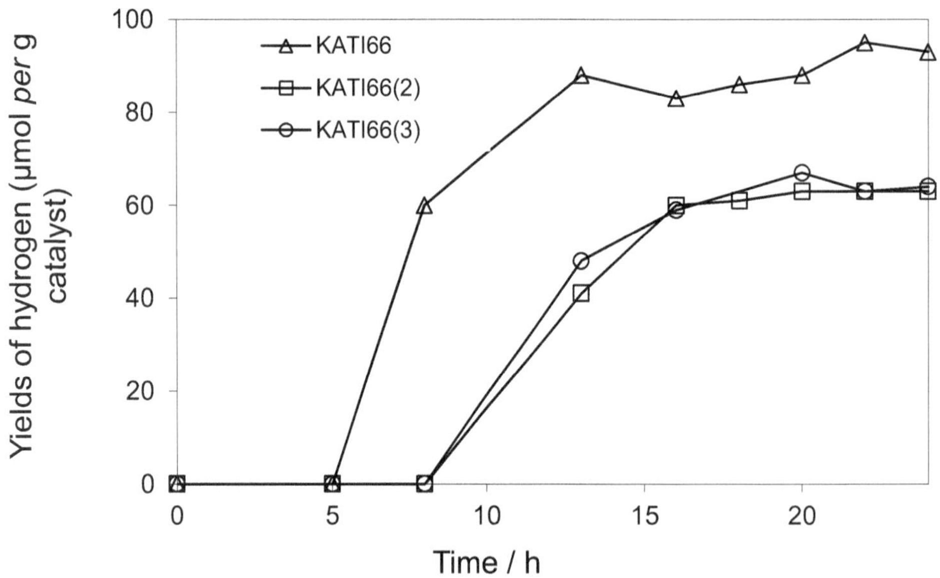

A

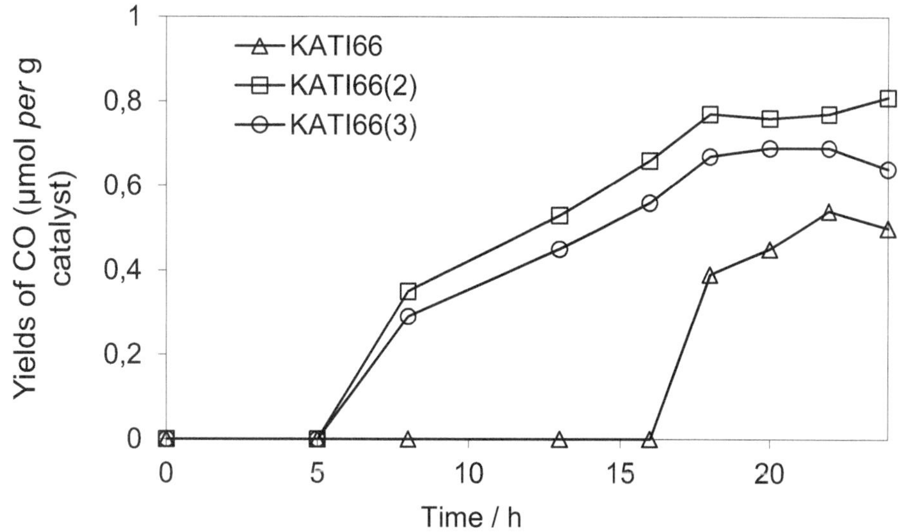

B

Fig. 7 Time dependence of the hydrogen (A) and the carbon monoxide (B) yields over the kaolinite/ TiO_2 catalysts calcined for different calcination time.

Generally, all studied physicochemical characteristics; phase composition, specific surface area, crystallite size and sulphur impurity contents can influence photocatalytic properties of kaolinite/TiO_2 catalysts. In our case, different conditions of calcination did not change the TiO_2 phase composition. Specific surface areas of calcinated composites were lower in comparison to a dried one. However, different calcination conditions (temperature, duration) did not change the specific surface area substantially; their deviations were close to the experimental error (± 10%). For the above mentioned reasons, the specific surface area and phase composition of kaolinite/TiO_2 composites calcined at different conditions were not the main parameters that determined their photocatalytic activity.

For examination of the effect of crystallite size on CO_2 photocatalytic reduction efficiency, the yields of CH_4, CH_3OH and CO after 24 hours irradiation for all tested catalysts were compared (Fig. 8). The data from the longest reaction time span were chosen for the photoactivity comparison because the yields of all products were the highest and their GC analysis was the most accurate. It is obvious from the Fig. 8 that the photoreactivity of kaolinite/TiO_2 increases when the crystallite size is increased from 6 nm to 22 nm, but decreases when it is further increased to 34 nm. The yields of all products were the highest over the catalysts with a crystallite size of 18-23 nm; these samples were calcined at 600°C for different times (1h, 2h a 3 h). The crystallite size influenced the dynamic of e^-/h^+ recombination, and so influenced the efficiency of photocatalytic reactions.

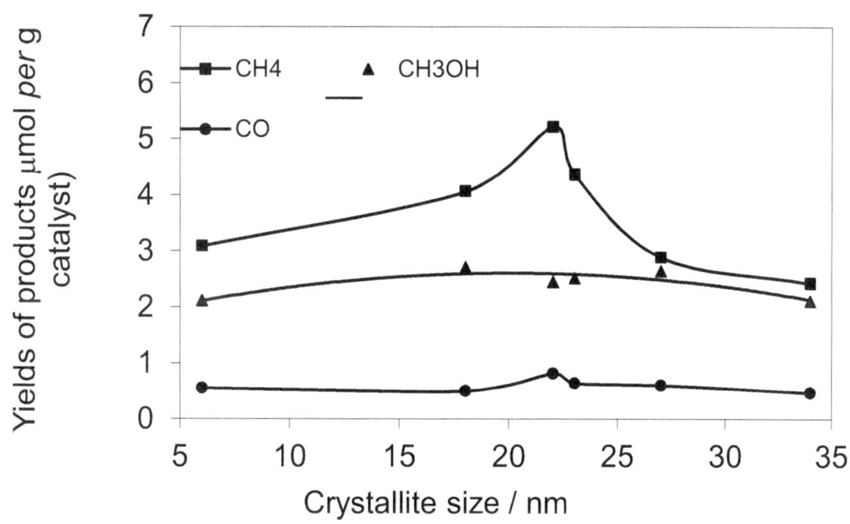

Fig 8. Dependence of the product yields (after 24 hours of irradiation) on different catalysts.

The optimal value of the primary particle size for pure TiO_2 (14 nm) was found in our recent work dealing with the CO_2 photocatalytic reduction [18]. An optimum crystallite size corresponding to the highest yields was also found for other photocatalytic reactions which depend on the oxidation potential of photogenerated holes, e.g. the degradation of trichloroethylene (with an optimal crystallite size of 7 nm) [25], phenol (with an optimal crystallite size of 25-40 nm) [26], trichlormethane (with an optimal crystallite size of 10 nm) [27], congo red (with an optimal crystallite size of 10 nm) [28]. From the results in Tab. 1 and Tab. 2, it is evident that the sulphur impurity contents decreased with an increase of temperature and duration of calcination. However, this parameter probably was not predominant for photocatalytic activity of kaolinite/TiO_2 composites.

4 CONCLUSIONS

The influence of calcination conditions (the temperature and duration) of the kaolinite/TiO_2 composite on the physicochemical properties and on the yields of CO_2 photocatalytic reduction was studied. It was determined that different calcination conditions did not influence TiO_2 phase composition; only anatase diffraction lines were found in all kaolinite/TiO_2 samples. A slight decrease of specific surface area was observed for the calcinated kaolinite/TiO_2 composite compared to that of a noncalcinated kaolinite/TiO_2 composite. The crystallite size was closely related to the temperature as well as to the duration of the calcination. The crystallite size increased with an increase of the calcination temperature and duration. The photocatalytic results showed that the crystallite size plays an important role in the nanocrystalline kaolinite/TiO_2 composite. The yields of all products were the highest over the catalysts with a crystallite size of 18-23 nm; these samples were calcined at 600°C for different times (1 h, 2 h a 3 h).

ACKNOWLEDGEMENTS

This paper has been elaborated in the framework of the Nanotechnology – the basis for international cooperation project, reg. no. CZ.1.07/2.3.00/20.0074 supported by Operational Programme Education for Competitiveness funded by the European Union Structural Funds and state budget of the Czech Republic, and the EU project ENET no. CZ.1.05/2.1.00/03.0069, and by the Ministry of Education, Youth and Sports of the Czech Republic – SP2012/196 and SP2012/25. The authors would like to thank Daniel Casten for controlling and correcting the use of English in this paper.

REFERENCES

[1] ADACHI, K.; OHTA, K.; MIZUNO, M. Photocatalytic reduction of carbon dioxide to hydrocarbon using copper-loaded titanium dioxide. *Solar Energy.* 1994, LIII. Nr. 2, pp. 187-190. ISSN 0038-092X.

[2] MIZUNO, T.; ADACHI, K.; OHTA, K.; SAJI, A. Effect of CO_2 pressure on photocatalytic reduction of CO_2 using TiO_2 in aqueous solutions. *Journal of Photochemistry and Photobiology A: Chemistry.* 1996, IIC, Nr. 1-2, pp. 87-90. ISSN 1010-6030.

[3] TSENG, I.-H.; CHENG, W.-C.; WU, J. C. S. Photoreduction of CO_2 using sol–gel derived titania and titania-supported copper catalysts. *Applied Catalysis B: Environmental.* 2002, XXXVII, Nr. 1, pp. 37-48. ISSN 0926-3373.

[4] SASIREKHA, N. ; BASHA, S. J. S. ; SHANTHI, K. Photocatalytic performance of Ru doped anatase mounted on silica for reduction of carbon dioxide. *Applied Catalysis B: Environmental.* 2006, LXII, Nr. 1-2, pp. 169-180. ISSN 0926-3373.

[5] ZHANG, Q.-H.; HAN, W.-D.; HONG, Y.-J.; YU, J.-G. Photocatalytic reduction of CO_2 with H_2O on Pt-loaded TiO_2 catalyst. *Catalysis Today.* 2009, CXLVIII, Nr. 3-4, pp. 335-340. ISSN: 0920-5861.

[6] ZHAO, Z.; FAN, J.; XIE, M.; WANG, Z. Photo-catalytic reduction of carbon dioxide with in-situ synthesized CoPc/TiO_2 under visible light irradiation. *Journal of Cleaner Production.* 2009, XVII, Nr. 11, pp. 1025-1029. ISSN 0959-6526.

[7] VARGHESE, O. K.; PAULOSE, M.; LATEMPA, T. J.; GRIMES, C. A. High-Rate Solar Photocatalytic Conversion of CO_2 and Water Vapor to Hydrocarbon Fuels. *Nano Letters.* 2009, IX, Nr. 2, pp. 731-737. ISSN 1530-6992.

[8] WU, J. C. S. Photocatalytice reduction of greenhouse gas CO_2 to fuel. *Catatalysis Surveys from Asia.* 2009, XIII, pp. 30-40. ISSN 1574-9266.

[9] KIBANOVA, D.; TREJO, M.; DESTAILLATS, H.; CERVINI-SILVA, J. Synthesis of hectorite–TiO_2 and kaolinite–TiO_2 nanocomposites with photocatalytic activity for the degradation of model air pollutants. *Applied Clay Science.* 2009, XLII, Nr. 3-4, pp. 563-568. ISSN 0169-1317.

[10] KOČÍ, K.; MATĚJKA, V.; KOVÁŘ, P.; LACNÝ, Z.; OBALOVÁ, L. Comparison of the pure TiO_2 and kaolinite/TiO_2 composite as catalyst for CO_2 photocatalytic reduction. *Catalysis Today.* 2011, CLXI, Nr. 1, pp. 105-109. ISSN 0920-5861.

[11] HAMADANIAN, M.; REISI-VANANI, A.; MAJEDI, A. Synthesis, characterization and effect of calcination temperature on phase transformation and photocatalytic activity of Cu,S-codoped TiO_2 nanoparticles. *Applied Surface Science*. 2010, CCLVI, Nr. 6, pp. 1837-1844. ISSN 0169-4332.

[12] MAHDJOUB, N.; ALLEN, N.; KELLY, P.; VISHNYAKOV, V. SEM and Raman study of thermally treated TiO_2 anatase nanopowders: Influence of calcination on photocatalytic activity. *Journal of Photochemistry and Photobiology A: Chemistry*. 2010, CCXI, Nr. 1, pp. 59-64. ISSN 1010-6030.

[13] JIA, A.; LIANG, X.; SU, Z.; ZHU, T.; LIU, S. Synthesis and the effect of calcination temperature on the physical–chemical properties and photocatalytic activities of Ni,La codoped $SrTiO_3$. *Journal of Hazardous Materials*. 2010, CLXXVIII, Nr. 1-3, pp. 233-242. ISSN 0304-3894.

[14] KUBACKA, A.; COLÓN, G.; FERNÁNDEZ-GARCÍA, M. N- and/or W-(co)doped TiO_2-anatase catalysts: Effect of the calcination treatment on photoactivity. *Applied Catalysis B: Environmental*. 2010, XCV, Nr. 3-4, pp. 238-244. ISSN 0926-3373.

[15] ZHENG, R.; GUO, Y.; JIN, C.; XIE, J.; ZHU, Y.; XIE, Y.; *J. Mol. Catal. A: Chem.* **2010**, *319*, 46.

[16] MAMULOVÁ-KUTLÁKOVÁ, K.; TOKARSKÝ, J.; KOVÁŘ, P.; VOJTĚŠKOVÁ, S.; KOVÁŘOVÁ, A.; SMETANA, B.; KUKUTSCHOVÁ, J.; ČAPKOVÁ, P.; MATĚJKA, V. Preparation and characterization of photoactive composite kaolinite/TiO_2. Journal of Hazardous Materials. 2011, CLXXXVIII, Nr. 1-3, pp. 212-220. ISSN 0304-3894.

[17] SCHERRER, P. Estimation of the size and internal structure of colloidal particles by means of röntgen. *Nachrichten von der Akademie der Wissenschaften in Göttingen*. 1918, XXVI, pp. 96-100.

[18] KOČÍ, K.; OBALOVÁ, L.; MATĚJOVÁ, L.; PLACHÁ, D.; LACNÝ, Z.; JIRKOVSKÝ, J.; ŠOLCOVÁ, O. Effect of TiO_2 particle size on the photocatalytic reduction of CO_2. Applied Catalysis B: Environmental . 2009, LXXXIX, Nr. 3-4, pp. 494-502. ISSN 0926-3373.

[19] BALLARI, M. M.; BRANDI, R.; ALFANO, O.; CASSANO, A. Mass transfer limitations in photocatalytic reactors employing titanium dioxide suspensions: I. Concentration profiles in the bulk. *Chemical Engineering Journal*. 2008, CXXXVI, Nr. 1, pp. 50-65. ISSN 1385-8947.

[20] BALLARI, M. M.; BRANDI, R.; ALFANO, O.; CASSANO, A. Mass transfer limitations in photocatalytic reactors employing titanium dioxide suspensions: II. External and internal particle constrains for the reaction. *Chemical Engineering Journal*. 2008, CXXXVI, Nr. 2-3, pp. 242-255. ISSN 1385-8947.

[21] KOČÍ, K.; OBALOVÁ, L.; PLACHÁ, D.; LACNÝ, Z. Effect of temperature, pressure and volume of reacting phase on photocatalytic CO_2 reduction on suspended nanocrystalline TiO_2. *Collection of czechoslovak chemical communications*. 2008, LXXIII, Nr. 8-9, pp. 1192-1204. ISSN 1212-6950.

[22] CHONG, M. N.; VIMONSES, V.; LEI, S.; JIN, B.; CHOW, C.; SAINT, C. Synthesis and characterisation of novel titania impregnated kaolinite nano-photocatalyst. *Microporous and Mesoporous Materials*. 2009, CXVII, pp. 233-242. ISSN 1387-1811.

[23] PATSOURA, A.; KONDARIDES, D. I.; VERYKIOS, X. E. Photocatalytic degradation of organic pollutants with simultaneous production of hydrogen. *Catalysis Today*. 2007, CXXIV, Nr. 3-4, pp. 94-102. ISSN 0920-5861.

[24] KONDARIDES, D. I.; DASKALAKI, V. M.; PATSOURA, A.; VERYKIOS, X. E. Hydrogen production by photo-induced reforming of biomass components and derivatives at ambient conditions. *Catalysis Letters*. 2008, CXXII, pp. 26-32. ISSN 1572-879X.

[25] MAIRA, A. J.; YEUNG, K. L.; LEE, C. Y.; YUE, P. L.; CHAN, C. K. Size Effects in Gas-Phase Photo-oxidation of Trichloroethylene Using Nanometer-Sized TiO_2 Catalysts. *Journal of Catalysis*. 2000, CXCII, Nr. 1, pp. 185-196. ISSN 0021-9517.

[26] ALMQUIST, C. B.; BISWAS, P. Role of Synthesis Method and Particle Size of Nanostructured TiO_2 on Its Photoactivity. Journal of Catalysis. 2002, CCXII, **Nr. 2, pp.** 145-156. ISSN 0021-9517.

[27] ZHANG, Z.; WANG, C.-C.; ZAKARIA, R.; YING, J. Y. Role of particle size in nanocrystalline TiO_2-based photocatalyst. *The Journal of Physical Chemistry B*. 1998, CII, Nr. 52, pp. 10871-10878. ISSN 1520-5207.

[28] WAHI, R. K.; YU, W. W.; LIU, Y.; MEJIA, M. L.; FALKNER, J. C.; NOLTE, W.; COLVIN, V. L. Photodegradation of Congo Red catalyzed by nanosized TiO_2. *Journal of Molecular Catalysis A: Chemical*. 2005, CCXLII, Nr. 1-2, pp. 48-56. ISSN 1381-1169.

RESUMÉ

Kompozit kaolinit/TiO_2 (60 hm% TiO_2) byl připraven termální hydrolýzou suspenze surového kaolinu v síranu titanylu a kalcinován při různých teplotách (600, 650 a 700°C) a po různou dobu (1, 2 a 3 h). Získaný vzorek byl charakterizován pomocí XRPD, N_2 fyzikální adsorpcí a SEM, a testován na fotokatalytickou redukci CO_2. Rozdíl kalcinačních podmínek neovlivnil složení fáze TiO_2, pouze se mírně pozměnila specifická povrchová plocha a výrazně byla ovlivněna velikost krystalitu kompozitu kaolinit/TiO_2. Vyšší teplota a delší doba kalcinace vedly k vyšší krystalitě prášku. Fotokatalytické výsledky ukázaly, že velikost krystality určuje účinnost fotokatalyzítoru kaolinit/TiO_2.

4

MINING OF COAL SEAMS UNDER THE MAIN HAULAGE LEVEL

Josef CHOVANEC[1], Petr URBAN[2], Ivo PĚGŘÍMEK[3], Vitězslav KOŠNOVSKÝ[4]

[1] *Ing. PhD. Institute of Mining Engineering and Safety, Faculty of Mining and Geology, VSB-TU Ostrava, 17. listopadu 15, Ostrava-Poruba*
e-mail:josef.chovanec@vsb.cz

[2] *Ing. Ph.D. Institute of Mining Engineering and Safety, Faculty of Mining and Geology, VSB-TU Ostrava, 17. listopadu 15, Ostrava-Poruba*
e-mail: petr.urban@vsb.cz

[3] *Ing. Czech Mining Authority, Kozi 4, Praha 1-Staré Město*
e-mail: pegrimek@cbusbs.cz

[4] *Ing. OKR, Paskov Mine, Staříč 528*
e-mail: vitezslav.kosnovsky@okd.cz

Abstract

The question of mining under the main haulage level as a way by which in some cases coal reserves can be effectively extracted is still discussed. The paper introduces advantages and disadvantages of this method. It presents examples of the bedding of a seam in a limited depth under level, for example 50 m, when the opening by the under level mining method is obviously more advantageous. Special attention is given to risks of ventilation short circuit which are higher in this method compared to classical opening. The paper also gives a brief economical reflection allowing to determine the efficiency of under level mining.

Abstrakt

Otázka podpatrového dobývání, jako způsobu, kterým lze v určitých případech efektivně vytěžit uhelné zásoby, je stále diskutována. V uvedeném příspěvku se uvádějí výhody i nevýhody této metody. Jsou uvedeny příklady uložení slojí do omezené hloubky pod patrem, například 50 m, kdy je příprava sloje podpatrovým dobýváním zřejmě výhodnější. Upozorňuje se také na rizika větrního zkratu, které je u této metody, oproti klasické otvírce, vyšší. V práci je také uvedena stručná ekonomická úvaha, podle které lze určit efektivnost podpatrového dobývání.

Key words: Gas, methane, seam, under the main haulage level, costs of coal mining.

1 INTRODUCTION

The presented work starts from findings, as published in [3], and supplement them by certain facts which were not given in [3] for various reasons, especially the possibility of scope of the article. We draw attention especially to the danger of a ventilation short circuit, which may cause serious complications when working under the main haulage level, and suggest some opportunities for its improvement. We dealt with the comparison, when working under the main haulage level is effective from an economic point of view. We give an example, when it seems preferable to use a typical way by deepening pits for opening a deposit.

2 UNDER LEVEL MINING DEPENDING ON SEAM BEDDING

Figure 1 and 2 show fairly frequent cases of mining a seam under the main haulage level on different conditions of its bedding. These are the situations of individual seams (or a seam), whose typical opening would require a relatively high costs. In the literature [1] this method is described as "occasional, random".

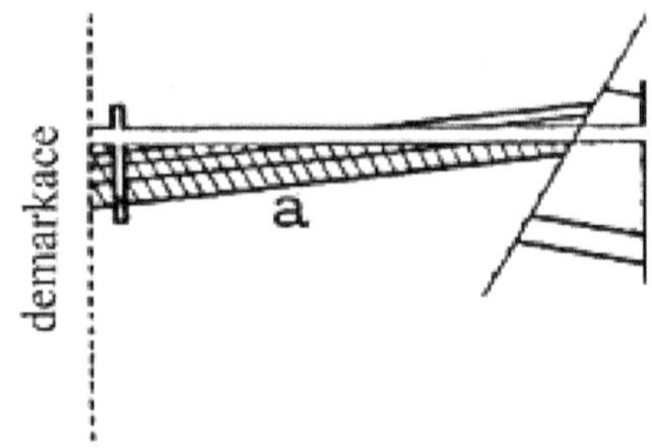

Fig. 1 Possible ways of under level mining depending on seam bedding. (Case "a")

In the case "a" as in Figure 1 it is a group of coal seams, which is directed to the boundary of the mining field. The development of these coal seams would require driving out a long cross-cut plus a staple pit from the bottom at the level of 595 m. Therefore, the under level mining method is more preferable.

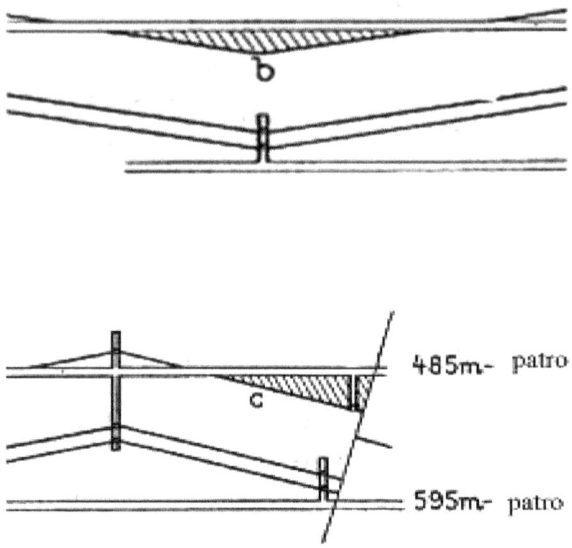

Fig. 2 Possible ways of under level mining depending on seam bedding. (Cases "b, c")

The case "b" in Figure 2 represents a flat syncline, whose greatest depth is close to the upper level at 485 m, thus it is in a long distance from the bottom level of 595 m.

The case "c" represents the throwing of the seam (thrust fault), also in a small depth below the top level at 485 meters.

In all these cases it is so-called "outstanding" (incomplete) mining under the main haulage level. It is characterized by the fact that fresh air is fed from a higher (mining) level (the level at 485 m in Figure 1 and Figure 2).

In contrast, in so-called solved (full) mining under the main haulage level the intake air is fed from the bottom level (the level at 595 m in Figure 1 and Figure 2).

The fresh intake in the cases shown in Figures 1, 2 is fed from the main mining level through a staple pit, or dip working to the lowest point, the location of the seam, and from here then it progresses up through appropriate mine workings. The exhaust air returns to the main mining level and must be diverted to the return level through the shortest path. However, in this way the risk of a ventilation short circuit occurs. This must be prevented by air doors, or other appropriate means.

3 INCOMPLETE UNDER LEVEL MINING

Figure 3 shows according to [2] a simple air diagram of the method of incomplete under level mining. Since this is primarily the representation of ventilation, it is rather simplified. For instance, there is no pass-by in the outline, which should be built for the filling point at the intake level. (This is shown in Figure 9).

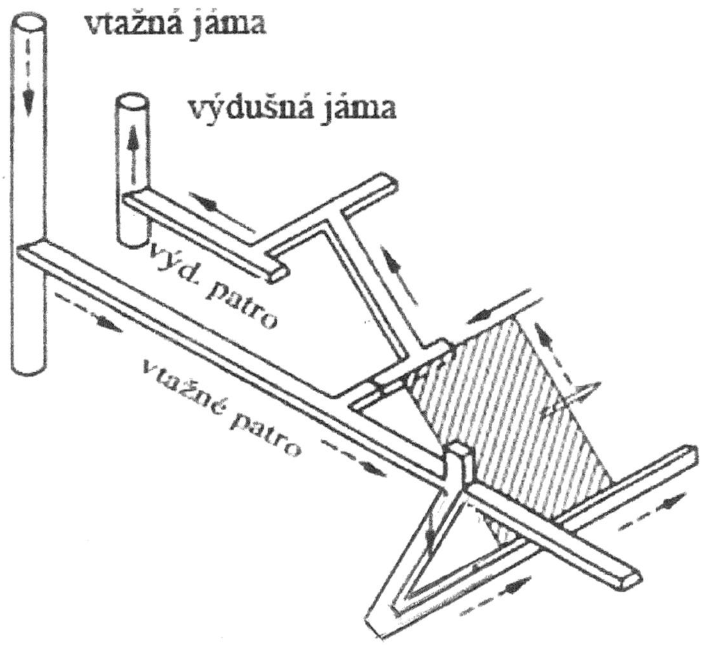

Intake shaft

Upcast shaft

Upcast level

Intake level

Fig. 3 Isometric diagram of incomplete under level mining

From Figure 3 we can deduce how the air from the intake level is drained through an incline to the bottom road under the coal face. Thereout then it proceeds across the face through the upraise to the upcast level. So as the air flows at the intake level through the incline under the face, an explosion control object must be installed behind the incline - see Fig. 4.

And just the possibility of its opening represents the risk of volume flow rate depletion for the active face. Another sensitive point is also a permanent stopping between the return airway of the face and the upcast level.

4 PLANNED UNDER LEVEL MINING

Planned mining under the main haulage level lies in that the entire coal reserves under the mining level will be mined out on this level upwardly. Usually, the interval for mining the bottom seam managed to increase and thus the development of the seam could start later. But due to certain disadvantages of under level mining and in the meantime achieved progress for mining out between seams, the planned under level mining is generally not recommended. Especially chutes and drag conveyors from the new means have proved themselves in mining out between seams.

Fig. 4 The ventilation control object in a coal mine

5 ADVANTAGES AND DISADVANTAGES OF UNDER LEVEL MINING

The major advantage of mining under the main haulage level lies in savings of costs on development work, but at least in saving of time to carry out extensive development work. Therefore, possible under level mining, as an alternative, cannot be completely excluded for instance when transiting to a deeper level. When even the development of the bottom level is in default and there are enough reserves for mining under the main haulage level at the mining level, the planned under level mining is very desirable.

On the other hand, however, the under level mining have many disadvantages that cannot be neglected. In addition to complex ventilation management, it is necessary to pay special attention to water regime. The costs of capturing and pumping water is the larger, the more water is contained in the mountain range. Also the remaining water in the abandoned mining area of the initial under level mining after completing the extraction must be taken into account.

The affecting of the main mining level by undermining by the coal seam (seams) being mined is however also very unpleasant. Especially the sites of filling stations, whose shifters and chain tracks are very sensitive to dip changes, are quite critical. Also the transport by means of mining large-volume cars is getting difficult.

6 VERTICAL AND DIP WORKINGS IN UNDER LEVEL MINING

For developing a seam (or a group of seams) either vertical or dip workings are used.

For dip workings the method of transport and excavation, which will take place on them, must always be considered. Figure 5 shows several possibilities.

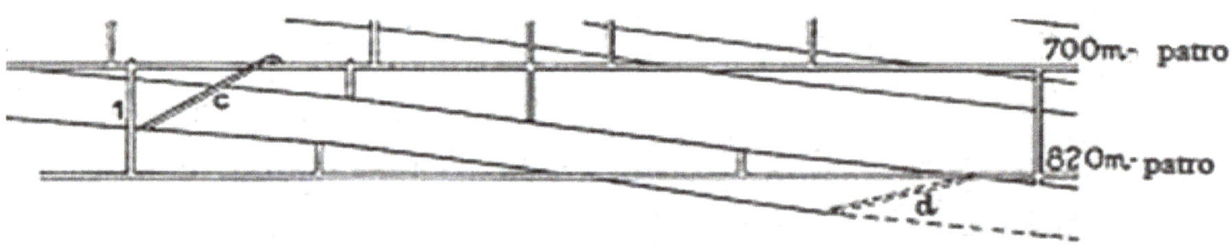

Fig. 5 Vertical and dip workings in under level mining. (The cases „c, d")

In the case "c" it concerns the development of seams under the main haulage level by a dip working (incline) with a dip of 30°. Selecting a higher inclination means shortening the working, but it also requires a more intensive means of transport for the extraction. In this case, a special conveyor should be chosen. For details, see [2].

When developing according to "d" when the incline has the dip of 13° to 15° a conveyor belt could be used.

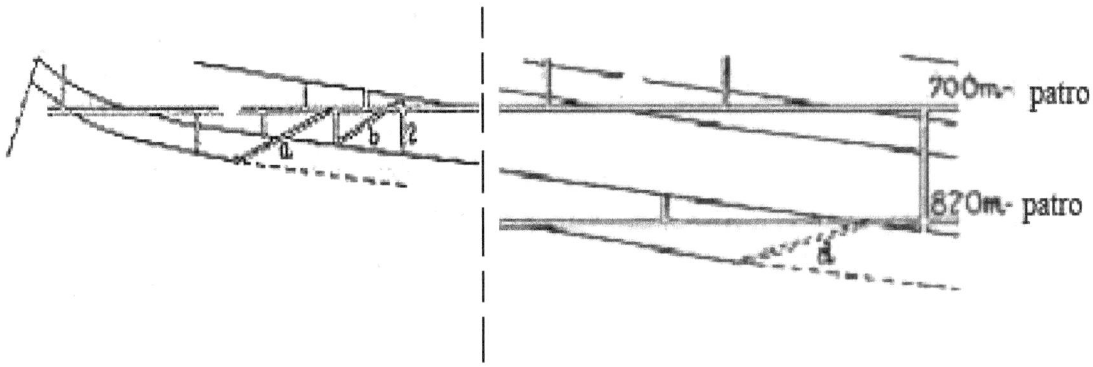

Fig. 6 Vertical and dip workings in under level mining. (Case „a, b“)

As a transport means on inclines (Fig. 6) according to "a" or "b" with a dip of 19° to 21° a plate conveyor should be used.

The inclines "b" (Fig. 6), or "c" are led out over the mining level - 700 m and leads into the bin over the filling station.

For the "complete" under level mining, fresh intake is led between blind pits from the bottom level at 820 m. The upper level at 700 m is also intake one, namely up to the ventilation gate at the cross-cut, through which the under level mining is open under this level at 700 m.

A relatively large distance of about 400 m between the filling point and a blind air pit creates a space large enough for empty cars, which come from one side, and on the other side of the blind pit a ventilation gate can be installed, where long trains can be placed just as well.

7 DANGER OF VENTILATION SHORT CIRCUIT

One of the serious risks in under level mining is a possible short circuit when opening the ventilation door. Today, most of doors, especially those through which traffic is intense, or due to which unclosing a dangerous short circuit can occur, have implemented a function of automatic opening and closing.

Most often the automatic opening function is provided by a pneumatic or hydraulic cylinder - see figure 7, which is activated by motion of a lever mechanism hung in front of the door. The lever mechanism and opening cylinders are connected by pulleys and a cable. Closing the door is controlled by signalisation that informs on closing at the control room.

Fig.7 Control of ventilation and dam doors

The risk of a ventilation short circuit often threatens on air bases of coalfaces, where the intake and upcast are separated by a wooden door. The ventilation door is still without mechanized closing. They are inclined to close themselves after release. Today, only sensors indicating opening the door are implemented. The signals are led out to the mine control room. But also this way of signalling represents a clear increase in security. The dispatcher may warn the nearest staff or technical supervision on the signal of opening by a loud intercom.

8 DOOR DESIGN ACCORDING TO FOREIGN SOURCES

Due to the importance of ventilation doors in case of a possible short circuit in ventilation network, we were also interested in possibilities occurring in the door design over the world.

According to [4] there are several companies in the USA specializing in the manufacture of doors for mining operations.

For example, the U.S. company American Mine Door, [5] offers 2 types of doors: High-and low-pressure ones. Figure 8 shows an example of the low-pressure doors design.

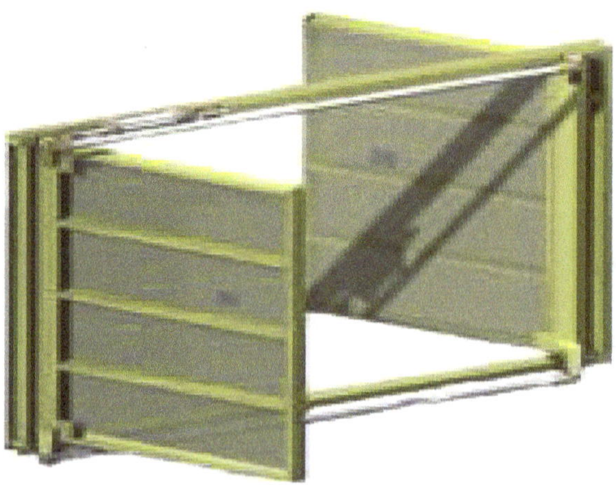

Fig. 8 The mine doors design of the company American Mine Door

The low-pressure doors, i.e. apparently a similar type that is in our mines used to control the air at less frequented workings, are made of lighter materials, narrower wings and a smaller frame. Therefore, they are lighter and cheaper.

The doors, during their opening, use opposing wings and thus cancel the effect of static pressure. Also at higher pressure in surroundings they are relatively easy to open, because the pressure against one wing is abolished by the wing on the opposite side.

The same pressure helping in opening the doors helps also in their closing. The pressure created by the ventilation system is evenly distributed. This feature is particularly preferable in building the doors in dip workings. The doors can be used several times. If they lose their importance in the original working, they can be disassembled and relocated.

The mining doors are mechanically controlled and their position is indicated by light and sound signals. They have an internal lock, which ensures the door not to open when another door is open too. They have an open-door sensor that secures any object not to be ajar; remote opening and closing, sound sensors.

9 INTERIM BALANCE TO ASSESS THE EFFECTIVENESS OF UNDER LEVEL MINING

In order to assess, whether it is economically advantageous to choose the method of under level mining or proceed to the typical development, it is possible to use an interim economic balance. As an example of its implementation we used the following procedure.

In [3] there is given a rough overview of the investment costs of opening out workings required for the development of the block for under level mining. That block is schematically depicted in Figure 9. The amount of investment costs on opening out workings required for obtaining reserves under the level - 436 m is listed in Table 1.

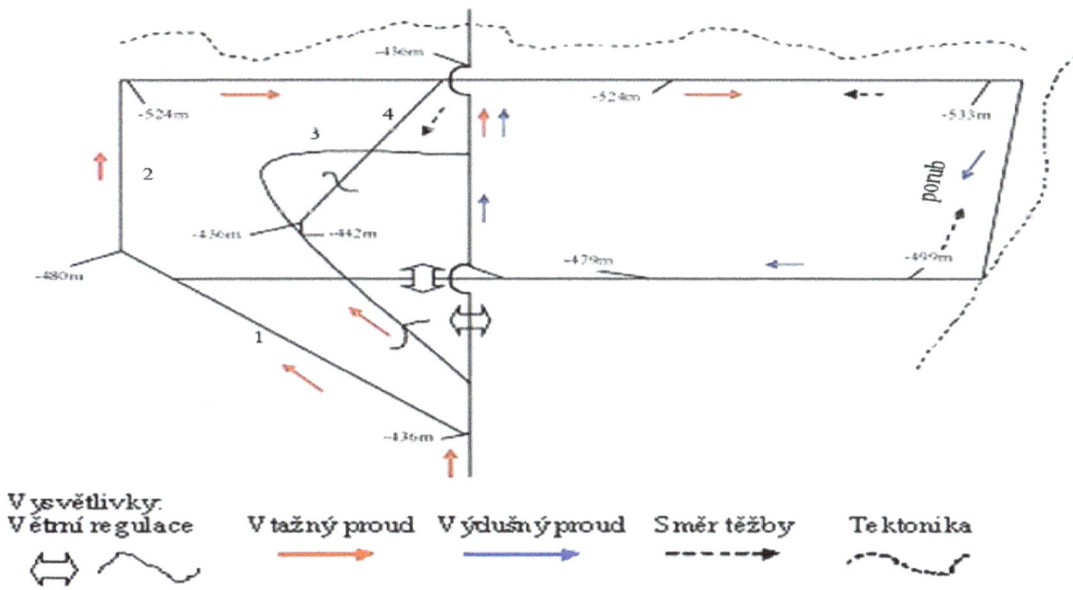

Legend:

Air control Intake air current Upcast air current Direction of mining Tectonics

Fig. 9 Schematic outline for assessing the balance of under level mining

1., 2. dip working for making the reserves under the level accessible

3. pass-by from cross-cut at the level of 436 m

4. upraise for leading the mining output into the bin

Tab. 1 Costs of developing the coal seam for under level mining.

Name of working	**Number**	**Profile**	**Length**	**Costs of erection (CZK)**
Pass-by from cross-cut at -436m	"3"	K14-P28-0,5 m 1 m̄ 81 735,-CZK	300m	24 520 500
Upcast crosscut	"4"	K 14-P28-0,5 m 1 m̄ 81,735,-CZK	110m	8 990 850
Vertical binbetween upcast cross-cut 2436 3 and pass-by at -436 m	"4"	1 m about CZK 180,000	10m	1 800 000
Incline	"1"	K12-P20-0,7 m 1 m̄ CZK 36,816,-	175m	6 442 800
Incline	"2"	K12-P20-0,7 m 1 m̄ CZK 36,816,-	90m	3 313 440
Total costs on developing the workings for under level mining				45 613 893

From the face 44,974 tons were extracted. From the bottom block under the face further 48,610 tons were extracted (i.e. 93,684 tons in total), so the work required for developing the seam for under level mining in the given area would burden a ton of coal by the sum **487,- CZK/t.**

For our own consideration on the profitability of applying the method of under level mining, we chose the solution of using the typical way for opening out and preparing a certain quantity of reserves. This means deepening pits and driving out the cross-cut at a lower level. In order the consideration to be based on the same level heights for obtaining reserves, as those showed in Fig. 9, we assume therefore deepening pits by 45 m.

If the investment costs of deepening a pit by 1 m are CZK 500,000, then the required total investment for both the pits will reach CZK 45 million. We also assumed that for making the reserves accessible up to the development stage (when only the workings in the seam will be open out), it will be necessary to open a 200m cross-cut. When the costs per 1 m of the cross-cut are CZK 85,000,-, it would be CZK 17 million in total.

Thus the development phase, in case we theoretically make the reserves accessible by deepening the pits, will require the costs of deepening the pits and opening out the workings **CZK 62 million**.

From this very simple consideration the costs of opening out by the method of under level mining (which amount to CZK 45.6 million, see Table 1) are more preferable.

However, if in opening out using the method of deepening the pits and crosscutting also the reserves at the deepened level are made available, which is very likely, then the consideration of economic efficiency would be changed. In order the way of developing by deepening the pits to achieve the same amount of costs per 1 ton of reserves, as in opening by the under level mining method, it should achieve according to the equation (1) at least

$$\frac{I_n}{T} = 487 \qquad (1)$$

I_n investment costs of deepening the pits and crosscutting (CZK)

T weight of open reserves (t),

which results in **T = 127,310 t.**

This very simplified consideration implies that if the manner of developing resting on deepening the pits opens the reserves, whose weight exceeds 127,310 tons, this way would be more preferable compared to under level opening.

The consideration, though simplified, as neglects e.g. the costs required for mining equipment (plate conveyors, etc.), may be very appropriate for an indicative decision. For the given hypothetical example a graph can be compiled that would express how the costs of opening by deepening the pits burdens the costs per 1 ton of open reserves. (Fig. 10).

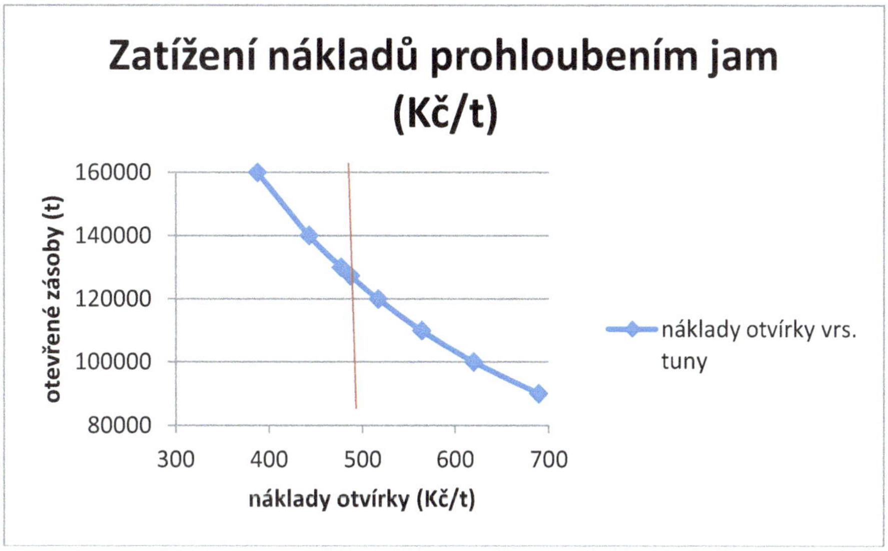

Burdening costs by deepening pits
(CZK t)
Open reserves (t)
Costs of development vs. tons
Costs of development (Czk t)

Fig. 10. The dependence of the total costs per 1 t of open reserves in the given situation according to Fig. 9 on the costs per one ton of the reserves opened by deepening pits and crosscutting

For the situation according to Figure 9 and Table 1, on the curve left from the red line there is the weight of reserves when opened by deepening the pits, the typical development would burden the costs per one ton of the open reserves less than the system of under level opening.

10 CONCLUSION

The presented article shows the possibilities of mining coal reserves by means of mining under the main haulage level. From extensive problems associated with this method of opening reserves only basic systems of development have been selected.

Furthermore, we pointed out the danger of an air short circuit, which is in under level mining specially significant, and indicated opportunities to improve its handling. The brief economic balance should indicate the procedure allowing evaluating the way of under level mining in comparison with the typical development by deepening pits.

REFERENCES

[1] FRITZSCHE C. G.: *Bergbaukunde. Springer Verlag Berlin*, Göttingen, Heidelberg 1958.

[2] WALTER BISCHOFF, HEINZ BRAMANN,:*Westfälische Berggewerkschaftskasse Bochum: Das kleine Bergbaulexikon. 7.* Auflage, Verlag Glückauf GmbH, Essen, 1988, ISBN 3-7739-0501-7.

[3] URBAN, P.: *Řešení větrání a bezpečnosti u podpatrového dobývání v oblasti 112sloje na Dole Stařič,OKR - ČR. [Solution of ventilation and safety in under level mining in the area of the 112 seam of the Stařič Mine, OKR - CR.]* Sborník mezinárodní konference *[Proceedings of the International Conference]* "Systemy wspomagania w zarzadzaniu środowiskia" Politechniky Śląske, Harrachov 2009.

[4] http:. www.google.com (mining door).

[5] info@minedoor.com.

RESUMÉ

Současné poznatky a vědomosti o podpatrovém dobývání se jeví pro uhelné hornictví ne- zcela dostačující.

Vznikají nesrovnalosti v definování podpatrového dobývání uhlí a v odborné hornické veřejnosti bývá tento termín vykládán s mírnými odlišnostmi.

Část báňských odborníků považuje za podpatrové dobývání pouze hornickou činnost dobývání pod posledním patrem z přístupněným vtažnou jámou, jiná část pak i dobývání pod jakýmkoliv patrem, pokud se porub dostane pod výškovou úroveň těžního patra.

Z porovnání výhod a nevýhod podpatrového a nadpatrového dobývání vyplývá kromě jiného zřejmá nevýhoda a nebezpečnost podpatrového dobývání.Představuje problém především v oblasti větrání a havarijních stavů, zejména protipožární prevenci.

Na tyto problematiky je z části zaměřen presentovaný článek "Podpatrové dobývání uhelných slojí".

5

MONITORING AND ANALYSIS OF SURFACE CHANGES FROM UNDERMINING

Vlastimil KAJZAR [1], *Hana DOLEŽALOVÁ* [2]

[1] *Ing., Ph.D., Institute of Geonics AS CR, v. v. i.*
Studentska 1768 Ostrava-Poruba
e-mail vlastimil.kajzar@ugn.cas.cz

[2] *Ing., Ph.D., Institute of Geonics AS CR, v. v. i.*
Drobneho 28, Brno
e-mail hana.dolezalova@ugn.cas.cz

Abstract

It is well known that the exploitation of mineral deposits negatively affects the environment. In the case of underground mining, there are usually the most serious effects which are associated with mining activities: movements and surface deformations lead to the creation of subsidence depressions.

The determination of the actual state of surface changes by conventional surveying methods is technically exacting, time-consuming and very expensive. With the development of geo-information technologies, new approaches arise and they can be applied to resolving this issue. These resources seem to be suitable not only for providing the necessary spatial information on ongoing processes, but also for their subsequent processing and comprehensive evaluation.

The presented paper deals with the possibilities of applying geo-information technologies for monitoring and evaluating the effects of underground mining on surface, summarizes the actual state of this research at ICT, and offers alternatives for their use in the future.

Abstrakt

Je známo, že exploatace ložisek nerostných surovin negativně ovlivňuje životní prostředí. V případě hlubinného dobývání jsou bezpochyby nejzávažnějšími důsledky spojenými s těžbou pohyby a deformace povrchu vedoucí k vytvoření poklesové kotliny.

Stanovení skutečného stavu povrchových změn je konvenčními měřickými metodami technicky, časově i finančně velmi náročné. S rozvojem geoinformačních technologií dnes přicházejí nové přístupy, jež mohou být aplikovány při řešení této problematiky. Tyto prostředky se zdají být vhodné nejen k zajišťování potřebných prostorových informací o probíhajících procesech, ale rovněž k jejich následnému komplexnímu zpracování a vyhodnocení.

Předkládaný příspěvek se věnuje možnostem uplatnění geoinformačních technologií pro sledování a hodnocení vlivů hlubinného dobývání na povrch. V jeho rámci jsou shrnuty poznatky zjištěné dosavadním výzkumem a nabízeny alternativy jejich využití do budoucna.

Key words: mining effects, subsidence depression, surveying and monitoring methods, GNSS, spatio-temporal data analysis

1 INTRODUCTION

The underground mining of raw material deposits leads to the formation of vacant underground spaces. The spaces are then filled with the mass of overlying strata which bend into them and break through. In the case of the extraction of a large deposit, the ongoing displacements of materials show up on the surface, typically by creating a subsidence depression (a basin). Movements and deformations occurring during its formation negatively affect both the surface and the objects associated with it. Although the theory for the assessment and prediction of the surface movements and deformations above the exploited deposits have been systematically developed for more than a century, it is not yet possible to perfectly describe the expected surface changes with certainty.

At present, the extent of the affected surface is numerically determined mainly by relatively simple prediction models based on generally accepted theories. The data obtained by seldom observations *in situ* are only used for local validation of predictive calculations. However, the real state of the surface changes is not always entirely consistent with the predictions. Therefore, there is a need to design a new universally applicable method for detecting surface changes in undermined areas to complement and or replace the existing survey methods used so far.

On an example of newly forming subsidence depressions, our effort is to demonstrate the possibilities of using modern geo-information technologies for capturing real spatio-temporal surface changes associated with the creation of subsidence depressions. It is based on a combination of direct and indirect surveying techniques, particularly the methods using global positioning navigation systems, supplemented by other geodetic and geophysical techniques and remote sensing. The collected data are then evaluated in the context of geological and geomechanical situation, advance of longwall mining and predicted surface changes.

2 AREAS OF INTEREST

The proposed procedures were subsequently applied to the selected pair of active mining areas. Both localities are situated in the Upper Silesian Coal Basin near the town of Karviná in the north-eastern part of the Czech Republic which has a long history of the hard coal deposits mining.

The rock mass consists of typical for the Upper Silesian Coal Basin upper carboniferous molasse sediments consisting mostly of coal-bearing siliciclastic continental deposits. The Upper Silesian Coal Basin is divided into tectonic blocks by a set of normal faults of tens to hundreds of meters amplitude (Dopita, 1997). The longwall method with controlled caving was used as the mining method.

The local environment can be characterized as a mining landscape with no permanent buildings. Vegetation begins to flourish on the former mine spoil deposits. The effects of undermining may cause significant changes mainly in local roads and rail tracks, infrastructure projects, or in the stability of sewage storage basin dams (Kajzar et al. 2012). The surface manifestations of mining do not include only surface movements and deformations, but also the manifestations caused by mining reclamation, i.e. creating mine spoil deposits on the surface, flooding parts of the affected surface, etc.

2.1 Louky locality

The first locality, called Louky, is situated in the mining area of the ČSM colliery. The selected northern part of this locality occupies an area of c. 6 km^2. The rock massif of this area belongs to the 0 and 1 mining blocks and was exploited in the past. Due to the significant tectonic faults in the massif, the coal seams were mined in an insular way. The first exploitation started in 1991. In the following years, there were several horizontal mining panels exploited. The new exploitation period started in 2006. In this period (until 2012), there were four horizontal coal mining panels gradually exploited at the depth of around 1000 m under the surface. The average exploited thickness of the mining panels varied from 2.0 to 3.6 m. For the detail information please see Tab. 1.

Tab. 1 Mining panels data

Mining panel	a	b	c	d
Start of exploitation	10 2006	05 2007	01 2009	07 2009
End of exploitation	06 2007	04 2008	12 2009	06 2010
Face length	180 m	190 m	160 m	180 m
Lateral length of longwall advance	480 m	880 m	600 m	670 m
Depth under the surface	945 m	995 m	960 m	1025 m
Average exploited thickness	2.0 m	3.6 m	2.4 m	2.5 m
Mining panel area	78.600 m^2	152.600 m^2	87.000 m^2	99.000 m^2

Source: OKD, a.s.

2.2 Gabriela locality

The second one, Gabriela locality, is situated in the Darkov mining area. The locality occupies an area of c. 3 km^2. The mining history goes back to the 19th century in this locality. The tectonic faults called "Gabriela", "Eliška" and "Ležatá" are main geological structures which substantially influence the stress state of the rock mass in this area. There also occur some tectonic faults of a minor importance in the vicinity of the exploited mining panel.

In the Gabriela locality, the exploitation of the last mining panel started in July 2011. The exploited mining panel is located east of former safety pillars of the Gabriela mine. This mining panel is situated beneath the previously extracted mining panels and its exploited thickness is very high as for local conditions. It is situated in the coal seam No. 40 called "Prokop" that has a total thickness of about 6-7 m. The depth of the coal seam Prokop varied from c. 790 m to 820 m. The face length of the mining panel is 200 m and the lateral length of longwall advance is up to 900 m. The average thickness of extraction is 5.5 m. More than 20 coal seams were gradually exploited above the studied mining panel in the past (Kajzar et al. 2012).

3 GNSS – MAIN SURVEYING METHOD

Observation networks of fixed points were built in the selected areas to observe the process of creation of a subsidence depression above exploited coal mining panels. Building the Louky observation network with around 100 points started in 2007. Based on the gained experience, another observation network was built in the framework of ICT in the Gabriela locality in 2011. This network consists of 36 fixed points covering the whole studied area.

GNSS, specifically the GPS system, was used to record the spatial changes in the fixed points of both observation networks not only during the actual mining, but also in the periods between each longwall mining activity (in case of the Louky locality) and during the final period after the termination of all mining activities.

GNSS is a method that can be successfully used in an undermined area. Its advantage over other common geodetic methods lies not only in providing space coordinates of a surveyed point, but also in the ability of a quick connection to the stable area not affected by undermining. This ability is very important because common geodetic methods of surveying may take time to connect the large undermined area to the non-affected stable area which means that these methods are not only time consuming, but they may also bring inaccuracy to the surveying as the points in the undermined area may move meanwhile due to undermining. The ability to provide space coordinates of points is also important. As a subsidence is the main after-effect from undermining, repeated geodetic levelling is the most common way to survey the undermined areas. But it only provides the information about vertical changes of the surface points or profiles, while the information about horizontal displacements, i.e. the direction of the surface movements is absent. But GNSS provides both the information. Our previous research of the undermined areas proved that the analysis of both vertical and horizontal movements enables more complex and precise evaluation of surface changes and their causes in the rock massif (Doležalová et al. 2009, Doležalová et al. 2010, Doležalová et al. 2011, Doležalová et al. 2012, Kajzar et al. 2011, Staš et al. 2009).

For surveying in studied localities, a static way of GNSS geodetic surveying method was chosen, with the observation for at least 10 minutes on each point with the Leica GPS system 1200. While a GNSS field receiver was surveying in the observation network, a GNSS reference station was surveying on a trigonometric point situated outside the assumed undermining effects, at a distance of several kilometres from the observation network. This point belongs to the national network.

For the used GNSS equipment, the stated accuracy for static surveying with subsequent post-processing is 0.005 m ± 0.5 ppm in the horizontal position of a point and 0.010 m ± 0.5 ppm in the vertical position of a point. Since the surveyed points were only few kilometres far from the reference point, constellation geometry was controlled during the whole observation and precise ephemeris were input into post-processing, the obtained accuracy of the spatial position of surveyed points may be estimated within an interval from 0.01 to 0.03 m (Kajzar et al. 2012).

In order to record an incremental development of undermining effects, the interval of roughly 1 or 1.5 month was chosen for repeated GNSS surveying action at first. Afterwards, the data from field surveying were processed on a computer. The spatial coordinates of the surveyed points in the WGS-84 and S-JTSK systems are the results obtained from the post-processing by the Leica GeoOffice software application.

4 DATA EVALUATION AND ANALYSIS

The data obtained by this surveying method are used as a suitable source of input information for the analysis of surface changes in the observed areas. Thanks to the re-use of this method and also thanks to the fixation of the points of the observation network in the form of lines and scattered points, it is possible to evaluate the continuous development of the subsidence and horizontal displacements of individual points as well as over the entire observed area.

The effective work with the obtained data was allowed by modern software geo-information technologies. Where necessary, their functionality was extended with newly created specialized software tools (scripts, extensions, applications) for the new procedures of processing and evaluation. The tools then allowed performing very interesting spatial analyses with the acquired data.

4.1 Analysis of point subsidence development in time

By performing this analysis, it was verified that the development of subsidence in the undermined area corresponds to the theoretical subsidence curve and to the theory of the full effective area. It was confirmed that a large part of the total subsidence and deformations of the surface caused by undermining occur during the next few months. After that, the subsidence desisting period follows characterized by a gradual reduction in the size of subsidence which may take several years. But it was also proved that the behaviour of the surface points as a reaction to the exploitation of the mining panels depends not only on the distance of the point from the mining panel, i.e. approaching the coalface, but also on the geo-tectonic conditions in the given locality (Doležalová et al. 2009, Doležalová et al. 2012).

4.2 Evaluation of line profiles

Many of the fixed points are parts of line profiles. Thanks to them, it is possible to evaluate the development of subsidence in different directions and in the context of surrounding points at the same time.

The results of the analysis of two selected profiles in the Louky locality (Fig. 1) show that while the subsidence in the profiles that are situated in parallel (the left graph) with the tectonic faults reported almost regular and smooth development throughout their length (in accordance with the theory of the creation of a subsidence depression; Neset, 1984), in the case of profiles intersecting these tectonic faults (the right graph), the subsidence development is usually quite different, irregular and very complicated. The major tectonic fault caused an irregular course of the profile subsidence where the difference in subsidence of close points above the fault was marked (Doležalová et al. 2009).

4.3 Analysis of horizontal displacements

A horizontal movement of a surface point can be defined as a change in its position in a horizontal plane. The size of the horizontal displacement from undermining is much smaller than the subsidence. However, its diagnosis effectively complements the evaluation of movements and surface deformation processes in the area of interest. Based on graphical evaluation of horizontal displacements, we are able to define the influence of a mining activity on the resulting change in position of surveyed points.

Determining the directions of point movements is a great benefit to the evaluation of the process of subsidence depression formation. We can thus detect the interface caused by a tectonic fault or interface of the various mining effects. The results of the analysis of the horizontal displacements of all points in the Louky locality confirmed the significant influence of the major tectonic fault. The horizontal movements of individual points corresponded to the expected movements in the context of position of points in the observation network towards the exploited mining panels. However, by a close-up view, it was possible to locate subareas, in which the points behave partly or totally out of the said presumption. We may conclude that the major tectonic fault makes a natural barrier (Doležalová et al. 2010, Doležalová et al. 2012).

4.4 Analysis of the time-dependence of surface changes on the progress of exploitation

The dynamics of the development of subsidence and displacements is not uniform and it is closely linked with mining activities. The change of a spatial position is always the most marked in the period of active mining. The main reaction of surface points to the changes in the rock massif and the movements of the points were different, according to their surface position, local geo-mechanical conditions etc. (Doležalová et al. 2012).

4.5 Modelling of spatio-temporal surface development

The acquired spatial data are supplemented by knowledge of mining activity, geological conditions and structural geology in the areas. Then they are processed with sophisticated mathematical statistics, geo-statistics methods and interpolation functions. In the modeling process, the values of height changes were calculated from irregularly located points into a regular network of points (the grid) using a suitable interpolation method. In the process of spatio-temporal modeling, it was appropriate to divide the evaluation period into sub-intervals, and to process them separately. The final model is then presented by a series of created grids corresponding to the individual time intervals, which represents the spatio-temporal development of surface height changes in the studied area during the evaluated period.

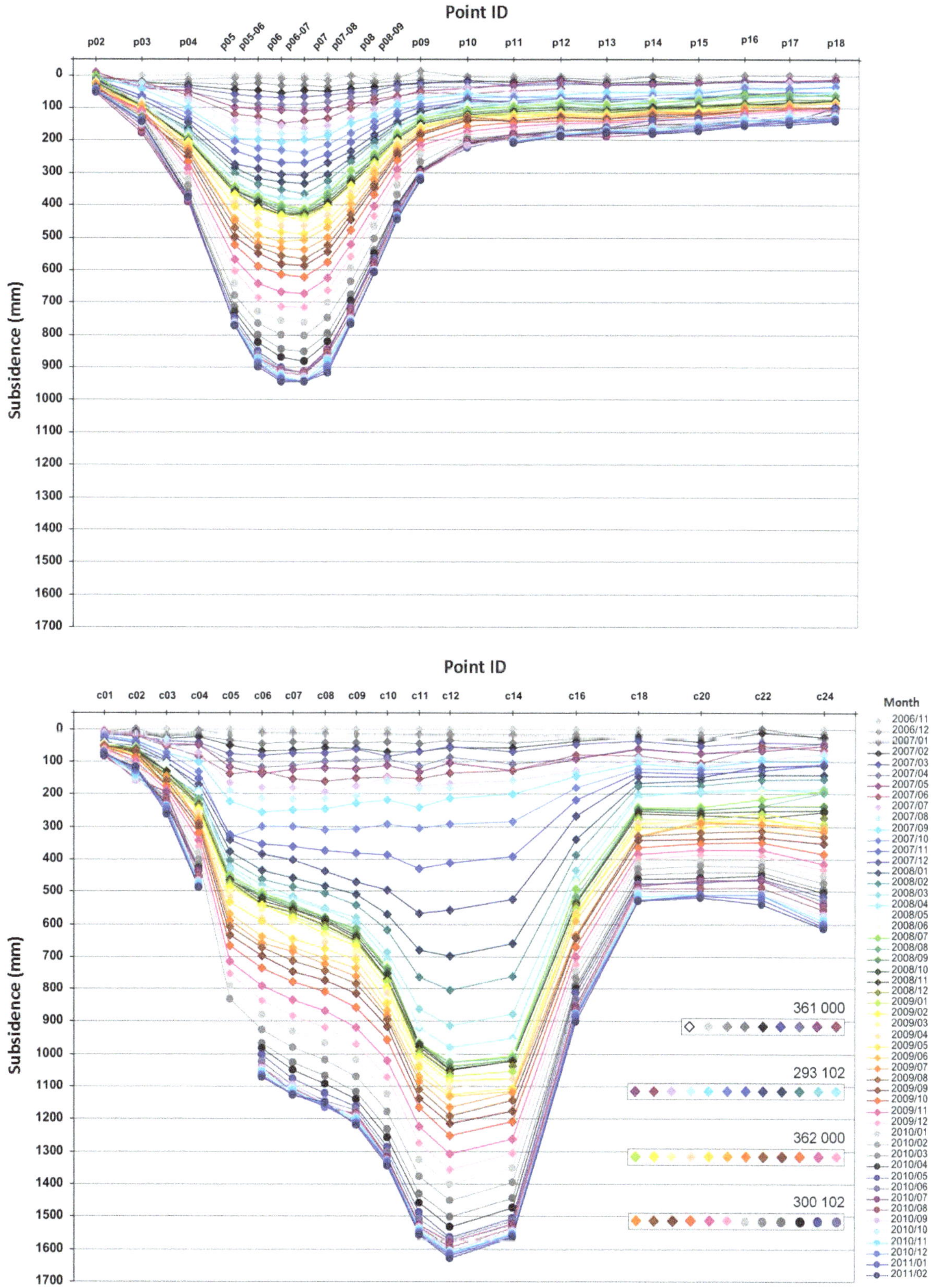

Fig. 1. Subsidence development of two selected line profiles in Louky locality. (Kajzar, 2011)

The results of comprehensive assessment are the spatio-temporal models describing the evolution of real subsidence above the exploited coal deposit which occurred in the selected area (Fig. 2). It was proved that the shape of the subsidence depression markedly depends among others on the structure, geometry of tectonic zoning and the massive failure of the overburden massif (Doležalová et al. 2010).

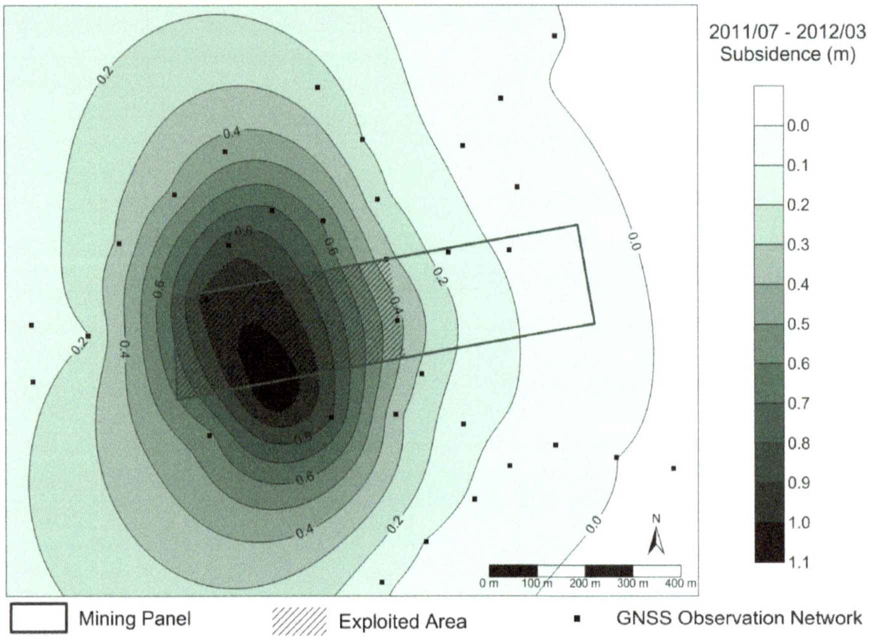

Fig. 2 Contour map of surface subsidence and mined out area of the mining panel in the Gabriela locality 8 months after the start of exploitation (Jiránková et al. 2013)

The results of processing and analysis of the obtained spatial data together with the created spatio-temporal models demonstrate the actual development of the studied subsidence depression, i.e. the landscape relief, during the observed period, which can be used to be compared with the initial predictions of the undermining influence. Based on the data, it is also possible to make a prediction of the development of the studied area in the future.

4.6 Comparison with official calculated predictions

The obtained surface changes were confronted with the results of official calculated predictions (provided by a mining company). Thanks to the created spatio-temporal models, the credibility of conventional predictive models generated by the mining company was validated, particularly in the areas with homogeneous structure of the rock massif. Noted deviations from the predictions were recorded in the areas with complicated structural-geological situation that is not considered in the calculation of the prediction models (Doležalová et al. 2009). In the future, it is therefore recommended to seek more suitable solutions of predictions for such complicated cases.

4.7 Prognosis of future subsidence development

The phase of subsidence desisting, which is characterized by a very gradual reduction in size of subsidence, may take even several years. For this phase, an outline of possible future subsidence development in the study areas by means of a regression analysis was suggested (Kajzar, 2011).

5 ADITIONAL SURVEYING AND MONITORING METHODS

To obtain additional knowledge about the deformation processes, verify the accuracy of the surveyed data, and obtain realistic ideas of the ongoing surface changes in a wider context of the studied area, methods of remote sensing were used together with the data from the geophysical, geomechanical and geodetic monitoring.

5.1 Remote sensing method - aerial photogrammetry

Aerial photogrammetry is a remote sensing method that allows determining the spatial position of studied elements with a surveying precision and its application may be suitable for monitoring the surface changes caused by underground mining and related reclamation. Repeated aerial photogrammetry at Louky locality showed the range and size of the surface changes (Fig. 3). The results from the aerial photogrammetry show that a large part of the scanned area is affected by mining. The method also enabled to observe the surface changes even in the areas of active reclamation (surface covered with mine spoils, flooded areas...) where land surveying is not possible (Kajzar et al. 2011).

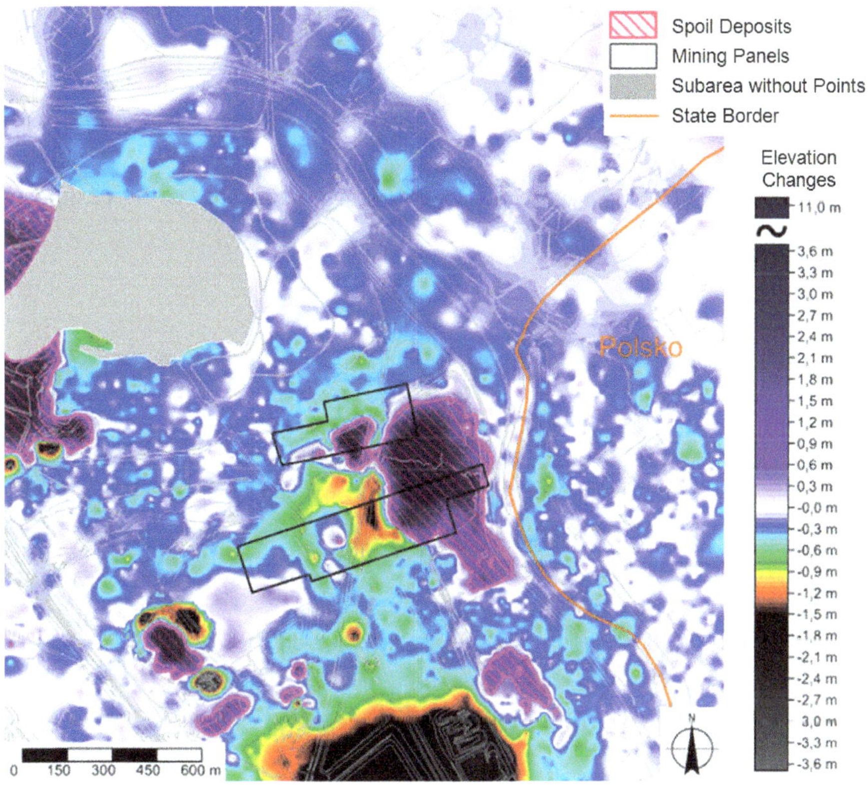

Fig. 3 Vertical changes between 2007 and 2009 generated from the aerial photogrammetry data in the Louky locality (Kajzar et al. 2012b)

5.2 Remote sensing method - radar interferometry

The method of satellite radar interferometry (Interferometry SAR – InSAR) was applied for the detection of subsidence for specific time periods. Using the differential interferometry SAR (DInSAR), it is possible to determine the extent of subsiding areas for the processed time period (up to few months). DInSAR results for consecutive time-periods display the spatio-temporal evolution of the mining induced subsidence.

In addition, the comparison of DInSAR and GNSS methods helped us to improve the values of subsidence and the delimitation of the area affected by subsidence in the studied locality (Kadlečík et al., 2012).

5.3 Geophysical methods - seismological observations

During coal extraction, induced seismic events are observed underground and or on the surface as well. This seismic activity can significantly affect the progress of subsidence in the undermined areas, as was in the case of the seismological event in the Gabriela locality (Jiránková et al., 2013).

Therefore, seismological observations, which could continuously record and determine absolute units of seismic energy or magnitude were recommended to be, simultaneously, part of the subsidence depression research. On the other hand, seismology offers instantaneous dynamic and time dependent image of seismic regime in the investigated area.

5.4 Geomechanical methods - stress changes measurements

The stress detection using the CCBM probe (Compact Conical-ended Borehole Monitoring) is based on an accurate measurement of the deformations of a borehole conical bottom using an oriented strain gauge measuring probe. Then, using laboratory tests of mechanical properties of materials from the probe installation place, the deformations can be transformed into the shape of a stress tensor (Staš et al., 2011, Sugawara and Obara, 1999).

5.5 Geodetic method - precise levelling

It is a method of height surveying which uses the level instrument to determine the elevation between two points. By comparing the available data from the precise leveling with the GNSS data, the accuracy of the GNSS surveying method was verified. This method is also helpful for surveying small height changes that happen near the margins of the subsidence depression and also in the phase of the subsidence desisting. These small changes cannot be detected by less precise surveying methods (Doležalová, 2009).

Within the ICT project, the precise digital level instrument Sokkia SDL1X was acquired for the implementation of precise leveling combined with two three-meter and one-meter invar rods. At present, this instrument is used to perform more accurate measurements on the point profiles and also to verify the stability of slopes (Doležalová and Kajzar, 2012).

6 CONCLUSION

The purpose of this research is the creation of monitoring scheme using the combination of direct and non-direct surface surveying methods with other suitable additional methods and then the application of the proposed scheme to a real mining situation in selected part of the Upper Silesian Coal Basin. To reach this, GNSS and other geodetic, geophysics and remote sensing methods are combined.

At present, periodic surveying is repeated in the Gabriela locality (8 times per year) for the purpose of detailed research of the prolonged phase of surface subsidence desisting. In the Louky locality, only annual periodic surveying is currently repeated, in order to maintain the continuity of data series. The possibility of using a 3D laser scanner for the research of the problems of undermining is within our considerations now.

AKNOWLEDGEMENT

This paper has been prepared in connection with the project ICT CZ.1.05 2.1.00 03.0082 (Institute of Clean Technologies for Mining and Utilization of Raw Materials for Energy Use) supported by the European Union.

REFERENCES

[1] DOLEŽALOVÁ, H. *Analysis of Geodetic and Geophysical Monitoring of the Undermined Area. Ph.D. Thesis*. Institute of Geonics of the AS CR, Ostrava, 2009. 127 pp.

[2] DOLEŽALOVÁ, H., KAJZAR, V. *Deformometric and levelling measurements in Dětmarovice in 2012*. Report 790 22 10. Ostrava: OKD, 2012. 61 pp.

[3] DOLEŽALOVÁ, H., KAJZAR, V., SOUČEK, K., STAŠ, L. Analysis of geodetic measurements of the undermined area. *EGRSE. Exploration Geophysics, Remote Sensing and Environment*. Vol. 18, No. 3, 2011, 27-36, ISSN 1803-1447 (in Czech).

[4] DOLEŽALOVÁ, H., KAJZAR, V., SOUČEK, K., STAŠ, L. Analysis of surface movements from undermining in time. *Acta geodynamica et geomaterialia*, Vol. 9, No. 3 (167), 2012, 389–400, ISSN 1214-9705.

[5] DOLEŽALOVÁ, H., KAJZAR, V., SOUČEK, K., STAŠ, L. Evaluation of mining subsidence using GPS data. *Acta geodynamica et geomaterialia*, Vol. 6, No. 3 (155), 2009, 359-367, ISSN 1214-9705.

[6] DOLEŽALOVÁ, H., KAJZAR, V., SOUČEK, K., STAŠ, L. Evaluation of vertical and horizontal movements in the subsidence depression near Karviná. *Acta geodynamica et geomaterialia*, Vol. 7, No. 3 (159), 2010, 355-361, ISSN 1214-9705.

[7] DOPITA, M. ET AL. *Geology of the Czech part of Upper Silesian Basin*. Praha, MŽP ČR, 1997, 278 pp., ISBN 80-7212-011-5, (in Czech).

[8] JIRÁNKOVÁ, E., STAŠ, L, KAJZAR, V., DOLEŽALOVÁ, H., SOUČEK, K. Mechanism of Rigid Overlaying Strata Failure in Face Mining in Conditions of Multiseams Carboniferous Deposits. *Acta geodynamica et geomaterialia*. 2013, ISSN 1214-9705. – in print.

[9] KADLEČÍK, P., KAJZAR, V., MAREK, T. Determination of extent of the subsiding areas in the Czech part of Upper Silesian Basin. *Proceedings of International Multidisciplinary Scientific GeoConference & EXPO SGEM. Part 2,* 2012, 715-721. ISSN 1314-2704.

[10] KAJZAR, V. *Modelling the effects of mining of mineral deposits. Ph.D. Thesis*. Institute of Geonics of the AS CR, VSB – TU Ostrava, 2011. 137 pp.

[11] KAJZAR, V., DOLEŽALOVÁ, H., SOUČEK, K., STAŠ, L. Aerial Photogrammetry observation of the subsidence depression near Karviná. *Acta geodynamica et geomaterialia.* 8, No. 3 (163), 2011, 309–317, ISSN 1214-9705.

[12] KAJZAR, V., DOLEŽALOVÁ, H., SOUČEK, K., STAŠ, L. Gabriela locality: Starting geodetic observations to detect the surface manifestations from undermining. *Acta geodynamica et geomaterialia.* Vol. 9, No. 3 (167), 2012, 401-407, ISSN 1214-9705.

[13] KAJZAR, V., DOLEŽALOVÁ, H., SOUČEK, K., STAŠ, L. Using aerial photogrammetry to observe surface undermining processes. *EGRSE. Exploration Geophysics, Remote Sensing and Environment.* Vol. 19, No. 2, 2012. ISSN 1803-1447.

[14] NESET, K. 1984, *Effects of undermining*. SNTL Praha 1984, 344 pp. (in Czech). OKD, a.s.: Internal memo, 2010.

[15] STAŠ, L., KAJZAR, V., DOLEŽALOVÁ, H., SOUČEK, K., SMETANOVÁ, L., GEORGIOVSKÁ, L. GPS monitoring of subsidence depression progress on undermined area. *Nowoczesne metody eksploatacji węgla i skał zwięzłych.* Kraków: AGH, 2009, 99-106, ISBN 83-915742-10.

[16] STAŠ, L., KNEJZLÍK, J., PALLA, L., SOUČEK, K., WACLAWIK, P. Measurement of stress changes using compact conical-ended borehole monitoring. *Geotechnical Testing Journal*, 2011, Vol. 34, Issue 6, 685-693, ISSN 0149-6115.

[17] SUGAWARA, K. AND OBARA, Y. Draft ISRM suggested method for in-situ stress measurement using the compact conical – ended borehole overcoring (CCBO) technique. *International Journal of Rock Mechanics and Mining Sciences*, 1999, No. 1, 905-912, ISSN 1365-1609.

RESUMÉ

Při hlubinném dobývání ložisek nerostných surovin dochází k vytváření volných podzemních prostor. Tyto prostory jsou následně vyplňovány masou nadložních vrstev, které se do nich prohýbají a prolamují. Probíhající přesuny materiálu se v případě vydobytí rozsáhlé plochy projevují až na povrchu, obvykle vytvořením kotliny poklesového charakteru. Pohyby a deformace, ke kterým při její tvorbě dochází, negativně působí jak na povrch, tak na objekty s ním spojené. Ačkoliv jsou teorie odpovídající potřebám hodnocení a prognózování pohybů a deformací povrchu nad dobývanými ložisky systematicky rozpracovány již déle než jedno století, stále není možné očekávané změny s jistotou dokonale popsat.

Snahou představeného výzkumu je na příkladu nově vznikajících poklesových kotlin demonstrovat možnosti využití moderních geoinformačních technologií pro zachycení skutečných časoprostorových změn povrchu spojených s tvorbou poklesové kotliny. To vychází z kombinace přímých i nepřímých měřických technik, zvláště metod využívajících globálních polohových navigačních systémů, doplněných o další geodetické, geofyzikální a geomechanické techniky a techniky dálkového průzkumu Země. Zjištěné údaje jsou následně vyhodnocovány v kontextu s geologickou a geomechanickou situací., s postupem těžebních prací a s očekávanými změnami povrchu.

Navržené postupy byly postupně aplikovány na dvojici vybraných těžebně aktivních oblastí. Obě lokality se nacházejí v Hornoslezské uhelné pánvi u Karviné v severovýchodní části České republiky. První lokalita, Louky, se nachází v dobývacím prostoru Dolu ČSM. Vybraná severní část této lokality se rozkládá na ploše cca. 6 km^2. Poslední období těžby zde začalo v roce 2006. V tomto období byly postupně do roku 2012 vydobyty čtyři horizontální poruby. Druhá lokalita, Gabriela, se nachází v dobývacím prostoru Dolu Darkov. Lokalita se rozkládá na ploše cca. 3 km^2. V lokalitě Gabriela bylo v červenci 2011 zahájeno dobývání posledního porubu. Ten se nachází ve sloji č. 40 s názvem "Prokop", která má celkovou mocnost asi 6 – 7 m. Pro sledování procesu vytváření poklesové kotliny nad exploatovanými uhelnými ložisky byly ve vybraných lokalitách vybudovány pozorovací stanice.

Pro zaznamenání prostorové změny stabilizovaných bodů obou pozorovacích stanic nejen během samotné těžby, ale také v obdobích mezi těžbou jednotlivých porubů (v případě lokality Louky) a během konečného období po ukončení všech těžebních aktivit, bylo využito GNSS, konkrétně systému GPS. GNSS je metoda, která může být na poddolovaném území úspěšně používána. Její výhoda oproti jiným běžným geodetickým metodám spočívá nejen v poskytování prostorových souřadnic zaměřovaných bodů, ale také ve schopnosti rychle připojit poddolované území na stabilní oblast, která není ovlivněna poddolováním. Náš výzkum na sledovaných poddolovaných oblastech ukázal, že analýza vertikální i horizontálních pohybů umožňuje komplexnější a přesnější vyhodnocení povrchových změn a jejich příčin v horninovém masivu.

Data získaná metodou GNSS slouží jako vhodný zdroj vstupních informací pro analýzu povrchových změn ve sledovaných oblastech. Díky opakovanému použití této metody a také díky stabilizaci bodů pozorovací stanice ve formě profilů a roztroušených bodů je možné vyhodnotit průběžný vývoj poklesů a horizontálních posunů jednotlivých bodů, stejně jako celé sledované oblasti.

Efektivní práci se získanými údaji umožnily moderní programové prostředky geoinformatiky doplněné o účelově vytvořené programové nástroje (skripty, nádstavby, aplikace). Tyto nástroje pak umožnily s pořízenými daty provádět velmi zajímavé prostorové analýzy - analýza časového vývoje poklesu bodů, hodnocení liniových profilů, analýza horizontálních posunů, analýza časového vývoje poklesů a posunů bodů v závislosti na postupu porubní fronty, modelování časoprostorového vývoje povrchu, srovnání predikce poklesů se skutečným stavem, prognóza budoucího vývoje poklesu bodů.

Pro zajištění doplňujících poznatků o přetvárných procesech, ověření přesnosti naměřených dat, získání reálnější představy o probíhajících povrchových změnách celé oblasti v širším kontextu studované oblasti bylo využito metod dálkového průzkumu Země – letecké fotogrammetrie a radarové interferometrie a rovněž údajů z geofyzikálního, geomechanického a geodetického monitoringu.

V současné době jsou prováděna pravidelná měření na lokalitě Gabriela (8x ročně) za účelem detailního postihnutí dlouhotrvající fáze doznívání poklesů. Na lokalitě Louky se provádí v současnosti pouze roční periodické měření, za účelem zachování kontinuity datové řady. V rámci úvah zůstává možnost využití laserového 3D skeneru pro řešení zvolené problematiky.

6

CHANGES IN MOLLUSC SPECIES COMPOSITION IN THE MINING LANDSCAPE ON THE EXAMPLE OF THE EXTINCT POND SYSTEM TERRITORY

Jiří KUPKA [1], Hana ŠVEHLÁKOVÁ [2], Kamila KAŠOVSKÁ [3]

[1] *Ing. Ph.D., Institute of Environmental Engineering , Faculty of Mining and Geology, VŠB –Technical University of Ostrava, 17. listopadu 15, Ostrava*
e-mail jiri.kupka@vsb.cz

[2] *Ing., Institute of Environmental Engineering, Faculty of Mining and Geology, VŠB –Technical University of Ostrava, 17. listopadu 15, Ostrava*
e-mail hana.svehlakova@vsb.cz

[3] *Ing. Ph.D., Dean´s Office, Faculty of Mechanical Engineering, VŠB –Technical University of Ostrava, 17. listopadu 15, Ostrava*
e-mail kamila.kasovska@vsb.cz

Abstract

The development of the industry in last two centuries, and especially intensive coal mining activities definitely changed the local landscape character. It led to the destruction of indigenous biotopes with natural flora and fauna communities. Many ecological studies use a group of molluscs as indicators of changes in the landscape. The advantage of this group is a relatively small number of taxa, relatively simple determination and very good knowledge of ecological requirements and distribution of individual species. The territory of Loucké ponds represents an interesting area in the industrial landscape of Karvinsko, where based on the above aspects of malacocenosis, it is possible to use and compare historical data of their incidence as well as to compare the data with current research. The pond system affected by the declension has an important function in the landscape today. The first malacological researches in the Loucké ponds territory were conducted in 1954 and 1955 (Brabenec, 1954, Ložek, 1964). Detailed survey was conducted by Macha (1978, 1979) who found 29 aquatic species (12 Bivalvia and 17 Gastropoda) and 19 land species of molluscs. This research was followed by the authors' survey in 2006, 2007 and 2013 which confirmed the presence of 48 molluscs species altogether (42 Gastropoda and 6 Bivalvia), from which 21 species were aquatic molluscs and 27 species were land molluscs.

Abstrakt

Vývoj průmyslu v posledních dvou staletích, především pak intenzivní těžba uhlí, definitivně změnil místní krajinný systém. To vedlo k destrukci původních biotopů s přirozenými společenstvy rostlin a živočichů. V mnoha ekologických studiích se jako indikátory změn v krajině používá skupina měkkýšů. Výhodou této skupiny je poměrně nízký počet taxonů, poměrně jednoduchá determinacene a velmi dobré znalosti ekologických nároků a rozšíření jednotlivých druhů. Území Louckých rybníků představuje zajímavou oblast v průmyslové krajině Karvinska, kde lze uvedených vlastností malakofauny využít a také srovnat historická data o jejich výskytu a srovnat je se současnými průzkumy. První malakologické výzkumy na Louckých rybnících byly provedeny v roce 1954 a 1955 (Brabenec, 1954, Ložek, 1964). Podrobný průzkum provedl Mácha (1978, 1979), který nalezl 29 druhů vodních (12 druhů mlžů a 17 druhů plžů) a 19 druhů suchozemských měkkýšů. Na tento výzkum, navazuje práce autorů v letech 2006, 2007 a 2013, která potvrdila přítomnost 48 druhů měkkýšů (42 Gastropoda, 6 Bivalvia), z toho 21 vodních druhů a 27 suchozemských druhů.

Key words: molluscs, vegetation, bioindication, mining landscape

1 INTRODUCTION

Karvinsko is one of the regions with the most disturbed environment in the Czech Republic. The development of the industry in last two centuries and especially intensive mining activities accompanied by the rise of anthropogenic forms of the relief definitely changed the local landscape character. It led to the destruction of indigenous biotopes.

Changes in these outstanding site conditions lead regularly to the species structure changes. There is a vegetation (the appearance of phytocoenosis, the dominance and the abundance of plant species), traditionally used for the ecosystem characterization and their relevant changes. However, also the animal components are very important in an ecosystem. Many animals, thanks to their mobility or close ecological valence, react to the human erosion of the site conditions much faster than the vegetation.

Molluscs are applied in many ecological studies evaluating the environment quality and thus they can be used for evaluating changes in site conditions of selected landscape parts. Molluscs are an animal group explored in detail, with its advantage of a quite low number of taxa, relatively simple determination and especially very good knowledge of ecological requirements of the individual species and their distribution. Their sensitivity to environmental changes, low vagility and the presence of receptacle on their bodies make molluscs a very interesting group of bioindicators. The indication of biotope erosion can be very often performed just only on the basis of the presence and the absence of important bioindicator species.

2 HABITAT CONDITIONS OF TERRITORY

Geomorphologically the territory belongs to a geomorphological district Ostravská niva, which consists of quarternary fluvial sediments. The southern part of the territory belongs to Hornožukovská pahorkatina with flysh sandstones and claystones, teschenites and quarternary sediments. A consequence of deep mining is the deformation of the earth's surface (Demek et al, 1987).

Geologically the territory belongs to the Bohemian massif of the carboniferous age, which consists of bedrock of the whole Karviná district. There are significant quarternary sediments with a bad depth of up to 20 m (Weissmannová, 2004).

Soils are represented by fluvisols, pseudogleys, luvisols, and antroposols. There is the disruption of the natural soil mantle in many places caused by mining operations (Weissmannová, 2004).

Climate of the territory belongs to the climatic region MT 10 with slightly warm, dry winter with frequent inversions and long, slightly warm, dry summer (Quitt, 1971).

Hydrologically the territory belongs to the Olza partial basin and the Mlýnka stream flows through. There are ponds supplied by the Mlýnka stream as important hydrological and ecological elements (Kupka et Kašovská, 2011).

Vegetation of the territory belongs to the Ostrava basin area phytogeographical district with monotonous flora and prevailing oak – a beech vegetation zone (Skalický, 1988). The potential natural vegetation includes especially waterlogged oak – beech forests (association *Carici brizoides – Quercetum)* that are maintained insularly in the west Silesia only. There are alluvial bird – cherry – ash forests (association *Pruno Fraxinetum*) and wetland alder woods (alliance *Alnion glutinoseae)* in river floodplains. Hhardwood floodplain forests are represented by elm – oak forests (association *Querco – Ulmetum)* (Neuhäuslová, 1998). There are saved lime - oak – hornbeam forests (association *Tilio – Carpinetum*) on slopes and also herb rich beech forests (suballiance *Eu – Fagenion*) and acidophilous beech forests (alliance *Luzulo – Fagion*).

The construction of ponds in 17th century led to an increase in diversity of habitats and plant species. New plant communities as reed swamps or tall sedges and communities of natural eutrophic and mesotrophic lentic water appeared in the landscape. Kantorek, Simanov (1982) show a rare communities occurrence as vegetation eutrophic and mesotrophic lentic water with *Utricularia sp.* (alliance *Utricularion vulgaris)* and vegetation of a water table with *Salvinia sp.* and *Spirodela sp.* (alliance *Salvinio – Spirodeletum*). These communities are not here today any more. Recent phytocoenosis of the territory are represented by reed beds and tall sedges of alliances *Phragmition australis, Phalaridion arundinaceae, Magnocaricion elatae*, and the vegetation of annual nitrophilous hygrophytes (alliance *Bidention tripartita).* Plant communities of mesotrophic and eutrophic water are represented by the alliance *Lemnion minoris*. Lentic water especially in shallow depressions is colonized by species from the alliance *Ranunculion aquatilis* or the alliance *Oenanthion aquaticeae.* We can find also wetland willow woods (alliance *Salicion cinereae*) and wetland alder woods (alliance *Alnion glutinoseae*).

Zoogeographically the territory belongs to a Polish province characterised by not very rich fauna and consequently geological and geomorphological conditions as well (Culek, 1995). However, aquatic animals are abundantly represented, which is used in this study.

In 1979, a research was conducted (Mácha, 1982) which proved there the occurrence of 48 species of molluscs (19 land species, 29 aquatic species). There were very prominent taxa between them, e.g. *Viviparus contectus*, *Aplexa hypnorum*, *Musculinum lacustre* and 9 species of genus *Pisidium*. This is why the Loucké ponds represent an interesting area in the Karvinsko industrial landscape where is possible to do the comparison with present conditions and to evaluate mining activity consequences on malacocenosis. The ponds system affected by the declension has its important function in the landscape today.

History of land use and fundamental habitats: There was the biggest development of fish-pond cultivation in Těšínsko in 16th century. At that time, the Loucké ponds were established in the alluvial plain of the Olza River. In spite of the industry development as well as the urbanization (e.g. in 1781 a railway was built near the pond), this area saved the pure nature character. Because of that, the Loucké ponds were proclaimed the State Nature Reserve (SPR) (in the land register of the Louky nad Olší community, on the area of 32, 98 ha in 1970). The reserve consisted of a flood-plain forest and the system of 15 ponds in the alluvial plain of the Olza River. As a result of the area devastation (caused by mining activities), the reserve was vitiated in 1987.

2 MATERIAL AND METHODS

Watter molluscs were acquired by metal sieve (washing submerged vegetation and sediment). When emerging, the found species including the empty receptacles were (if necessary) taken away and determined under the binocular magnifying glass. Collecting the aquatic molluscs was completed by collecting land mollusc species. The land molluscs were collected by hand.

Ecological classification: 1: strictly forest species (SI), 2: predominantly forest species (SIMS – mesophilic, Sith – shrubwood) , 3: species of alluvial and wetland forests (Sih), 5: silvifobic species, species of open areas (PT – transitional), 7: mesohygrophilous and mostly euryecious species (MS - mesicolae), 8: species with high moisture demands, but not confined to wetlands (HG - hydricolae), 9: species with high moisture demands which confined to wetlands (PD - paludicolae), 10: aquatic species (PD, PDt – periodic wetlands, RV – rivicolae, SG stagnicolae, SG(RV), SG-PD, SG-PDt, SG-RV, SGRV- transitional biotopes).

Exposure to danger: NE – Not Evaluated, NT – Near Threatened, LC – Least Concern, VU – Vulnerable, EN – Endangered.

Categories of the relative strength populations: VO – very sporadic occurrence, O – sporadic occurrence, R – scattered occurrence, H – plentiful occurrence, VH – very plentiful occurrence. Results of collecting in 1954 – 1955 and 1979: ¯ (presence), ~ (absence).

3 RESULTS

In the territory of the previous SPR of the Loucké ponds, during the research period (June and October 2006, March 2007 and May 2013), 48 molluscs species altogether were found (42 Gastropoda and 6 Bivalvia), from which 21 species were aquatic molluscs and 27 species were land molluscs. The summary of all the founded species, including the ecological division and exposure to danger is stated in Tab. 1.

Tab.1: List of species found in the Loucké ponds, the last natural reserve, their ecological classification (according to Ložek 1964 and Lisický 1991, adapted), conservation status (Beran et al., 2005), and relative abundance.

Ecological classification		Species	Exposure to danger	1954	1955	1979	2006	2007	2013
1.	SI	*Daudebardia brevipes* (Draparnaud, 1805)	EN	~	~	-	~	~	~
		Daudebardia rufa (Draparnaud, 1805)	NT	~	~	~	~	~	O
2.	SI (MS)	*Fruticicola fruticum* (O. F. Müller, 1774)	LC	~	~	~	H	H	H
		Arion subfuscus (Draparnaud, 1805)	LC	~	~	-	O	O	O
		Alinda biplicata (Montagu, 1803)	LC	~	~	-	H	H	R
		Monachoides incarnatus (O. F. Müller, 1774)	LC	~	-	-	VH	VH	VH
	Sith	*Helix pomatia* (Linnaeus, 1758)	LC	~	~	-	H	H	H
3.	Sih	*Deroceras precox* (Wiktor, 1966)	NT	~	~	~	~	O	O
5.	PT	*Vallonia pulchella* (O. F. Müller, 1774)	LC	~	~	~	O	O	O
7.	MS	*Euconulus fulvus* (O. F. Müller, 1774)	LC	~	~	-	~	~	~
		Boetgerilla pallens (Simroth, 1912)	LC	~	~	-	~	~	O
		Deroceras reticulatum (O. F. Müller, 1774)	LC	~	~	-	~	~	R
		Plicuteria lubomirskii (Slósarskii, 1881)	LC	~	~	-	O	O	~
		Trochulus hispidus (Linnaeus, 1758)	LC	~	~	~	VH	VH	VH
		Oxychillus celarius (O. F. Müller, 1774)	LC	~	~	-	~	O	R
		Oxychilus draparnaudi (Beck, 1837)	LC	~	~	~	~	~	O
		Vitrina pelucida (O. F. Müller, 1774)	LC	~	-	~	~	O	O
		Cochlicopa lubrica (O. F. Müller, 1774)	LC	~	-	-	H	H	R
		Arion distinctus (Mabille, 1868)	LC	~	~	-	~	R	R
		Arion vulgaris (Moquin-Tandon, 1855)	LC	~	~	~	~	~	H
		Perpolita hammonis (Ström, 1765)	LC	~	-	-	~	O	R
		Punctum pygmaeum (Draparnaud, 1801)	LC	~	~	~	~	O	R
8.	HG	*Succinella oblonga* (Draparnaud, 1801)	LC	~	-	~	~	~	~
		Deroceras laeve (O. F. Müller, 1774)	LC	~	~	-	~	~	R
		Perpolita petronella (L. Pfeiffer, 1853)	EN	~	~	-	~	~	~
		Carychium tridentatum (Risso, 1826)	LC	~	~	~	O	O	R
Ecological classification		Species	Exposure to danger	1954	1955	1979	2006	2007	2013
9.	PD	*Carychium minimum* (O. F. Müller, 1774)	LC	~	-	-	~	R	H
		Euconulus praticola (Reinhardt, 1883)	VU	~	~	~	~	R	R
		Oxyloma elegans (Risso, 1826)	NT	~	-	-	VH	H	H
		Zonitoides nitidus (O. F. Müller, 1774)	LC	~	-	-	VH	VH	VH
		Succinea putris (Linnaeus, 1758)	LC	-	-	-	H	H	VH
10.		*Radix ampla* (Hartmann, 1821)	NT	~	~	-	~	~	~
	PD	*Segmentina nitida* (O. F. Müller, 1774)	VU	-	-	-	~	~	~
		Musculinum lacustre (O. F. Müller, 1774)	NT	-	-	-	O	O	O
	PDt	*Aplexa hypnorum* (Linnaeus, 1758)	NT	~	-	-	R	H	R
		Anisus leucostoma (Millet, 1813)	LC	-	-	-	O	O	O
		Pisidium obtusale (Lamarck, 1818)	NT	~	-	-	~	O	~
	RV	*Pisidium amnicum* (O. F. Müller, 1774)	EN	~	-	-	~	~	~
	RV(SG)	*Anodonta anatina* (Linnaeus, 1758)	LC	~	~	~	VH	VH	VH
		Anodonta cygnea (Linnaeus, 1758)	VU	~	~	-	~	~	~
		Pisidium henslowanum (Sheppard, 1823)	LC	-	-	-	~	~	~
	RV-PDt	*Pisidium nitidum* (Jenyns, 1832)	LC	~	~	-	~	~	~
		Pisidium casertanum (Poli, 1791)	LC	~	~	-	~	~	R
	SG	*Pisidium personatum* (Malm, 1855)	LC	~	~	-	~	~	R
		Acroloxus lacustris (Linnaeus, 1758)	LC	-	-	-	~	~	O
		Ferrissia clessiniana (Jickeli, 1882)	NE	~	~	~	~	~	O
		Stagnicola corvus (Gmelin, 1791)	LC	-	~	-	O	O	O
		Radix auricularia (Linnaeus, 1758)	LC	-	~	-	O	O	O
		Lymnea stagnalis (Linnaeus, 1758)	LC	-	-	-	VH	VH	H
		Physa fontinalis (Linnaeus, 1758)	NT	~	-	-	~	~	~
		Gyraulus albus (O. F. Müller, 1774)	LC	-	-	-	O	O	R
		Gyraulus crista (Linnaeus, 1758)	LC	~	-	-	~	O	O
		Hippeutis complanatus (Linnaeus, 1758	LC	-	-	-	R	R	R
		Planorbarius corneus (Linnaeus, 1758)	LC	-	-	-	VH	VH	VH
		Pisidium hibernicum (Westerlund, 1894)	EN	-	-	-	~	~	~
	SG(RV)	*Sphaerium corneum* (Linnaeus, 1758)	LC	~	-	-	O	O	VO
	SG-PD	*Viviparus contectus* (Millet, 1813)	NT	~	-	-	~	~	~
		Anisus vortex (Linnaeus, 1758)	LC	-	-	-	VH	H	H
	SG-PD(-t)	*Galba truncatula* (O. F. Müller, 1774)	LC	-	~	-	~	~	R
	SG-RV	*Pisidium milium* (Held, 1836)	VU	~	~	-	~	~	~
		Pisidium substruncatum (Malm, 1855)	LC	-	-	-	H	H	R
	SGRV	*Radix peregra* (O. F. Müller, 1774)	LC	-	~	-	H	H	H
		Physella acuta (Draparnaud, 1805)	NE	~	~	~	VH	VH	VH

From the species of aquatic molluscs, common representatives such as *Anisus vortex*, *Gyraulus albus*, *Hippeutis complanatus*, *Lymnaea stagnalis*, *Musculinum lacustre*, *Planorbis planorbis* were mostly found. According to the exposure to danger of particular species (Beran et al., 2005), the occurrence of only 3 aquatic molluscs from the category Near Threatened: *Aplexa hypnorum*, *Pisidium obtusale* and *Musculinum lacustre* was found.

In terms of the use of molluscs as indicators of change in the mining landscape, we can consider the alliances *Phragmition australis*, *Phalaridion arundinaceae*, *Lemnion minoris*, *Ranunculion aquatilis*, *Oenanthion aquaticeae*, possibly *Bidention tripartita* as fundamental habitats. Phytocoenosis of these habitats consist of submersed and emersed forms of plants with a frequent occurrence of the observed mollusc species. From this perspective, among the most important plants especially Common cattail (*Typha latifolia*), Canebrake (*Phragmites communis*), Reed canary grass (*Phallaris arundinacea*), Canadian pondweed (*Elodea canadensis*), Flowering rush (*Butomus umbellatus*), and Arrowhead (*Sagittaria sagittifolia*) can be included.

4 CONCLUSIONS

Biotopes vitiated by the previous State Nature Reservation (SPR) Loucké ponds are mostly influenced by an extensive declension. This proved, among others, also in the malacofauna composition, especially in a decrease of sensitive species.

The first malacological research in the Loucké ponds territory was conducted by Brabenec (1954) and Ložek in 1955 (Ložek, 1964). At that time, there was found 32 species of molluscs, 9 of them were land species and 23 were aquatic species. This research was focused on aquatic species of molluscs. Later in 1978-1979, Mácha dealt with the locality. This research described the situation of the Loucké ponds malacofauna before its cancellation and destruction at 17 localities, 13 of which were various character stretches of water (watercourse and stationary ditchwater). This research was unambiguously focused on aquatic species of molluscs. Mácha found the occurrence of 19 land species and 27 aquatic species (12 Bivalvia and 15 Gastropoda). We can expect that the difference between the previous researches and collections and Mácha research (24 years later) is caused by a more detailed research performed mainly due to selecting a wider territory.

The present malacozoological research proved an expressive decrease of all mollusc species. Except some usual species, the recent research has not proved the occurrence of this sensitive molluscs species yet: *Daudebardia brevipes, Perpolita petronela, Viviparus contectus, Physa fontinalis, P. henslowanum, P. hibernicum, P. amnicum, P. milium* and *P. nitidum*. These species are mostly aquatic species, which are quite exacting on water cleanness and which are very sensitive to negative interferences to biotopes. All previously found species of molluscs from the category endangered (4 species) probably died out there (*Daudebardia brevipes, Perpolita petronella, Pisidium amnicum, Pisidium hibernicum*). On the contrary, following from the results that in the investigated territory, the species took root very well, which live in human-influenced and changed areas, and species with a wide ecological valence, such as *Helix pomatia* or *Fruticicola fruticum*, and the introduced species (*Physella acuta* and *Ferrissia clessiniana*). From rare species of molluscs, there lives a quite strong population of *Alexa hypnorum* in the territory which inhabits exclusively periodic pools and wet grounds, and *Pisidium obtusale*, inhabiting minute ditchwaters. On the contrary, there were new species, imported.

In conclusion we can say that mollusc fauna in the Loucké ponds is the consequence of the destruction of previous site conditions considerably impoverished. Despite of all the extreme load of environment, there lives an interesting and also rich guild of molluscs (42 Gastropoda and 6 Bivalvia) in the area.

The research results suggest the influence of changes in ecosystems caused by mining activities on molluscs, along with other current research that will be used in the characterization of the mollusc mining landscape. A comparison of the results, especially with the results by Macha will offer us interesting information about the development and changes in the character of the landscape in the last 50 years.

REFERENCES

[1] BERAN L., JUŘIČKOVÁ L. & HORSÁK M., 2005: Mollusca (Měkkýši). In *Červený seznam ohrožených druhů České republiky. Bezobratlí.* FARKAČ J., KRÁL D. & ŠKORPÍK M. (eds), AOPK ČR, pp. 67–69.

[2] BRABENEC, J. Malakozoologický výzkum Slezska a některých částí Západních Karpat. *Přírodovědecký sborník Ostravského kraje.* Opava. 1954. 14 (3-4), pp. 428-469.

[3] CULEK, M. *Biogeografické členění České republiky: II. Díl.* Praha : Agentura ochrany přírody a krajiny ČR, 1995. 589 pp. ISBN 8086064824

[4] DEMEK, J. *Zeměpisný lexikon ČSR. Hory a nížiny.* Brno : Academia, 1987. 584 pp.

[5] KANTOREK, J., SIMONOV. Taxon Urticularia vulgaris. In *Virtuální floristická kartotéka.* DANIHELKA J., PETŘÍK P. & WILD J. (2010). Dostupné: http: florabase.cz kartoteka

[6] KUPKA, J, KAŠOVSKÁ, K. Měkkýši zrušené Státní přírodní rezervace Loucké rybníky (Slezsko, Česká republika) [Molluscs of the abolished reserve of the Loucké Rybníky ponds (Silesia, Czech Republic)]. In *Malacologica Bohemoslovaca*, 2011. 10, pp. 68–72. Online serial at http: mollusca.sav.sk 11-Nov-2011.

[7] LISICKÝ, M. J. *Mollusca Slovenska*. Bratislava : Veda, SAV, 1991. 244 pp. ISBN 80-224-0232-X.

[8] LOŽEK, V. *Quartärmollusken der Tschechoslowakei.* Praha : Nakladatelství ČSAV, 1964, sv. 31. 374 pp.

[9] MÁCHA, S. Revizní výzkum měkkýšů Louckých rybníků. *Přírodovědecký sborník Ostravského muzea* 26, 1982. pp. 41- 50.

[10] NEUHÄSLOVÁ, Z., NOVOTNÁ, Z., ŠTURSA, J., ELCOVÁ, V. *Mapa přirozené potenciální vegetace České republiky.* 1st ed. Praha : Academia, 2001.

[11] SKALICKÝ, V. Regionálně fytogeografické členění. In Hejný S. & Slavík B. *Květena ČSR I.* Praha : Academia, 1988. 103-121pp.

[12] QUITT, E. *Klimatické oblasti ČSR.* Studia Geographica 16. Brno : Academia, geografický ústav ČSAV, 1971. 73pp.

[13] WEISSMANNOVÁ, H. *Ostravsko – chráněná území ČR X.* Brno : AOPK ČR, 2004. 454 pp. ISBN 80-86064-67-0.

RESUMÉ

Karvinsko patří k regionům s nejvíce poškozeným životním prostředím v ČR, zejména v důsledku hornické a hutnické činnosti. Vážným důsledkem hornické činnosti je především destrukce původního reliéfu a jeho náhrada antropogenními tvary. S tím úzce souvisí i zásadní změna ekologických podmínek a narušení původních ekosystémů a společenstev, které je osidlují.

Ekologické podmínky a jejich změny jsou charakterizovány jednak vegetací (výskyt společenstev, početnost druhů a jejich dominance, druhová pestrost), avšak velmi důležitá je rovněž živočišná složka ekosystému. Mnoho živočichů reaguje na změny v ekosystému rychleji než rostliny díky své pohyblivosti a nižší schopnosti tolerance. Jako významný bioindikátor je z tohoto důvodu používána skupina měkkýšů, kteří se vyznačují vysokou senzitivitou ke změnám a nízkou vagilitou.

Jako modelové území byl zvolen systém Louckých rybníků – bývalé státní přírodní rezervace, která je od 70. let 20. století silně narušena poddolováním. Sběry měkkýšů byly provedeny v červnu a říjnu 2006 a březnu 2007 a v květnu 2013. Pozornost byla věnována zejména vodním druhům, zemní druhy byly determinovány jako doplňkové. Druhové spektrum bylo srovnáno s nálezy z let 1954, 1955 a 1979 a rovněž byla sledována vazba na základní biotopy, rostlinná společenstva a konkrétní rostlinné druhy.

Na daném území bylo nalezeno celkem 48 druhů měkkýšů (42 plžů, 6 mlžů). Mezi vodní měkkýše náleží 21 druhů, 27 druhů bylo suchozemských. Seznam všech druhů včetně jejich zařazení do ekologické skupiny a stupně ohrožení (dle Beran et al., 2005) je uveden v tab. 1. Z druhů vodních měkkýšů, které byly v území nejčastěji nalezeny, uvádíme např. *Anisus vortex*, *Gyraulus albus*, *Hippeutis complanatus*, *Lymnaea stagnalis*, *Musculinum lacustre*, *Planorbis planorbis*. Z hlediska ohrožení byly nalezeny 3 druhy z kategorie téměř ohrožený (NT): *Aplexa hypnorum*, *Pisidium obtusale* a *Musculinum lacustre*.

A BRIEF OVERVIEW OF DEVELOPMENT IN THE USE OF INDICATOR GASES FOR COAL SPONTANEOUS COMBUSTION AND PROSPECTS FOR FURTHER SOLUTION

Lukáš SNOPEK [1], Alois ADAMUS [2]

[1] *Ing., Institute of Mining Engineering and Safety, Faculty of Mining and Geology, VŠB - Technical University of Ostrava, 17. listopadu 15, 708 33 Ostrava - Poruba, Czech Republic,*

e-mail: lukas.snopek@vsbcz

[2] *Prof. Ing., PhD., Institute of Mining Engineering and Safety, Faculty of Mining and Geology, VŠB – Technical University of Ostrava, 17. listopadu 15, 708 33 Ostrava – Poruba, Czech Republic,*

e-mail: alois.adamus@vsbcz

Abstract

One of the most serious risks associated with mining using mainly underground methods is the risk of mine fires. Causes of these fires can be exogenous (external) or endogenous (internal). The authors of the article pay attention to endogenous fires in coal mines, especially to the timely indication of them. As already mentioned in many contributions, spontaneous combustion is a complicated process that passes, in the final stage, into a mine open fire. The localization of a place of such danger can be done by means of so-called indicator gases that are desorbed from the coal substance at a point of the originating seat of spontaneous combustion. The article deals with a development trend in the use of indicator gases for spontaneous combustion in coal mining. The objective of the authors was to arrange chronically and briefly knowledge of the use of indicator gases both abroad and in the Czech Republic and to provide input information about a research project being dealt with at present.

Abstrakt

Při hornické činnosti prováděné především hlubinným dobýváním je jedním z nejzávažnějších rizik vznik důlních požárů. Příčiny vzniku těchto požárů mohou být, buď exogenního (vnějšího) anebo endogenního (vnitřního) původu. Autor článku věnuje pozornost požárům endogenním v uhelných dolech a to především jejich včasné indikaci. Jak už bylo v mnoha příspěvcích zmíněno, samovznícení je složitý proces, který ve své konečné fázi přechází v otevřený důlní požár. Odhalit místo takového nebezpečí lze pomocí tzv. indikačních plynů, které desorbují z uhelné hmoty v místě vznikajícího ohniska samovznícení. Článek se zabývá trendem vývojem využití indikačních plynů samovznícení v uhelném hornictví. Cílem autora bylo chronologický a stručně setřídit poznatky o využití indikačních plynů jak v zahraničí, tak v ČR a podat vstupní informaci o současně řešeném výzkumném projektu.

Keywords: spontaneous combustion, indicator gases, coal, higher hydrocarbons

1 INTRODUCTION

Indicator gases for coal spontaneous combustion can be divided into two groups, namely majority gases (CO, CO_2, O_2, CH_4), continuously observed in the mine atmosphere, and minority gases (C_2H_6, C_2H_4, C_2H_2, C_3H_8, C_3H_6, C_4H_{10}), analysed in cases of indications of spontaneous combustion. In mines of the Czech Republic, these gases are sampled in different ways. The majority gases are sampled using a so-called "wet sampling" method to glass gas collecting tubes filled with a sealing liquid. The minority gases are sampled using a so-called "dry sampling" method by means of vacuum ejectors.

2 FOREIGN KNOWLEDGE

The first who started to investigate these problems was a British researcher J.I. Graham (1920). He focused mainly on the release of CO and CO_2. During his research he found that in the course of oxidation of coal, CO was released and oxygen concentration decreased depending on the temperature of oxidized coal. His research led to the development of a method based on the determination of two basic indexes (Graham's ratios), i.e. CO O_2 deficiency and CO_2 O_2 deficiency.

A Japanese author Kitagawa (1959) noticed in the course of observation of a mine fire that, in addition to CO and CO_2, ethylene and other gases were released. On the basis of this finding, an impulse was given all over the world to examine gaseous higher hydrocarbons as possible gases for the indication of a spontaneous combustion hazard. Because ethylene occurred in very low concentrations and at the given time it was difficult to measure, Kitagawa designed his own detection tubes (colorimetric tube for CO detection and linear tube for C_2H_4).

Pursall and Banerjee (1961) were other authors who were concerned with the determination of higher hydrocarbons, above all ethylene and propylene in the mine atmosphere. In their researches they used chromatography with argon as carrier gas. For gas detection, a thermal conductivity detector was used; however, for low concentrations of these gases it turned out to be a failure. After experience thus the authors proposed to use a more sensitive detector for further observations. Subsequently, Pursall and Gosh (1963) used a modified argon ionization detector. Traces of ethylene began to appear before changes in values of Graham indexes were observed. They found that at a sevenfold increase in the basic CO ΔO_2 ratio, propylene began to appear. On the other hand, propylene disappeared at a threefold decrease in the basic ratio. At a ninefold increase in the $CO/\Delta O_2$ ratio, acetylene occurred. Based on these findings they concluded that the occurrence and increase in the amount of ethylene, propylene and acetylene in this order indicated a danger of heating of the coal substance. Furthermore, they arrived at the conclusion that with decreasing temperature, these hydrocarbons disappeared in the order from acetylene, propylene and ethylene.

In Poland, Muzyczuk (1966) was concerned with the problems of determination of higher saturated and unsaturated hydrocarbons. He carried out his research at the Central Mining Institute (GIG), a Polish research institute in Katowice. Thanks to his methods, the presence of aliphatic and aromatic hydrocarbons in very low concentrations could be detected. Saturated and unsaturated aliphatic hydrocarbons were generated during low-temperature oxidation. Muzyczuk also stated that unsaturated hydrocarbons, in which the prevailing component was ethylene, were formed merely during the oxidation process. In a case of aromatic hydrocarbons, these gases did not occur at temperatures less than 300°C. Muzyczuk was of the opinion that CO, the concentration of which was 100 to 1000 times higher than that of ethane or ethylene, seemed to be the most suitable indicator gas.

Chamberlain (1970) together with his research team was engaged in heating British coal samples at a constant flow of air or nitrogen. Gaseous hydrocarbons were analysed chromatographically. For the determination of aliphatic hydrocarbons (from methane to butane), a chromatograph with a FID detector was used; the detector was able to detect the presence of the observed gases even in low concentrations, e.g. the minimum detection limit for methane was about 0.5 ppm and that for butane was about 2 ppm. For the determination of hydrogen, nitrogen and oxygen, the other chromatograph with a molecular sieve, which was able to determine e.g. hydrogen even at a minimum detection limit of about 1 ppm, was used. Chamberlain came to the conclusion that at temperatures more than 40 °C, CO began to be released from the examined coal. Another released gas was hydrogen; it was released at a temperature of about 80°C with maximum release at about 200°C. Propylene was released at 137 °C with maximum release at about 230°C; but immediately after that a sharp drop in it followed. Ethylene was released at temperatures of about 155°C. From these conclusions it followed that CO seemed to be the most suitable indicator gas indicating the occurrence of spontaneous combustion. In addition to these measurements, the authors were also concerned with the problems of effects of the degree of coalification on coal oxidation. They arrived at the conclusion that the dynamics of oxidation of low-rank coals was higher (higher consumption of oxygen and greater formation of CO) than that of high-rank coals.

In the year 1973 Chamberlain and Hall (1973) continued the previous researches. This time they used a system of enrichment of indicator gases proposed by Novák and his team (1965) for the determination of indicator gases; samples being analysed were concentrated to measurable limits. Thanks to this method, Chamberlain and Hall were able to isolate 30 different hydrocarbons; 20 of them they were able to determine by experiment. They also stated that a ratio between iso-butane and n-butane could indicate the level of ongoing oxidation of the coal substance.

Chamberlain (1976) during his researches always drew attention to one serious problem, namely the fact that although CO seemed to be the most suitable indicator gas, its concentrations grew only after exceeding the critical temperature of coal spontaneous combustion moving in a temperature range of 30-70°C (according to the coal rank). For this reason, in the next tests he focused especially on gases released in the course of oxidation at temperatures below 120°C. For testing, three coal samples of different ranks were selected. It was found that at temperatures less than 10°C, none of chromatographically analysed gases with a concentration higher than the 0.01 ppm limit occurred. Measurable concentrations of chromatographically analysed gases began to appear at temperatures of 10-70°C and increased rapidly. However, at exceeding 70°C, the generation of the gases but CO became stable.

The problem of timely indication of spontaneous combustion was also solved by a research team of Aijperovič, as stated in Faster (1976), in the former U.S.S.R. Aijperovič together with the team found that in the course of initial oxidation, unsaturated hydrocarbons, namely ethylene and acetylene in trace concentrations were released. During their research they however tackled the problem of detection of these hydrocarbons owing to their low concentrations. They developed a special concentrator that was able to separate required components from a rather large volume of gaseous oxidation products. Subsequently, the components were desorbed in the laboratory and in concentrated form were analysed chromatographically. For the estimation of temperature of the seat of spontaneous combustion a table of values of the ethylene acetylene ratio was determined.

Authors Chakravorty and Feng were researchers who were concerned with indicator gases for the spontaneous combustion of Canadian coals. They performed their tests on the coal from the Sparwood coal field. This coal was heated at temperatures from 20°C to 250°C. By measurements they came to the conclusion that at temperatures below 74°C, mainly CO and CO_2 were released. Only at temperature of 110°C, perceptible traces of hydrogen and ethane began to appear, but their concentrations were much lower than the concentration of CO. Higher saturated and unsaturated hydrocarbons (propane, butane and ethylene) were released only at higher temperatures. In Fig. 1 a well-known graph showing the formation of indicator gases (above all CO, hydrogen, ethane and ethylene) of the cited authors is stated. As for the dynamics of release of indicator gases, it is obvious that CO predominates. For this reason, the authors proposed to measure continuously CO using modified IR analysers directly in mining conditions.

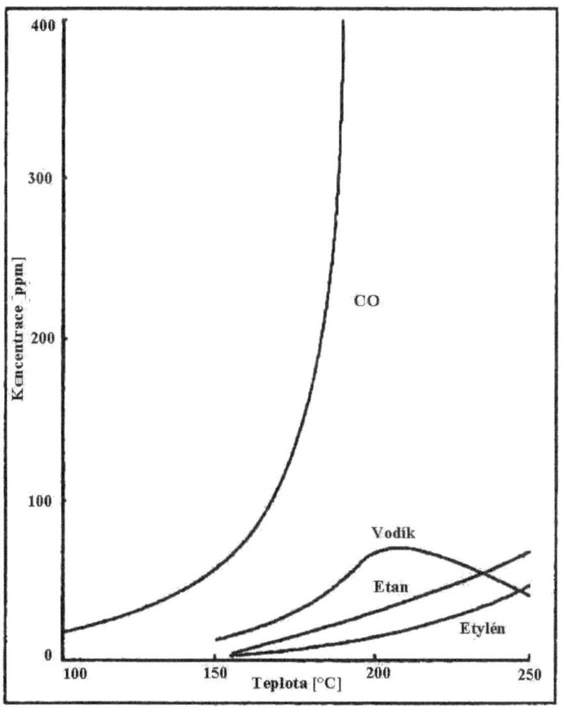

Fig. 1 Chakravorty's graph 1978

In the nineties, Cygankiewicz (1996) also began, on the basis of knowledge acquired during previous researches done at the coal research institute GIG in Katowice, to be concerned with the development of indicator gases by thermal oxidation. In samples examined at temperatures from 50 to 250°C, the formation of common indicator gases and simultaneously of C_2–C_4 saturated and unsaturated hydrocarbons was found. Based on these gases, the following six dimensionless indexes were taken as most suitable for the assessment of the process of spontaneous combustion:

- C_2H_4 C_2H_2 (ethane acetylene),
- C_3H_6 C_2H_2 (propylene acetylene),
- 100 * (C_2H_4 H_2),
- CO H_2,
- 100 * G (Graham's ratio),
- CnHm H_2.

From his research Cygankiewicz drew the conclusion that the value of all these dimensionless indexes increased with growing temperature. CnHm̧ H_2 and CO̧ H_2 indexes exhibited the highest tendency to grow; on the contrary, Graham's ratio showed the lowest. Later, this method was used successfully in practice in one of Polish mines, when the process of spontaneous combustion was revealed. According to the author, this method can be used for the assessment of the process of spontaneous combustion. The method has been offered by the research institute GIG for operational purposes up to now.

3 KNOWLEDGE OF UTILIZATION OF INDICATOR GASES IN THE CZECH REPUBLIC

At the beginning of the seventies, Lanková et al. (1975) as the first ones began carrying out oxidation tests on coal. This team began to determine the concentrations of methane, ethane, propane, n-butane, ethylene and propylene by means of a chromatograph at temperatures of 150, 200 and 240°C. Conclusions confirmed the above-mentioned information, i.e. the higher the temperature, the greater formation of hydrocarbons; the higher the hydrocarbon, the smaller formation of it; at the same number of carbons, a saturated hydrocarbon is formed in a higher degree. To distinguish hydrocarbons evolved by oxidation from hydrocarbons released by heat action, the tests in which nitrogen was used as a circulating gas were carried out. The result was the conclusion that ethylene and propylene were formed in low concentrations when the coal substance was exposed to the mere action of heat without the presence of oxygen.

Other authors concerned with these issues were Harašta, Čermák (1979). On the basis of research aimed at preparing a proposal for the method of determination of higher hydrocarbons to find indications of spontaneous combustion, to assess the extent of a space affected by spontaneous combustion in the early stage and temperature of spontaneous combustion in the early stage, they arrived, at the research institute VVUÚ in Ostrava-Radvanice, at the finding that higher hydrocarbons and their ratios (methane, ethane, ethylene, acetylene, propane, propylene and butane) were, under normal conditions of mining without indications of spontaneous combustion, with common ventilation systems and under stable pressure, constant and their ratios between each other did not change. By continuous decreasing the concentration ratios of methanę ethane, ethanę propane and propanę butane, it was possible to expect the increased oxidation of coal with accumulation of heat. The presence of higher unsaturated hydrocarbons, mainly ethylene and propylene, warned of spontaneous combustion in the early stage. The occurrence of acetylene in measurable concentrations indicated an increase in the temperature of coal and a situation when the fire almost originated.

In this period, researches based on the above-mentioned knowledge were also carried out at the Main Mining Rescue Station (henceforth referred to as MMRS) in Ostrava. The output of these researches was "A Temporary Method for the Evaluation of Chromatographic Determination of Hydrocarbons", Apfelthaler (1983), and subsequently "The Evaluation of Spontaneous Combustion Process of Hard Coals of Saddle Seams in the Ostrava-Karviná Coalfield Using Gas Chromatography – Method of MMRS in Ostrava", Hajník (1987). The researches concerned were based especially on the evaluation of numerous samples of the mine atmosphere. The mentioned method prescribes the way of dry sampling of the atmosphere. On the basis of detected concentrations of indicator gases, individual phases of spontaneous combustion process are assigned to particular gases. According to this method, in the early phase of spontaneous combustion, CO and gases of alkane series (ethane, propane, butane), which under normal conditions with the exception of ethane do not occur in the mine atmosphere, begin to appear. From the occurrence of other hydrocarbons, temperature at the seat of spontaneous combustion can be deduced as well; if propane and butane in concentrations of 0.01-1 ppm appear, temperatures at the seat of spontaneous combustion can be expected to be in the range of 30-60°C. Temperatures above 60°C indicate the presence of ethylene in concentrations of 0.01-1 ppm and a sharp increase in CO. By subsequent increase in temperature to 90-110°C, hydrogen appears in concentrations of 10-1 000 ppm and simultaneously the concentrations of hydrocarbons determined earlier grow to the values of 1-100 ppm. If the limit of 230°C is exceeded, the presence of acetylene is detected, when the fire originates. On the basis of these findings, a column graph, determining amounts of formed individual gases in litres in the given phases of spontaneous combustion was constructed (Fig. 2).

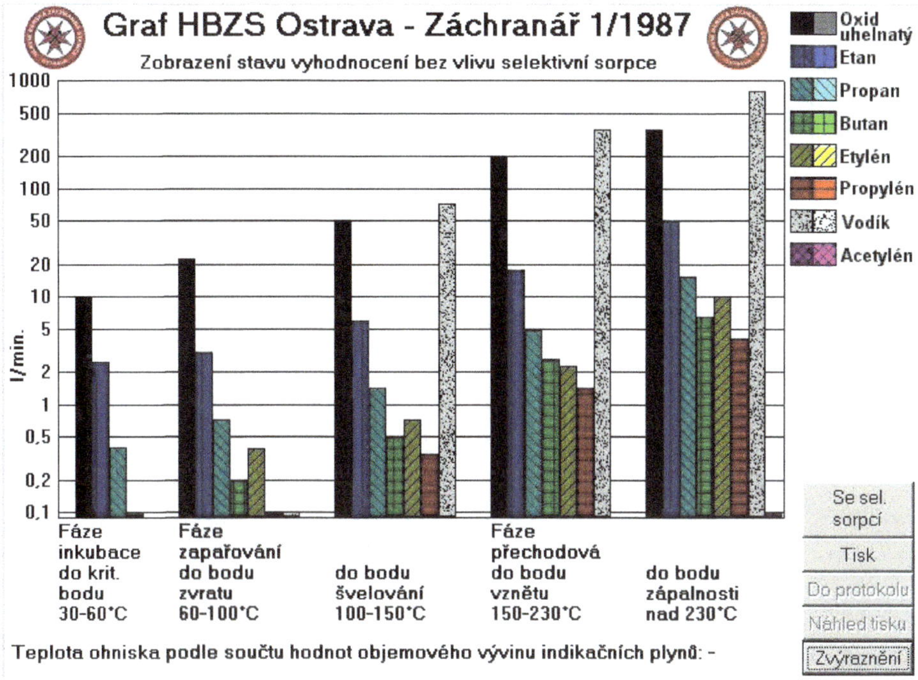

Fig. 2 Print screen from the program "Evaluation of Indicator Gases for Spontaneous Combustion of Coal Substance", version 2.3 (Stecker, 2005)

On the basis of chromatographic measurements of the amounts of formed hydrocarbons, Prof. Taraba (2003) developed a so-called gas thermometer, on which intervals for measurable evolved indicator gases were illustrated (Fig. 3). Thick lines represent the most frequently determined ranges of temperatures of indicator gases and thin lines represent the maximum spreads of temperature values.

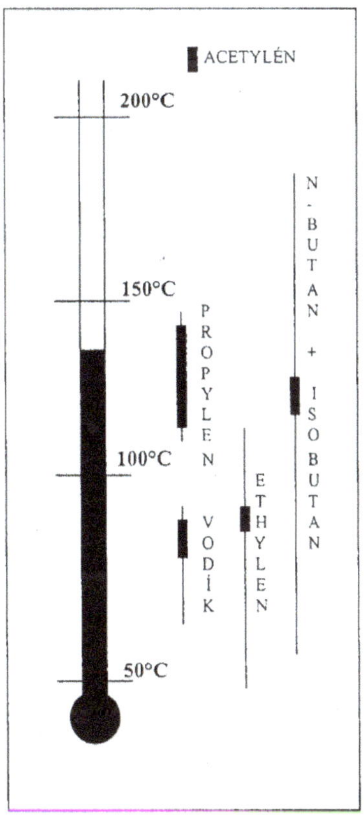

Fig. 3 Gas thermometer (Taraba, 2003)

Researches were done at VŠB – Technical University of Ostrava as well. One of researchers was Jindřich Šancer (2004), who in the course of his study for the doctoral degree participated in research dealing with the problems of indicator gases for spontaneous combustion of the coal substance. He stated acquired knowledge in his Ph.D. thesis. Above all, he was concerned with the formation of products of thermal oxidation of coal samples from the Ostrava-Karviná Coalfield (henceforth referred to as OKC) depending upon temperature (Fig. 4) and with their relations to the physical and chemical properties of coal and to the sampling locality. Based on knowledge obtained from previous researches, he tried to verify these properties in practice using mathematical statistical methods and data analysis.

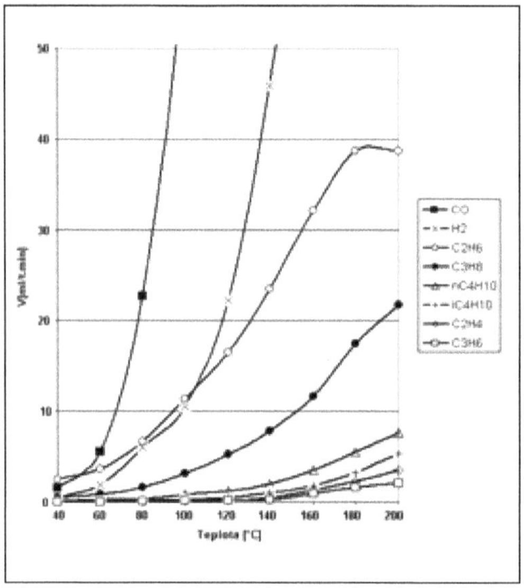

Fig. 4 Average volumes of formed products of thermal oxidation of OKC coal samples (Šancer, 2004)

Further research concerning these issues was carried out by a team of researchers led by Prof. Adamus (2002, 2004) in the framework of scientific research projects VaV ČBÚ No. 3 1999 (1999-2002) and No. 29 03 (2003-2005). One of subtasks of this research was timely indication of the early stage of spontaneous combustion of the coal substance. Here, findings from previous two researches were processed and a database making it possible to generate gas patterns of indicator gases for thermal oxidation of 63 verified coal samples was created. One of gas patterns of thermal oxidation from catalogues mentioned below is given in Fig. 5. In addition to the assessment of the released amounts of gases, ratios between particular oxidation products could be assessed here as well. As suitable dimensionless indexes of spontaneous combustion of coal, C_2H_6 C_3H_8, C_2H_6 C_2H_4, C_2H_4 C_3H_6 and CO_2 C_2H_4 indexes were determined besides the already used CO_2 CO ratio. Two of outputs of the concerned research were catalogues of coal seams prone to spontaneous combustion in the Czech Republic, Adamus (1999-2002), and a computer program for work with gas patterns of indicator gases for thermal oxidation of verified coal samples, Stecker (2005). At the Faculty of Mining and Geology of VŠB – Technical University of Ostrava, a system for the creation of gas etalons for partial developed faces in the OKC was designed and verified. The procedure for etalon preparation was based on the verification of gas patterns of indicator gases for spontaneous combustion using the method of thermal oxidation of three coal samples from a coal block prepared for mining. The concerned gas etalon was subsequently designed for use in practice in case of spontaneous combustion at the given face, e.g. "Indicator Gas Etalon for Coal Substance Spontaneous Combustion Valid for the ČSM-Sever Mine, Seam 32, Block 2, Face No. 320 203, Faculty of Mining and Geology of VŠB-Technical University of Ostrava, Ostrava 8 June 2005".

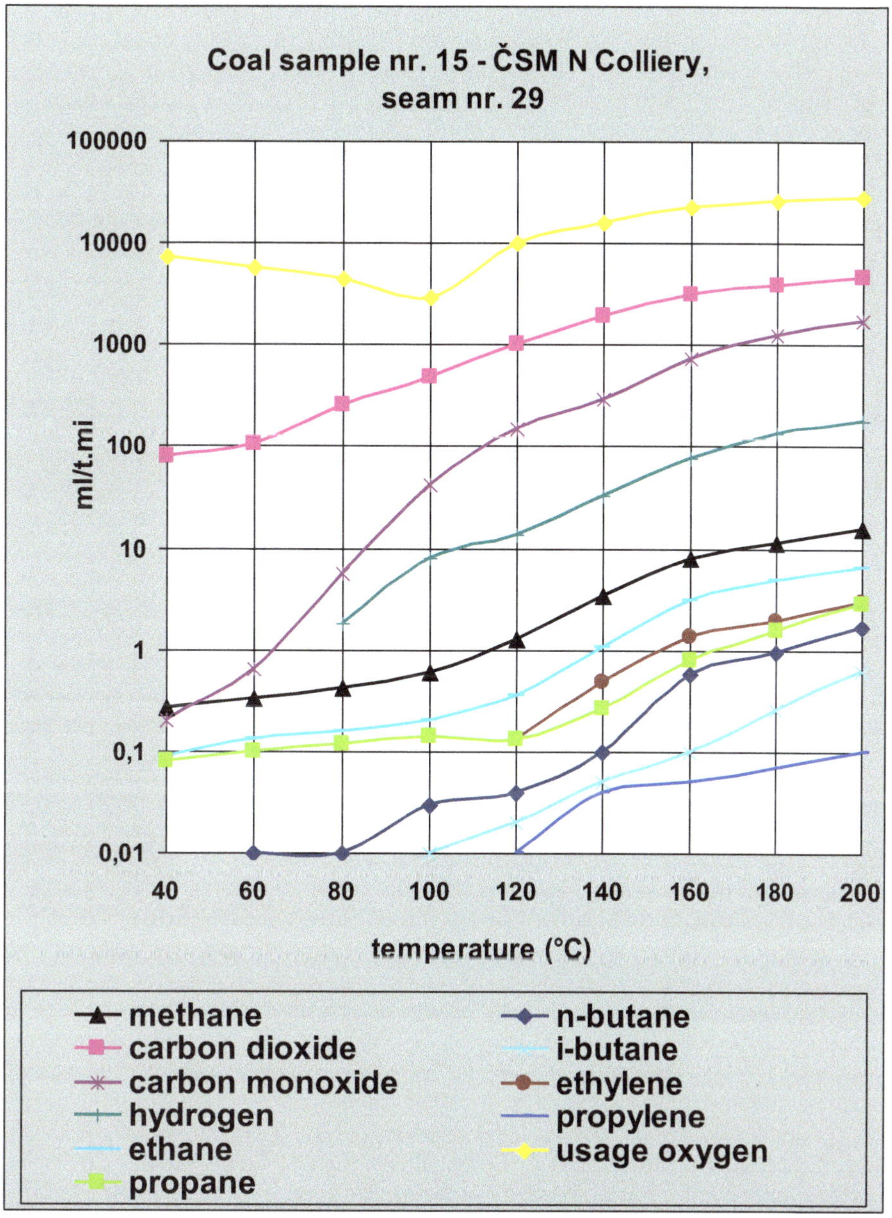

Fig. 5 Evolution of products of thermal oxidation of an OKC coal sample (Adamus, 1999-2002)

4 RESEARCH INTO INDICATOR GASES FOR SPONTANEOUS COMBUSTION OF BROWN COAL IN THE CZECH REPUBLIC

In the early seventies of the 20th century, Hrubý, Charamza and Kusý (Lanková, 1975) from the former Czechoslovak Socialist Republic were concerned with the problems of evolution of gaseous hydrocarbons from hard coals. These authors focused especially on the problems of distinguishing so-called primary methane, physically bound to the coal substance, from secondary methane. During tests they observed that at slight heating the coal (50-55°C), besides methane, iso-butane appeared; the other hydrocarbons, mainly ethene and n-butane were not identified. At temperatures by about 15°C higher (critical temperature of spontaneous combustion), ethane and ethene occurred. Other hydrocarbons began to appear at temperatures of 80-100°C.

The authors Hrubý, Trýzna, Kusý and Hautke (1990) examined, on the basis of research "Determination of a Boundary between the Oxidation Reduction Process and the Initial Stage of Spontaneous Combustion Process. Determination of Indicators of Processes of Coal Spontaneous Combustion", gaseous indicators of the process of spontaneous combustion in a caved zone of a longwall face. The research was carried out as laboratory research into coal oxidation and as measurement of air masses from the caved zone of the longwall face. In the conclusion, both these activities were evaluated and compared. After the evaluation of the result, the authors arrived at the conclusion that at the beginning of so-called "flushing" of the coal seam, in addition to the release of CO and CO_2, the desorption of methane (from 2 to 100 ppm) also took place. At heating the coal to temperatures of about 50-55°C, when water was released, even iso-butane in concentrations of 0.05-0.1 ppm began to appear besides CO, CO_2 and methane. After the vaporization of a large amount of water, temperature grew and methane content was small. Ethane and ethene began to appear in trace concentrations of about 0.1 ppm. At temperatures of 90-120°C, the proportion of methane increased and C_3 hydrocarbons began to appear. Up to the temperature of 180°C, an increase in the content of iso-butane was not too progressive. At temperatures above 180°C, a steep increase in the concentrations of individual components occurred.

Other findings concerning the indicator gases for the thermal oxidation of brown coal were obtained in the framework of preparation of catalogues of Czech Republic's coal seams prone to spontaneous combustion, Adamus (1999-2002).

5 CONCLUSION

Czech Republic's research findings in question are partially used within the meaning of Sections 187, 189 and 194 of Decree of the Czech Mining Authority No. 22 1989 Sb., on safety and health protection at work and operation safety in mining activity and in underground mining of non-reserved minerals, which was in the year 1990 amended for the OKC by the Decision of the District Mining Authority in Ostrava No. 10 1990, the currently valid Decision of the District Mining Authority in Ostrava No. S 0300 2008. At present, Article 27 of the Decision lays down e.g. the classification of coal seams into categories based on the proneness of the coal substance to spontaneous combustion; the localization of places that have to be checked regularly for possible spontaneous combustion of the coal substance; periods and places for CO concentration measurement; determination of frequency of taking air mass samples from the current of air for gas chromatographic analysis (determination of H_2, C_2H_6, C_2H_4, C_2H_2, C_3H_8, C_3H_6, n-C_4H_{10}, iso-C_4H_{10} concentrations); obligations in the case of repeated analysis of gaseous unsaturated hydrocarbons in the sense of closing an endangered area.

In spite of the fact that current technologies enable already improved prediction of spontaneous combustion in coal mines, this problem, especially in the area of determination of temperature condition of the seat of spontaneous combustion, is not solved sufficiently yet. The existing above-mentioned methods enable the estimation of temperature at the seat of spontaneous combustion but they do not enable its exact determination. For these reasons, it is suitable to continue to do the given research. One of activities that deal with the given problems is a project of the Technology Agency of the Czech Republic No. TA01020351, which is just in progress, "Research into Possibilities of Predicting the Occurrence of Initial Stage of Spontaneous Combustion and Subsequent Spontaneous Combustion of Brown Coal Fuels" with duration from 01 2011 to 12 2014, the recipient of which is the Brown Coal Research Institute, JSC in Most and a project participant is the Faculty of Mining and Geology of VŠB – Technical University of Ostrava. The project focuses on the use of thermometry and gasmetry in the framework of prediction of spontaneous combustion in brown coal dumping grounds.

The article was written with support from the Technology Agency of the Czech Republic in the framework of the project No. TA01020351.

REFERENCES

[1] ADAMUS, A., et al. 2002. *Snížení rizika vzniku samovznícení uhelné hmoty se zaměřením na indikační a prevenční metody*. Ostrava, listopad 2002. Závěrečná zpráva výzkumného projektu ČBU č. 3/99. VŠB-Technická univerzita Ostrava.

[2] ADAMUS, A., et al. 1999-2002. *Pasportizace uhelných slojí České republiky – náchylnost k samovznícení – I-IV*. Ostrava, 1999-2002. Dílčí zprávy k programu výzkumu a vývoje ČBÚ č. 2435/99III. VŠB-Technická univerzita Ostrava.

[3] ADAMUS, A., ŠANCER, J. 2005. *Zdokonalení metod včasného zjišťování samovznícení uhlí a kontroly průběhu jeho likvidace se zaměřením na indikační plyny*. Ostrava, listopad 2005. Závěrečná zpráva projektu VaV ČBU č. 29/03. VŠB-Technická univerzita Ostrava.

[4] APFELTHALER, R., et al. 1983. Prozatímní metodika vyhodnocování chromatografických stanovení uhlovodíků. *Záchranář*. 1983, listovka HBZS č.7-8.

[5] CYGANKIEWICZ, J. 1996. Ocena rozwoju ognisk samozagrzewania na podstawie precyzyjnej analýzy chemicznej prób powietrza kopalnianego. *Konference „Zagrozenia metanole i pozarowe w górnictwie, wykorzystanie metanu z pokladow wengla*. 2-4.10.1996. Ustroń.

[6] FASTER, P. 1976. Poměr C_2H_4:C_2H_2 kontroluje teplotu. *Záchranář*. 1976, listovka HBZS č.10.

[7] GRAHAM, J. I. 1920. The normal production of carbon monoxide in coal-mines. *Transaction - Institution of Mining Engineers*. 1920-21, vol.60, s. 222-234.

[8] HAJNÍK, M. 1987. Hodnocení samovzněcovacího procesu černých uhlí sedlových slojí v OKR s využitím plynové chromatografie – metodika HBZS Ostrava. *Záchranář*. 1987, listovka HBZS č.1.

[9] HARAŠTA, M., ČERMÁK, V. 1979. *Výzkum prostředků a metod pro včasné zjištění samovznícení a pro sledování a určení jeho průběhu*. Ostrava, 1979. Závěrečná výzkumná zpráva. VVUÚ Ostrava-Radvanice.

[10] HRUBÝ, M., TRÝZNA, P., KUSÝ, J., HAUTKE, P. 1990. *Stanovení hranice oxidačně redukčního a záparového procesu. Stanovení indikátorů procesů samovznícení uhlí*. Most, září 1990. Výzkumná zpráva. VÚHU o.z. SHD Most, s.p. Most.

[11] CHAMBERLAIN, E. A. C., HALL, D. A., THIRLAWAY, J. T. 1970. The Ambient Temperature Oxidation of Coal in Relation to the Early Detection of Spontaneous Heating – part 1. *The Mining Engineer*. October 1970, vol. 130.

[12] CHAMBERLAIN, E. A. C., HALL, D. A. 1973. The Ambient Temperature Oxidation of Coal in Relation to the Early Detection of Spontaneous Heating – part 2, *The Mining Engineer*. October 1973, vol. 132.

[13] CHAMBERLAIN, E. A. C., BARRAS, G., THIRLAWAY, J. T. 1976. Gases evolved and possible reactions dutiny low-temperature oxidation of coal. *Fuel*. July 1976, vol. 55, p. 217.

[14] KITAGAWA, T. 1959. Detection of underground spontaneous combustion in its early stage with detectors. *10th International Conference of Directors of Safety in Mines Research*. 1959. Pittsburgh, PA.

[15] LANKOVÁ, S., ČERNÝ, V., DOBAL, V. 1975. *Vývoj uhlovodíků při oxidaci černých uhlí*. Praha, 1975. Hornický ústav ČSAV.

[16] MUZYCZUK, J., KOLODZIEICZYK, B. 1966. *Wspólczesne kierunki rozwoju metod wczesnego wykrywania pozarów endogenicznych w kopalniach wegla kamiennego*, Przeglad górniczy, 1966, s. 364 – 370.

[17] PURSALL, B. R., BANERJEE, S. P. 1962. The Application of Gas Chromatography to the Analysis of Firedamp from various Sources in the Yorkshire Coalfield. *The Mining Engineer*. february 1962, s. 317-330.

[18] Rozhodnutí Obvodního báňského úřadu v Ostravě, spis. zn.:S 0300/2008-6-68 Ing.Kp/Pe, k zajištění jednotného plnění požadavků vyhlášky Českého báňského úřadu v Praze č. 22/1989 Sb., ve znění pozdějších předpisů, OBÚ Ostrava, Ostrava 2008.

[19] STECKER, J., ŠANCER, J., ADAMUS, A. *Hodnocení indikačních plynů samovznícení uhelné hmoty*. Ver. 2.3. Ostrava, listopad 2005. Výpočetní program s pracovním názvem „CnHm" na nosiči CD a tištěným manuálem. Realizační výstup projektu VaV ČBÚ č. 29/2003, VŠB-Technická univerzita Ostrava.

[20] ŠANCER, J. 2004. *Indikační plyny samovznícení uhelné hmoty*. Ostrava, 2004. Disertační práce (PhD). VŠB-Technická univerzita Ostrava. Fakulta hornicko-geologická.

[21] TARABA, B. 2003. *Nízkoteplotní oxidace a samovzněcování uhelné hmoty*. Ostrava: Ostravská univerzita, 2003. 112 s. ISBN 80-7042-832-5.

RESUMÉ

Předmětné poznatky výzkumu ČR jsou částečně uplatněny ve znění §187, §189 a §194 vyhlášky ČBÚ č. 22/1989 Sb., o bezpečnosti a ochraně zdraví při práci a bezpečnosti provozu při hornické činnosti a při dobývání nevyhrazených nerostů v podzemí, která byla v roce 1990 pro OKR doplněna rozhodnutím OBÚ č. 10/1990, nyní platné rozhodnutí OBÚ v Ostravě č. S 0300/2008. V článku č. 27 rozhodnutí je v současnosti stanoveno např. členění slojí do kategorií podle náchylností uhelné hmoty k samovznícení; stanovení míst, která musí být pravidelně kontrolována z důvodu možného vzniku samovznícení uhelné hmoty; lhůty a místa pro měření koncentrace CO; stanovení četností odběrů vzorků vzdušin ve větrním proudu pro plynovou chromatografickou analýzu (zjištění koncentrací H_2, C_2H_6, C_2H_4, C_2H_2, C_3H_8, C_3H_6, n-C_4H_{10}, iso-C_4H_{10}), povinnosti při opakované analýze nenasycených plynných uhlovodíků ve smyslu uzavření ohrožené oblasti.

Přesto, že současné technologie umožňují již zdokonalenou predikci samovznícení v uhelných dolech, není táto problematika dosud uspokojivě dořešená, především v oblasti určení teplotního stavu ohniska samovznícení. Dosavadní, výše zmíněné postupy umožňují odhad teploty ohniska samovznícení, nikoliv její exaktní určení. Z těchto důvodu je vhodné pokračovat v daném výzkumu. Jednou z aktivit, která se dané problematice věnuje je v současné době řešený projekt Technologické agentury České republiky č. TA01020351 „Výzkum možností predikce vzniku záparů a následného samovznícení hnědouhelných paliv“ s dobou řešení 01/ 2011 - 12/2014, jehož řešitelem je Výzkumný ústav hnědého uhlí v Mostu a spoluřešitelem Hornicko-geologická fakulta Vysoké školy báňské v Ostravě. Projekt je zaměřen na využití termometrie a plynometrie v rámci predikce samovznícení na deponiích hnědých uhlí.

Článek byl napsán s podporou projektu Technologické agentury České republiky č. TA01020351.

8

CONTRIBUTION TO A MAGNETITE CONSUMPTION IN THE KARVINÁ MINE'S PREPARATION PLANT OF THE LAZY PLANT

Radim PAVLÍK [1], *Vlastimil ŘEPKA* [2]

[1] *Ing, OKD, plc., Karviná Mine, Lazy Plant, Karviná doly, ZIP 73 506*
e-mail: radim.pavlik@okd.cz

[2] *Ing, Ph.D , Institute of Mining Engineering and Safety, Faculty of Mining and Geology ,*
VŠB – Technical University of Ostrava, 17. listopadu 15, Ostrava
e-mail: vlastimil.repka@vsb.cz

Abstract

In its introduction, this contribution deals with the usage of magnetite during the process of coal separation in the Karviná Mine's preparation plant of the Lazy Plant. Next it deals with the evaluation of quality of magnetite used in the preparation plant, losses of magnetite adhering to the products leaving the preparation plant, the recovery of a diluted suspension with focus on the efficiency of magnetite separation of the diluted suspension and the comparison of the magnetite consumption with the one in the ČSM preparation plant. In conclusions, the article presents options leading to the reduction of the magnetite consumption in the Karviná Mine's preparation plant of the Lazy plant. For comparison, the data of the ČSM Mine is stated here.

Abstrakt

Příspěvek se v úvodu zabývá použitím magnetitu při rozdružování těžného uhlí na úpravně Dolu Karviná, závodu Lazy, dále vyhodnocením kvality magnetitu používaného na úpravně, ztrátami magnetitu ulpívajícího na produktech odcházejících z úpravny, regenerací zředěné suspenze se zaměřením na účinnost separace magnetitu ze zředěné suspenze a porovnáním spotřeby magnetitu s úpravnou dolu ČSM. V závěru jsou uvedeny možnosti vedoucí ke snížení spotřeby magnetitu na úpravně Dolu Karviná, závodu Lazy. Pro porovnání jsou zde uvedeny hodnoty z Dolu ČSM.

Key words: magnetite, coal, separation.

1 INTRODUCTION

In the Kariná Mine's preparation plant of the Lazy Plant, magnetite is used to treat coal. Magnetite is used as a high density material from which together with water a dense medium is prepared. The high density material must meet at least several conditions, such as availability, low cost, sufficient density, abrasion resistance, and may not react with the treated raw material and may not be soluble in water as well. The dense medium must have at least several properties such as high stability, low viscosity, wholesomeness and easy recoverability. The dense medium suspension is used to create an environment conductive to the separation of products. The separation is a process in which materials are separated due to their different properties, in our case their different density. The separation in the preparation plant of the Lazy Plant takes place in a SM (Staatsmijnen) dense medium separator (Fig. 1) and in hydrocyclones where coal, dirt band and spoil are separated from one another. The shallow SM dense medium separator consists of a bath where a scraper conveyor is located. The dense medium and the material being treated are fed to the bath.

Through the upper branch of the discharge end floats are washed out (the fraction of lower density than the density of the dense medium) and through the lower branch of the discharge end sinks falling to the bottom of the bath are discharged (the fraction of higher density than the density of the dense medium). Dense medium cyclones are used to remove particles of smaller particle size where besides gravitational forces, as in a SM dense medium separator, also centrifugal field forces are applied.

The cyclone involves a tapered inlet through which the material is supplied together with the dense medium into a short cylindrical section followed by a conical section. The conical section ends with a tapered outlet through which the higher density fractions are discharged. In the axis of the dense medium cyclone there is an overflow outlet on top through which the lower density fractions pass out. During the separation, particles of the high density material adhere to the surface of the material being separated. If not removed, they can cause the deterioration of the quality of manufactured products and an increase in the consumption (losses) of the high density material. In order to minimize losses of the high density material (economically significant), the separation products are rinsed off, and so the adhered high density material removed from their surface. This way a mixture of particles of the high density material, fine particles of the material being separated and other foreign matters and water occurs which is referred to as a diluted suspension. For further use (re-use) of the high density material it is necessary to recapture its particles for the preparation of a new suspension. This technological operation is called dense medium recovery. [1] In the Lazy Plant, the recovery of magnetite particles is performed by multi-stage separation using magnetic separators. The principle of a uniflow magnetic separator is shown in Fig. 2. The choice of a recovery method for the diluted suspension depends on the type of the used high density material and its properties. As the high density material used by us has magnetic properties, we could use the properties to recover the diluted suspension. Regarding the price of magnetite has been increasing in recent years and the magnetite consumption in the Karviná Mine's preparation plant of the Lazy Plant amounts monthly to one million crowns on average, it is necessary to check regularly the magnetite consumption and simultaneously to seek ways to reduce it.

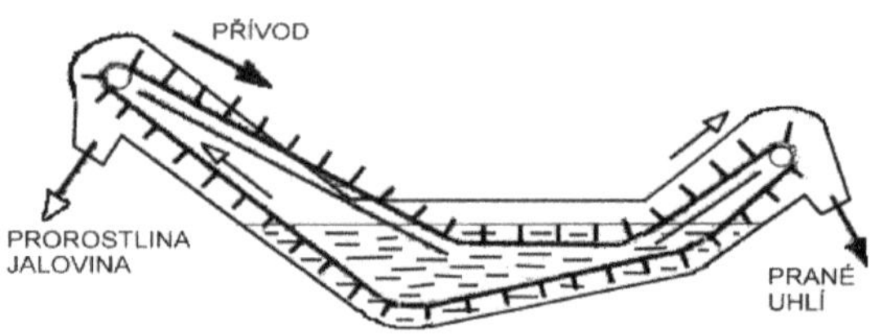

INLET
SPOIL WASHED COAL

Fig. 1 SM dense medium separator

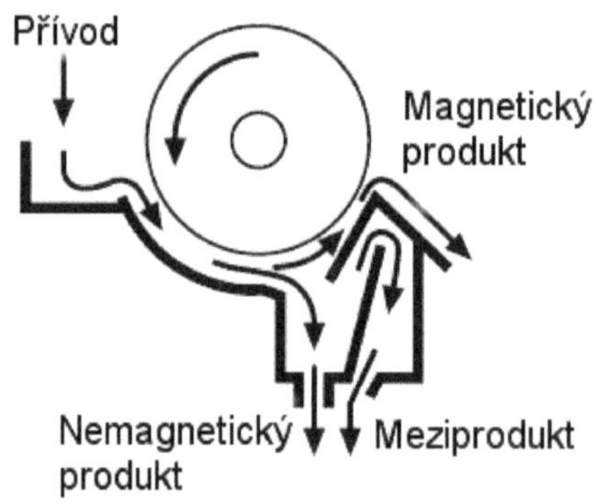

INLET MAGNETIC PRODUCT
NON-MAGNETIC PRODUCT INTERMEDIATE

Fig. 2 Uniflow magnetic separator

2 PROPERTIES OF MAGNETITE USED IN THE PREPARATION PLANT OF THE LAZY PLANT

Raw magnetite was taken from a landfill plant of the preparation plant, and then its properties were determined in laboratories at the VSB-Technical University of Ostrava. These were the density of magnetite, its grain-size composition, contents of magnetic particles in raw magnetite and properties of raw magnetite. The properties were found in accordance with the industry standard 44 7233 - *Evaluating high density materials and separating suspensions*.

Density of magnetite - magnetite was first dried in an oven at 105°C. The dried magnetite was screened at a mesh size of 0.5 mm. From the undersize the magnetite density of 4524 kg.m^{-3} was determined by pycnometry. The density of magnetite complies with the conditions determined for high density materials.

Grain-size composition of magnetite - was determined by dry distribution on screens with nominal lengths of mesh of 0.1 mm, 0.063 mm, 0.04 mm and 0.02 mm. The grain size distribution of magnetite and the recommended range of grades used in coal preparation plants are given in Tab. 1. A graphical representation is presented in Graph 1. It follows from the values shown in the table or in the graph that magnetite does not differ much from the recommended particle size composition.

Tab. 1 Particle size composition of magnetite and recommended range of grades. [2]

Grain size [mm]	Magnetite weight [g]	Reality Magnetite yield [%]	Yield according to ON 44 72 33 [%]
- 0.020	1.157	4.3	30
0.020 - 0.040	14.768	55.2	20
0.040 - 0.063	6.7887	25.3	30
0,063 - 0,100	1.6778	6.3	15
+ 0.100	2.378	8.9	5
Total	**26.7695**	**100**	**100**

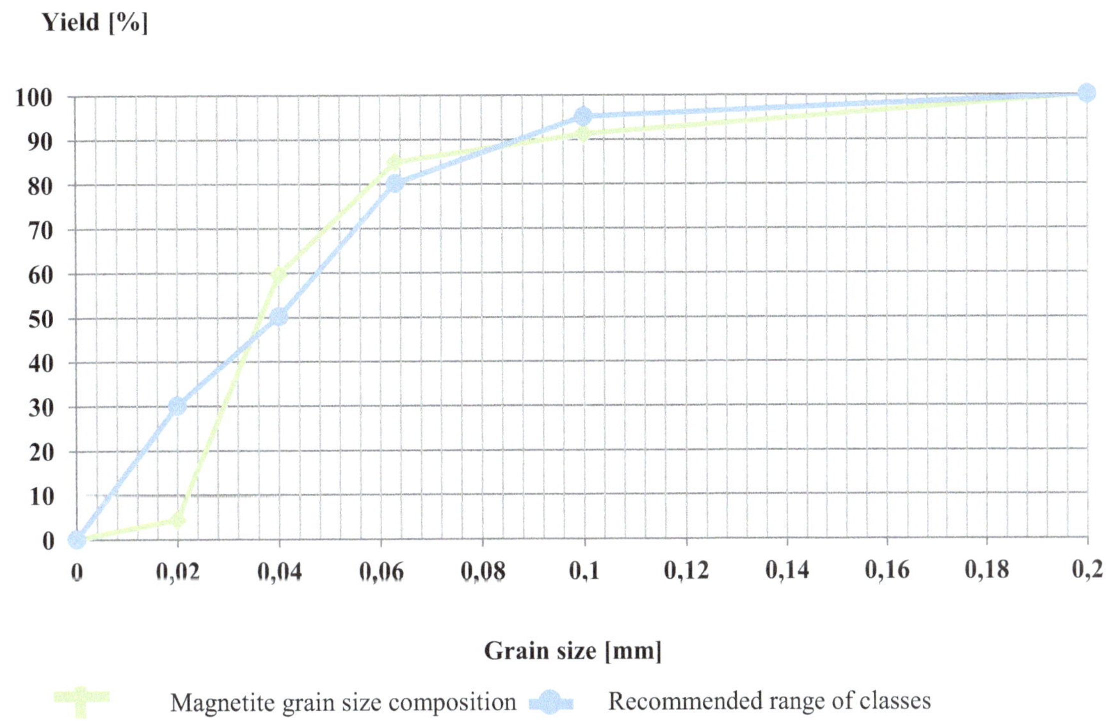

Graph 1 Particle size composition of magnetite and recommended range of classes. [2]

Determination of contents of magnetic particles and magnetic properties of raw magnetite - magnetic particle contents in raw magnetite was determined using the Mechanobr analyzer according to the industry standard 44 7233 which specifies the procedure for determining magnetic components. Magnetite contained on average 98.135 % of magnetic particles, which is a satisfactory value. The determination of magnetic properties of raw magnetite is voluntary. The procedure is imposed by the branch standard 44 7233. The observed average magnetic property of 0.76 is sufficient as it exceeds the value of 0.7 which is a recommended value according to the standard.

3 LOSS OF MAGNETIC FRACTIONS ON PRODUCTS

In the coal preparation plant, products already washed were taken after separation, i.e. those free of magnetic fractions. These were rinsed in the laboratory and the obtained mixture of slurry with remaining magnetite was subjected to control wet separation using the Mechanobr magnetic separator at the magnetic induction B = 0.1 T, level 2 and a voltage of 24 volts. The results are shown in Table 2. The reported losses are acceptable. The rinsing process can be considered effective if losses are not greater than 1 kg per ton of coal being processed, or per ton of charge. The different amounts of magnetite losses are caused by the different particle size of products and the associated size of the surface which the magnetite particles adhere to.

Tab. 2 Loss of magnetic fractions on products. [2]

Product	Loss [g/t]	Product	Loss [g/t]
Lumpy product	24.35	Dirt band	80.86
Coal from centrifuges	40.26	Fine stone	73.48
Washed coal	189.78	Rough stone	68.21
Dirt band from centrifuges	38.70	Spoil	45.16

For comparison, measurements at the ČSM Mine in 2010 under the same conditions showed higher losses of magnetite on products of dense medium cyclones at magnetic fraction contents of 97.94 % in dry magnetite. [3]

Losses of magnetic particles on DM HC products
Washed coal 322 g/t
Spoil 88 g/t
Losses of magnetic particles on DM HC products
Washed coal 391 g/t
Spoil 81 g/t
Losses of magnetite on DM HC products
Washed coal 328 g/t
Spoil 90 g/t
Losses of magnetite on DM HC products
Washed coal 399 g/t
Spoil 82 g/t

4 RECOVERY OF DILUTED SUSPENSION

Magnetite in the coal preparation plant of the Lazy Plant is recovered using drum magnetic separators. In order to ensure sufficient efficiency, a three-stage separation process is used. In the first stage of recovery, four separators are involved, in the second stage, six separators are included and in the third stage, four separators are used. The separator is a device consisting of a drum rotating in a bath which is fed with the diluted suspension. Non-magnetic particles are rinsed off through the opening of the environment being separated to the next stage of recovery or to a service water tank. Magnetic particles are attracted in the magnetic field to the drum surface, washed away from the diluted suspension into a chute area where they are rinsed off and drained into a tank of magnetic fractions. This way obtained magnetite is used for a new application. The efficiency of magnetic separation is shown in Table 3, and for comparison, the efficiency of magnetic separation at the ČSM Mine is given in Table 4, where the efficiency result is better than in the Lazy Plant for a lower value of magnetic particle contents in raw magnetite.

Tab. 3 Calculations of magnetic separation efficiency on separators. [2]

Indicator		Concentrate		Waste		Intake	
Contents of mag. fractions [%]	98.14	c	86.23	b	0.45	a	37.43
Yield [%]		vc	43.11	vb	56.89	va	100
Amount of magnetic portions [kg/10 t]		C	3717.64	B	25.36	a	3743
Contents of magnetite [%]		cr	87.87	br	0.45	ar	38.14
Contents of coal slurry [%]	slurry	cb	12.13	bb	99.55	ab	61.86
Sum			100		100		100
High density material		Cr	3788.28	Br	25.84	Ar	3814.11
Contents of slurry	slurry	Cb	523.04	Bb	5662.85	Ab	6185.89
Sum			4311.31		5688.69		10000
Yield [%]	Fe_3O_4	mc	99.23	mb	0.68	ma	100
Yield [%]	of slurry	wc	8.46	wb	91.54	wa	100
Efficiency [%]			90.87	+	-90.87	=	0.00

Tab. 4 Calculations of efficiency of magnetic separation on a ČSM Mine's separator in 2010

Indicator		Concentrate		Waste		Intake	
Contents of mag. fractions [%]	97.94	c	88	b	0.35	a	31.7
Yield [%]		vc	35.76	vb	64.24	va	100
The amount of magnetic portions [kg/10 t]		C	3147.52	B	22.48	a	3170
Contents of magnetite [%]		cr	89.85	br	0.36	ar	32.37
Contents of coal slurry [%]	slurry	cb	10.15	bb	99.64	ab	67.63
Sum			100		100		100
Yield [%]	Fe_3O_4	mc	99.29	mb	0.71	ma	100
Yield [%]	of slurry	wc	5.37	wb	94.63	wa	100
Efficiency [%]			93.92	+	-93.92	=	0

Tab. 5 Waste water

2011	Waste water [m^3]	Charge [t]	Loss of magnetite [kg]	Loss of magnetite [kg/t] of charge
JANUARY	46000	136300	13800	0.10
FEBRUARY	50670	147460	15201	0.10
MARCH	51570	155440	15471	0.10
APRIL	47290	183570	14187	0.08
MAY	41850	137430	12555	0.09
JUNE	43460	163419	13038	0.08
AVERAGE	46807	153937	14042	0.09
TOTAL	**280840**	**923619**	**84252**	**0.09**

5 CONCLUSIONS

The paper shows that the used magnetite is of good quality, losses of magnetite particles adhering to the surface of products being separated are minimal and the efficiency of magnetic separation is sufficient. In addition to technological subordination and regular inspections of the function of equipment in the preparation plant, a reduction of the magnetite consumption can be achieved by a limited discharge of waste water. In order to minimize the discharge of waste water, the water and slurry circuits in the preparation plant must work well. These circuits are interconnected. The waste water discharged from the Lazy preparation plant contains 0.3 g/l of magnetite on average. The magnetite contained in the waste water is irretrievably lost.

Table 5 presents the amount of waste water discharged from the preparation plant, the amount of processed coal, magnetite losses in kg and the expression of losses of magnetite in the waste water in relation to the currently processed amount of coal.

REFERENCES

[1] NOVÁČEK, J.: – *Technologie úpravy uhlí I*, VŠB – TU OSTRAVA, 2000

[2] TŘINECKÝ, D.: – *Vyhodnocení účinnosti regenerace magnetitu na dole Karviná, závod Lazy*, BAKALÁŘSKÁ PRÁCE, VŠB – TU OSTRAVA, 2011

[3] ŘEPKA, V. a kol.: - *Provedení garančních zkoušek hydrocyklonové linky B na OKD, a.s. VOJ Důl ČSM* HS 542003 RPS Ostrava, a.s. Ostrava

RESUMÉ

Magnetit slouží jako zatěžkávadlo, ze kterého se společně s vodou připravuje těžká suspenze. Důvodem použití suspenze je vytvořit prostředí sloužící k rozdružování produktů. Rozdužování je proces, kde dochází na základě rozdílných vlastností materiálů, v našem případě rozdílné hustoty materiálů, k jeho oddělování. Rozdružování na úpravně dolu Karviná, závodu Lazy probíhá v těžkosuspezním rozdružovači SM (Staatsmijnen) a v hydrocyklónech, kde jsou od sebe oddělovány uhlí, proplástek a hlušina. Při rozdružování ulpívají částice zatěžkávadla na povrchu rozdružovaného materiálu. Nejsou-li odstraněny, mohou způsobit zhoršení kvality vyrobených produktů a současně představují spotřebu (ztráty) zatěžkávadla. Za účelem minimalizace ztrát zatěžkavádla (ekonomicky významné) jsou produkty rozdružování oplachovány, a tím je z jejich povrchu odstraňováno ulpělé zatěžkavádlo. Vzniká tím směs částic zatěžkávadla, jemných částic rozdružovaného materiálu i ostatních příměsí a vody, kterou označujeme jako zředěnou suspenzi. Pro další (opakované) použití zatěžkavádla je potřebné, částice zatěžkávdla ze zředěné suspenze znovu získat pro přípravu nové suspenze. Příspěvek se vedle popisu použití magnetitu na úpravně dolu Karviná, závodu Lazy, zabývá také oblastmi, které výrazně ovlivňují množství spotřeby magnetitu. Jedná se o tyto oblasti:

- *vlastnosti magnetitu používaného na úpravně závodu Lazy,*
- *ztrátami magnetických podílů na produktech,*
- *regenerací zředěné suspenze na závodě Lazy a porovnáním regenerací zředěné suspenze s dolem*

ČSM. Příspěvek se také zabývá odběrem a zpracováním vzorků v laboratořích VŠB – TO OSTRAVA, jejich vyhodnocením a hledáním cest vedoucích ke snížení spotřeby magnetitu na dole Karviná, závodu Lazy.

MEMBRANE PROCESSES FOR INDUSTRIAL WASTEWATER TREATMENT – DEMINERALIZATION STATION

Jitka HAJDUKOVÁ[1], Jan THOMAS[2], Petra MALÍKOVÁ[3], Veronika MATÚŠKOVÁ[4]

[1]Ing., Ph.D., Institute of Clean Technologies for Mining and Utilization of Raw Materials for Energy Use, Faculty of Mining and Geology, VŠB – Technical University of Ostrava, 17. listopadu 15, Ostrava - Poruba
e-mail: jitka.hajdukova@vsb.cz

[2]Ing., Ph.D., Institute of Clean Technologies for Mining and Utilization of Raw Materials for Energy Use, Faculty of Mining and Geology, VŠB – Technical University of Ostrava, 17. listopadu 15, Ostrava - Poruba
e-mail: jan.thomas@vsb.cz

[3]Ing., Ph.D., Institute of Clean Technologies for Mining and Utilization of Raw Materials for Energy Use, Faculty of Mining and Geology, VŠB – Technical University of Ostrava, 17. listopadu 15, Ostrava - Poruba
e-mail: petra.malikova@vsb.cz

[4]Ing., Institute of Clean Technologies for Mining and Utilization of Raw Materials for Energy Use, Faculty of Mining and Geology, VŠB – Technical University of Ostrava, 17. listopadu 15, Ostrava - Poruba
e-mail: veronika.matuskova@vsb.cz

Abstract

The paper deals with treatment of wastewater produced by the energy industry using a membrane processes (electrodialysis). Prior to application of this technology, the influence of wastewater on the chosen type of membranes is solved. Specifically, it means eliminating the influence of formation soluble (insoluble) salts which could negatively affect water treatment by the chosen technology. The paper also summarizes the results of laboratory tests of electrodialysis with the aim to acquire treated wastewater of such quality meeting legislative limits for discharges to surface water (i.e. total dissolved solids below $1 g.l^{-1}$).

Abstrakt

Příspěvek se zabývá problematikou úpravy odpadních vod z energetiky s využitím membránových procesů (elektrodialýzy). Před samotnou aplikací této technologie je řešen vliv chemismu upravované vody na použitý typ membrán. Konkrétně jde o eliminaci vlivu tvorby málo rozpustných (nerozpustných) solí, které by úpravu vody zvolenou technologií negativně ovlivnily. Příspěvek dále shrnuje získané výsledky doposud provedených testů elektrodialýzy se zaměřením na získání upravené odpadní vody vyhovující legislativním limitům pro vypouštění do vod povrchových (tj. rozpuštěné látky pod $1 g.l^{-1}$)

Key words: wastewater, treatment, chemical precipitation, clarification

1 INTRODUCTION

Membrane technologies are the ones of the ways of wastewater treatment. In comparison with conventional technologies, they have considerable advantages. These are high quality of treated water and reduction of the amount of chemical agents. Other advantages are high selectivity and low energy consumption. The development of the membrane technology affects current increased demands for the quality of discharged wastewater and current needs of high-quality process water.

The scale of desalinated water usage is largely dependent on the cost effectiveness of the desalination process. When membrane methods are used, the basic costs for desalination include those of water pre-treatment, membranes and production equipment, electrical power consumption for desalination and auxiliary operations [1].

The application of membrane technologies is limited by the quality of feed water. The performance of membrane process may be limited by the concentration changes in the vicinity of the membrane or the deposition of dirt on the surface of the membrane material. Therefore, the first indispensable step in the design of water treatment using membrane technology is a suitable pre-treatment method.

This paper summarizes the results of laboratory experiments of wastewater treatment using the laboratory units EDR-Z10/0.8. Electrodialysis (ED) was developed first for the desalination of saline solutions, particularly brackish water. The production of potable water is still currently the most important industrial application of electrodialysis. Other applications, such as the treatment of industrial effluents, the production of boiler feed water are gaining increasing importance in large-scale industrial installations [2,3]. Another industrial application is the production of table salt in Japan and Kuwait. In this application, ED is used for the concentration of NaCl from seawater prior to evaporation [3]. For water with a relatively low degree of salinity (below 5 $g.l^{-1}$), ED is an economically advantageous process compared to reverse osmosis [4]. Also water with high salinity (30 $g.l^{-1}$) may be successfully treated by ED. Thanks to the high selectivity and low demands for chemical agents, ED may be a reliable and effective method in wastewater demineralization.

2 MATERIALS AND METHODS

2.1 Wastewater characteristics

The research is focused on the treatment of wastewater from a demineralization station generated during chemical water treatment for boilers, turbines and piping. The chemical treatment of a water demineralization station with a mixed bed (MB) assures consistent water quality of treated water required for each device. MB consists of a homogenous mixture of strongly acid and strongly basic anion-exchange resins. The chemical agents for water treatment are: HCl, $Fe_2(SO_4)_3$ and hydrate $Ca(OH)_2$.

Wastewater produced by cleaning ion-exchange resins. Wastewater discharged together with other wastewater from the drainage system on a physico-chemical treatment plant. The treated water is discharged to surface water; part of this water is recycled as operating water.

The aim of the research is to design and verify the effectiveness of water treatment technology using electrodialysis. The criterion for successful application is to obtain the treated water meeting legislative limits for its discharge into surface waters.

Characteristics of the quality of the wastewater are given in Tasb.1 below. The analyses were carried out in the laboratory of VŠB-TU Ostrava, at the Institute of Clean Technology. The wastewater contains significant quantities of ions that can decrease the performance and life-span of the membrane technology. The problematic ions are: Ca^{2+} and Mg^{2+} (formation of low soluble salts). Heavy metals and humic substances could also pose a risk for the electrodialysis process as they foul the active centres on the surface of the ion-exchange membrane and prevent entrance of salts to be converted into the membrane. Increase levels of COD and BOD may cause problems for the membrane technology (formation of gel layer) as well.

Tab 1: Industrial wastewater characteristics

INDICATOR	UNIT	SAMPLE		
		1	2	3
pH		7.81	6.66	6.97
K	[$mS.cm^{-1}$]	10.2	11.63	13.35
Turbidity	[ZF]	31.97	>40	>40
HCO_3^-	[$mg.l^{-1}$]	220.82	68.32	208.01
CO_2	[$mg.l^{-1}$]	25.52	47.96	70.84
Hardness	[$mmo.l^{-1}$]	16.46	17.35	29.15
Ca^{2+}	[$mg.l^{-1}$]	526.65	535.07	983.16
Mg^{2+}	[$mg.l^{-1}$]	80.69	91.77	112.56
Cl^-	[$mg.l^{-1}$]	2 907.15	3 261.68	4 148.00
SO_4^{2-}	[$mg.l^{-1}$]	720.00	820.00	1238.00
COD	[$mg.l^{-1}$]	33.60	47.84	52.20
TOC	[$mg.l^{-1}$]	41.56	-	-
BOD	[$mg.l^{-1}$]	4.97	5.29	-
Mn	[$mg.l^{-1}$]	<0.30	<0.30	<0.30
Fe	[$mg.l^{-1}$]	0.15	0.25	0.52
Al	[$mg.l^{-1}$]	<0.10	<0.10	<0.10
NO_3^-	[$mg.l^{-1}$]	193.00	115.00	150.00
TDS	[$mg.l^{-1}$]	7 640.00	8 036.00	10 778.00

In order to apply the membrane technology (electrodialysis), it is necessary to eliminate the influence of undesirable constituents. It is necessary to achieve such quality of feed water which corresponds to the ED standard (critical) limit for the type of used membranes and equipment.

2.2 Water treatment for applying membrane technology

Based on the observed chemical composition of wastewater, the need occurred to adapt the technology of water treatment by focusing on pre-treatment – elimination of the influence of problematic quality indicators on the membrane technology (see Tab. 2). In particular, it is the case of removal of Ca^{2+}, Mg^{2+} and SO_4^{2-} ions.

Tab. 2: Standard and critical indicators for ED [5]

INDICATOR	UNIT	STANDARD	CRITICAL
pH		2-8	>10
Turbidity	[NTU]	0.5	5
Mn	[mg.l^{-1}]	0.1	1.0
Fe	[mg.l^{-1}]	0.30	3.00
Al	[mg.l^{-1}]	0.10	1.00

In order to choose a suitable pre-treatment method prior to the application of membrane technology, several methods were tested:

- reducing the concentration of Ca^{2+} and Mg^{2+} ions
  - alkalization (NaOH), sedimentation, filtration, acidification;
  - precipitation (Na_2CO_3), sedimentation, filtration;
  - precipitation ($Na_3PO_4.12H_2O$), sedimentation, filtration;
- reducing the concentration of SO_4^{2-} ions
  - alkalization, sedimentation, filtration, acidification;
  - precipitation (metal salts of Al^{3+}), sedimentation, filtration, acidification;
  - precipitation according to Patent No. 290953 [6];
- reducing the overall pollution
  - clarification, sedimentation, filtration;
- antiscalant.

2.3 Electrodialysis equipment

The tests were carried out on the lab-scale units EDR-Z/10-0.8; the electrodialysis module is fitted with 10 pairs of ion-selective membranes AM(H)-PES and CM(H)-PES. The tests were conducted in a form of batch cycles (batch process). The time duration was chosen based on the desired quality of the treated water (diluate).

The first part of the research was focused on the determination of specifications and operating parameters for the maximum of desalination rate. Another part of the research was focused on the relevant performance parameters and the time required for reaching the total dissolved solids concentration in the treated water below 1 g.l^{-1} [7].

The procedure of the batch test is as follows:

- dosage relevant volume of wastewater;
- 250 ml electrode solution for electrode circuit;
- set desired voltage;
- switch on equipment;
- pH, conductivity, temperature, voltage and current measured at interval of 5 minutes;
- analysis of products.

3 RESULTS AND DISCUSION

3.1 Water treatment for applying membrane technology

Alkalization and acidification are effective methods for reducing the concentration of Mg^{2+} ions (Fig. 1). To reduce the concentration of Ca^{2+} ions is ineffective. After alkalization, it is necessary to introduce the stage of neutralization which ensures an optimum pH value for the course of membrane processes. This type of pre-treatment brings along high costs of both the equipment and the own operation. It is necessary to deal with the formed sludge.

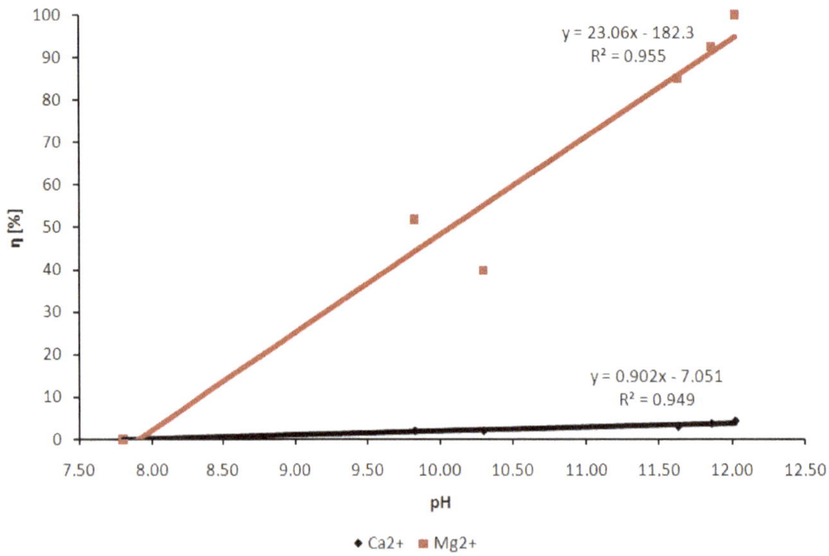

Fig. 1: Reduction of Ca^{2+} and Mg^{2+} ions – alkalization NaOH

Better results were obtained when using the precipitants Na_2CO_3 and $Na_3PO_4.12H_2O$. The main conditions for the tests were the optimum pH of treated water, i.e. less than 10 (the critical value of pH for the operation of ED). The tests results are shown in Fig. 2.

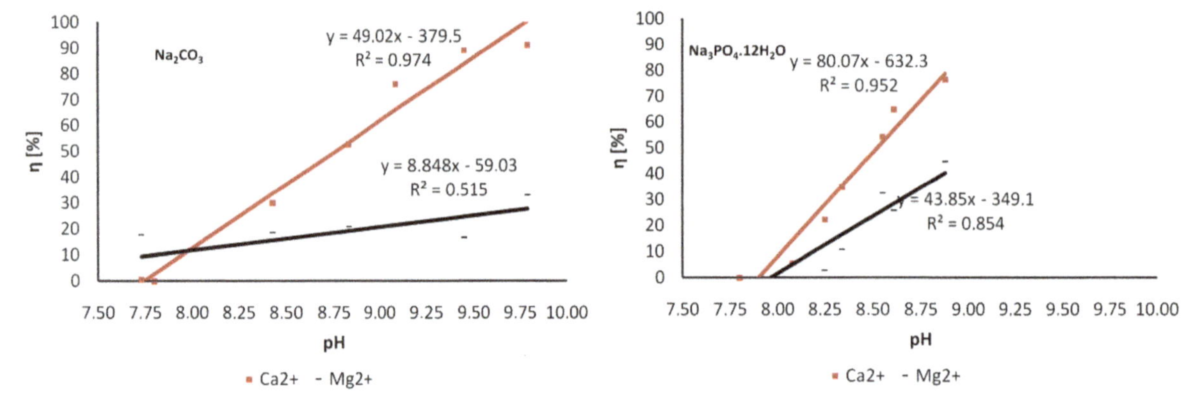

Fig. 2: Reducing the concentration of Ca^{2+} and Mg^{2+} ions – precipitation Na_2CO_3 and $Na_3PO_4.12H_2O$

The best results were achieved during clarification through the use of aluminium coagulants. Satisfactory results were obtained for both turbidity and total water pollution (COD). The indicator COD decreased by 53 %, turbidity reducing efficiency was around 98 %. During the tests, the emphasis was placed on keeping the residual coagulant Al^{3+} concentration below $0.1 g.l^{-1}$ and a minimal change of pH. The advantage of this method is the simultaneous reduction of sulphate ions.

The possibility of using antiscalant for wastewater treatment has not been confirmed. The problem is associated with the short functionality of antiscalant and its circulation in the concentrate circuit. It creates the membrane gel deposition. For the application of antiscalant, the optimization depending on the proper selection of a type and dosage of antiscalant is very important. When selecting a wrong type or non-optimal dosage, the effect of pre-treatment is rather the opposite.

3.2 Electrodialysis equipment

Fig. 3 and Fig. 4 show the course of the batch tests without pre-treatment and with pre-treatment of wastewater clarification. The desire degree of desalination is reached in the course of about 45 minutes and 14V (an input capacity of wastewater – 3 litters). The voltage regulating flow rates of the circuits are in a range of 60-70 $l.h^{-1}$.

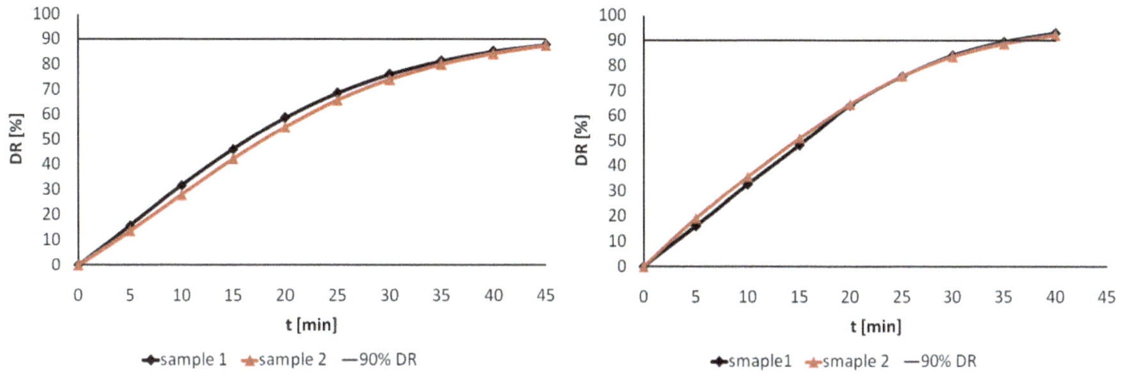

Fig. 3: Desalination rate of wastewater without pre-treatment (left) and with pre-treatment clarifier (right)

Without pre-treatment of wastewater, the process of desalination slows down, a precipitate is formed and the gradual introduction of input and output concentrate current happens. The desalination rate of 90 % without pre-treatment was achieved after 45 minutes of the test. The same effect with pre-treatment was achieved after 35 minutes of testing.

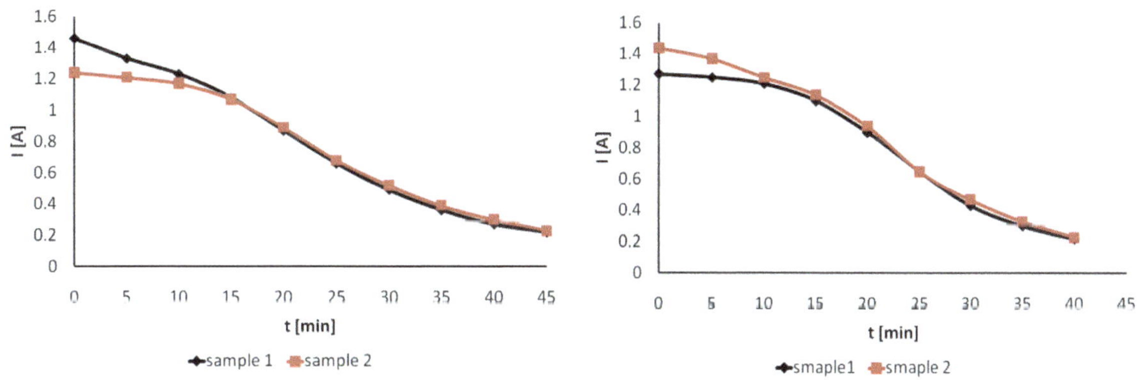

Fig. 4: Course of ED tests without pre-treatment (left) and with pre-treatment clarifier (right)

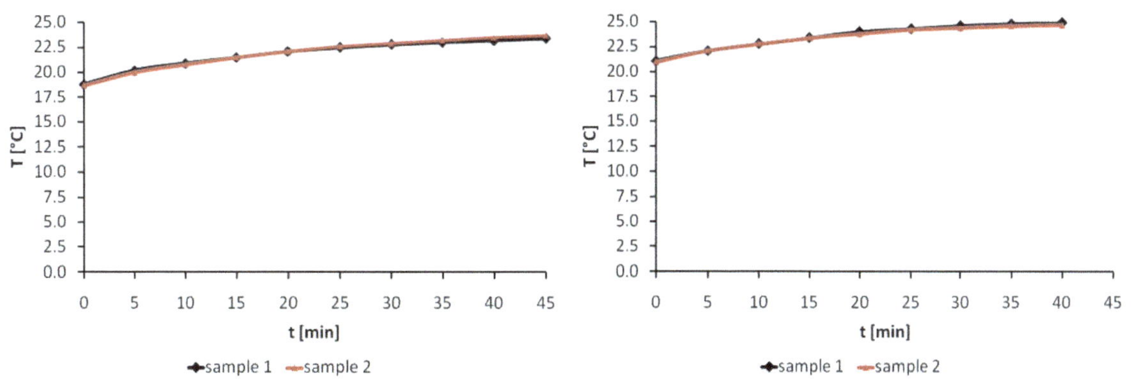

Fig. 5: Temperature of wastewater without pre-treatment (left) and with pre-treatment clarifier (right)

The quality of treated water with the application of electrodialysis corresponds to legislative limits for surface water in all indicators. Another product of desalination is a concentrated solution which is an area of our future research.

3 CONCLUSIONS

The results of previous research confirmed the possibility of using membrane technology, specifically electrodialysis, for treatment of wastewater from a demineralisation station. The successful application of the technology is dependent on many factors, especially on the choice of a suitable membrane type, associated with the pre-treatment of feed water for the application of membrane technology. The choice of an appropriate pre-treatment method is a complex process that requires the thorough comparison of the requirements for the quality of treated water, the requirements for operating conditions, and the associated economic cost of the process water treatment. The areas for further research are to verify the results obtained during feed-and-bleed tests and dealing with the treatment and further treatment of ED waste product.

ACKNOWLEDGEMENTS

This study was a part of the research project no. SP 2013/103 and has been done as well in connection with the project of the Institute of clean technologies for mining and utilization of raw materials for energy use, reg. no. CZ.1.05/2.1.00/03.0082 supported by the Research and Development for Innovations Operational Programme.

REFERENCES

[1] GREBENZUK, V.D., CHEBOTAREVA, R.D., PETERD, S., LINKOV, V. Surface modification of anion/exchange electrodialysis membranes to enhance anti/fouling characteristics. *Desalination*. 1998, vol. 115, p. 313-329. ISSN 0011-9164.

[2] R.D. Noble, S.A. Stern (Eds.), Membrane Separation Technology Principles and Applications, Elsevier, Amsterdam, 1995. ISBN 044481633X.

[3] Ghyselbrecht, K., et al. Desalination of industrial saline water with conventional and bipolar membrane electrodialysis. *Desalination*. vol. 2013, *318*, p. 9–18. ISSN 0011-9164.

[4] BANASIAK, L. J., KRUTTSCHNITT, T. W., SCHÄFER, A.I. Desalination using electrodialysis as a function of voltage and salt concentration, *Desalination*. 2007, 205, pp. 38-46. ISSN 0011-9164.

[5] MEGA a.s. Technická dokumentace – elektrodialyzační modul ED(R) –Z/10, Stráž pod Ralskem, 2009.

[6] CZ č.290953 „Způsob čištění důlních vod s nadlimitním obsahem síranů", 2002.

[7] Nařízení vlády č. 61/2003 Sb. o ukazatelích a hodnotách přípustného znečištění povrchových vod a odpadních vod, náležitostech povolení k vypouštění odpadních vod do vod povrchových a do kanalizací a o citlivých oblastech, ve znění nařízení vlády č. 229/2007 Sb. a nařízení vlády č. 23/2011 Sb.

RESUMÉ

Předložený příspěvek shrnuje dosavadní výsledky testů úpravy odpadních vod z demineralizační stanice s využitím laboratorní jednotky EDR-Z10/1.0. Tyto odpadní vody vznikají zpětným vymýváním směsného lože. Z pohledu možné aplikace membránové technologie odpadní voda obsahuje významné množství iontů, které by bez předúpravy mohly snižovat výkon a životnost membránové technologie. Jde zejména o kombinaci iontů Ca^{2+} a Mg^{2+} (tvorba málo rozpustných solí). Problematické by pro proces elektrodialýzy mohly být i těžké kovy a huminové látky (zanáší aktivní centra na povrchu iontovýměnné membrány a znemožňují vstup do membrány převáděným solím).

První část výzkumu se zabývá návrhem vhodného typu předúpravy před samotnou membránovou technologií. Další část výzkumu se zaměřuje již na testy elektorodialýzy ve vsádkovém režimu (batch test). Cílem těchto testů je ověřit účinnost zvolené předúpravy a samotné membránové technologie. Při laboratorních testech je kladen důraz na zjištění maximálního stupně odsolení testovaných vod a zjištění provozních parametrů potřebných pro dosažení požadované kvality upravené vody. Kritériem pro úpravu odpadních vod je dosažení koncentrace RL_{105} pod 1 $g.l^{-1}$.

Provedené testy potvrdily, že bez předúpravy odpadní vody dochází ke zpomalení procesu odsolení, k tvorbě sraženin a k postupnému zanesení vstupu a výstupu koncentrátového proudu. 90% účinnosti odsolení bylo u odpadní vody bez předúpravy dosaženo po 45 minutách testu. U předupravené odpadní vody čířením bylo stejné účinnosti dosaženo po 35 minutách testu. Upravená odpadní voda vyhovuje legislativním limitům pro vypouštění do vod povrchových ve všech námi sledovaných ukazatelích. Dalším produktem odsolení je zakoncentrovaný roztok (koncentrát), který je oblastí našeho dalšího výzkumu.

10

FILLING MIXTURES AND LEGISLATIVE ASPECTS OF THEIR APPLICATION FOR THE DISPOSAL OF MINE WORKINGS

Alena ORLÍKOVÁ [1]

[1] *Ing., Institute of Environmental Engineering, Faculty of Mining and Geology, VSB - Technical University of Ostrava, tř. 17. listopadu 15, 708 33 Ostrava-Poruba*
e-mail: alena.orlikova.st@vsb.cz

Abstract

The paper deals with the legislative aspects of the use of filling mixtures during the remediation of mining activities. The enterprises, carrying out the disposal of mine workings, buy filling mixtures within public procurement and need the quality of the products to be guaranteed. The requirements laid down in the technical guidelines for the certification of specified construction products raise doubts about the safety of products in terms of hygiene, protection of health and the environment. Currently, legislative development is underway in this area, striving for the implementation of the REACH Regulation into technical guidelines and thus to achieve the filling mixture to be produced and used in a safe manner.

Abstrakt

Příspěvek se zabývá legislativními aspekty používání základkových směsí při zahlazování následků hornické činnosti. Podniky, které provádějí likvidaci důlních děl, nakupují základkové směsi v režimu veřejných zakázek a potřebují záruku kvality výrobku. Požadavky zakotvené v technických návodech pro certifikaci stanovených stavebních výrobků vyvolávají pochybnosti o bezpečnosti výrobku z hlediska hygieny, ochrany zdraví a životního prostředí. V současné době probíhá v této oblasti legislativní vývoj, který směřuje k implementaci nařízení REACH do technických návodů a tím k dosažení cíle, aby se základkové směsi vyráběly a používaly bezpečným způsobem.

Key words: filling mixture, remediation of mining activities, technical guidelines, REACH Regulation

1 INTRODUCTION

The existence of mine workings has a significant impact on individual components of the environment in the long term. The major problem is a large number of old and abandoned mine workings, occurred from past mining operations and activities involving mining, which must be disposed under the applicable legislation in order not to become a risk in the future.

2 REMEDIATION OF MINING ACTIVITIES

The Czech Republic gives support to the coal industry for remediation of mining activities. The aid is compatible with proper functioning of the common market of the European Union. The funds are intended solely to cover inherited liabilities that arise in connection with the closure of coal mines. The Czech coal industry does not receive any grants for current coal mining operations. In this respect, the Czech Republic acts in accordance with Council Regulation (EC) No. 1407 2002 of 23 July 2002, on state aid to the coal industry.

In order to maintain the competitiveness of Czech private mining companies, an addressable system of providing state funds had to be created, exclusively for mining companies in decline that are owned by the state. Within the restructuring process of the coal mining industry in the Czech Republic, the sites, where mining had been completed, were separated from private mining companies.

The state enterprises that have adopted the subdued localities are provided with funds from the state budget for remediation of mining operations [1]. Thus addressing the consequences of ecological loads, incurred before the effective date of the amendment to Act No. 44 1988 Coll. on the protection and exploitation of mineral resources (Mining Act), was ensured in 1993. In that year, the Mining Act was amended by Act No. 168 1993 Coll. which imposes the obligation on mining companies to create reserve funds to ensure the settlement of mining damages.

The state enterprises DIAMO, s. p. and Palivový kombinát Ústí, s. p. (Fuel Combine) provide for the remediation of mining operations at two levels based on the resolution of the Government of the Czech Republic. On the one hand, they provide the disposal of mine workings, and on the other, the remediation and reclamation of the affected area. This is significant fact in terms of Act No. 100 2001 Coll., on environmental impact assessment and on amendments to certain related acts (Act on Environmental Impact Assessment).

Underground coal extraction becomes evident on the surface by two typical elements. These are rock (mine stone) dumps and subsidence due to undermining [2]. Uncontrolled outputs of mine gases, from which especially methane poses a threat to the environment, are concomitant phenomena of deep mining in OKR as well.

By nature of remediation and reclamation works, the remediation of mining damage on the surface subjects to the assessment under provisions of Article 4 of Act, Environmental Impact Assessment, according to Section 1.3 Water regulation, or other adjustments affecting drainage conditions on the area of 10-50 hectares, or in compliance with Section 2.10 Waste disposal, by depositing to natural or man-made rock structures and gaps, category II of Annex 1 of this Act [3].

The disposal of mine workings does not subject to the environment impact assessment of the project. The disposal of mine workings is performed by an organization under provisions of Act No. 61 1988 Coll., on mining activities, explosives and the State Mining Administration (Act on Mining Activities).

When the mine liquidation is prepared by an organization, a disposal plan for the mine working must be made in accordance with Article 32 (4) of Mining Act and Decree No. 104 1988 Coll., on efficient use of reserved deposits, on permits and notification of mining operations and activities involving mining. The application for the authorisation to safeguard or dispose mine workings must be submitted by the organization to the District Mining Office.

Mining companies may use mine waste for the disposal of their own mine workings being closed. The state-owned enterprises, which use state funds in the process of remediation of mining activities, shall procure needed services and materials for the mine disposal under provisions of Act No. 137 2006 Coll., on Public Contracts, as amended, as a certified product – a filling mixture. They need to guarantee the quality of the product in relation to the environment.

3 FILLING MIXTURES

In the submitted plan for the safeguard or disposal of mine, the method for securing mine workings must be specified. The organization is responsible for the correctness of the selected solution and shall demonstrate its expertise for mining operations. Therefore, the organization is also responsible for using the chosen filling mixture in compliance with the environmental legislation.

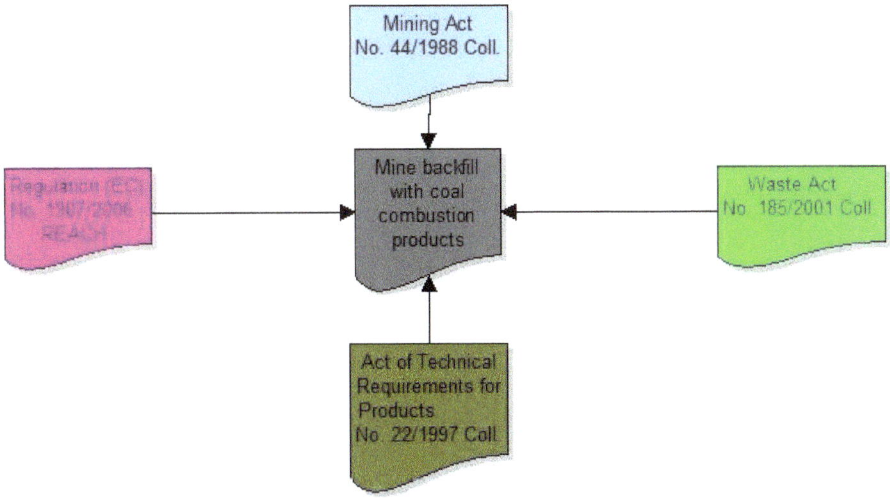

Fig. 1 Scheme of legal process control

In terms of mining legislation, the organization shall proceed in accordance with Decree No. 52 1997 Coll., on the requirements for health and safety at work and the safety of operations during the disposal of main mine workings. Provisions of Article 6 (1) impose: For the disposal of a main mine working, only non-flammable, insoluble, non-swellable and not turning slushy backfill materials can be used that may not pollute the environment with harmful vapours or leachates with toxic substances. In order to use another backfill material than natural stone, natural stone aggregates or secondary raw materials, the organization must request for the prior approval of competent environmental authorities [4]. Unfortunately, it is not clear which ones, as the decree refers to Article 4 of Act No. 238 1991 Coll., on waste, having been cancelled.

Fig. 2 Closing old mine (the photo taken by UNIGEO, a.s.)

These conditions are further substantially complicated when using a filling mixture made on the basis of energy by-products, until recently considered solely a waste. Manufacturers and users will thus get to a conflict with waste and product legislation requiremments. The organization is suddenly required to be well informed about its responsibilities imposed to the manufacturer of a filling mixture to comply with the requirements of responsible approach to the environment. These obligations must then be involved by the organization in terms and conditions for the safeguarding or disposal of a mine. The documentation submitted by the manufacturer of the filling mixture shall be then included in Annexes of the plan for safeguarding elaborated by the organization as part of the application for a mining permit to the extent specified in Article 6 of Decree No. 104 1988 Coll., as amended.

4 SPECIFIED PRODUCT CERTIFICATION

Filling mixtures are specified products under Act No. 22 1997 Coll., on Technical Requirements for Products. Therefore, before placing them on the market in the Czech Republic, they subject to mandatory conformity assessment and manufacturer's declaration of conformity according to the relevant Government Regulation.

In terms of conformity assessment, filling mixtures are by legislation included in the list of construction products which are assessed according to Government Regulation No. 163 2002 Coll., as amended by 312 2005 Coll. laying down technical requirements for selected construction products. They are included in Section 9 of Appendix 2 Special materials, products, structures and equipment under order No 13. The backfill material intended for disposal of main and old mine workings by backfilling operations. The certification under Article 5 of Government Regulation is a prescribed method for assessing the conformity of product [5, 6]. Such certificate or report of conformity assessment issued by an authorized person to the manufacturer for a specified product in relation to the requirements of Government Regulation No. 163 2002 Coll. forms the basis for issuing the declaration of conformity in accordance with Act No. 22 1997 Coll.

The identification of product properties of backfilling and filling materials must be provided by the manufacturer under Article 2 of Government Regulation No. 163 2002 Coll. through an authorized person. Based on these technical findings, the authorized person shall provide the manufacturer with the Building Technical Certificate with a limited duration for max 5 years, determining the technical characteristics of the product in relation to the basic requirements for the liquidation of mine, depending on the role the product has in the liquidation. This period may be extended by the authorized person.

5 REQUIRED PRODUCT PROPERTIES

Required characteristics, quality levels, end-use properties of materials intended for backfilling the mine workings being liquidated, are not fully identified in the defined standards. Therefore, the Technical Guidelines (TG) were elaborated which is a technical document prepared for construction products listed in Annex 2 to Government Regulation No. 163 2002 Coll. All authorized persons are obliged to follow the Technical Guidelines during the implementation of certification in accordance with the conditions for granting the authorization by the Office for Standards, Metrology and Testing (COSMT). The Technical Guidelines are constantly updated according to changes in the technical standardization and according to new knowledge in the field of science and technology [7]. Currently the following documents are valid for filling mixtures:

Name	**Number of Technical Guidelines**	**Date of Registration**
Backfill material intended for disposal of main and old mine workings by backfilling operations		
Consolidated backfill material – CBM	9. 13. 2001	1. 1. 2013
Unconsolidated backfill material – UBM	9. 13. 2002	1. 1. 2013

According to the Technical Guidelines, the UBM is rocky waste loose material from a grading, separating and washing treatment system of raw hard (black) coal composed of Carboniferous rocks – siltstones, sandstones and claystones. The CBM is, by a specific-recipe defined, optimal dense water composite slurry consisting of solid residues from the combustion of coal or solid residues from co-incineration (co-firing) of coal with biofuels.

Looking more closely at the Technical Guidelines, it can be found in which way one of the basic requirements is fulfilled, namely the requirement in terms of hygiene, health and the environment protection in accordance with Annex 1 to Government Regulation No. 163 2002 Coll., Article 3, letter (d), i.e. to prevent pollution of water or soil. This requirement is addressed by the reference to Decree No. 294 2005 Coll. of the

Ministry of the Environment, as amended, Annex 2, Article 1 and 2, i.e. only by specifying the sample treatment and analytical methods without links to the defined classes of leachability of wastes in Article 3. The material tested for certification will no longer be a waste, but it is interesting that limit values of the content of elements in water extract of the product given in the Technical Guidelines are higher for several monitored indicators (toxic metals) compared to the corresponding values of the Class I of leachability for landfilling the group S – inert waste. Similarly, it is also important that these limit values exceed the limits (for relevant indicators) of groundwater contamination according to the guidelines of the Ministry of the Environment „Contamination Indicators" established for making decisions on the necessity of rehabilitating interventions on sites affected by anthropogenic pollution. The guideline provides the values of indicators based on the values of the U.S. Environmental Protection Agency – USEPA valid in June 2011 [8]. Exceeding them should request further survey or removal of pollution.

Tab. 1 Comparison of the values of indicators imposed by individual regulations

Type of parameter	Indicator	Unit	Limit value Technical guidelines 09.13.01 19.13.02	Leachability of wastes - Class I Decree No. 294/2005 Coll.	Guideline by the Ministry of the Environment Indicators of pollution Ground water
Chemical elements in water extract	As[1)]	mg l	0.1	0.05	0.000045
	Ba	mg l	1.0	2	7.3
	Be	mg l	0.005	.	0.073
	Pb	mg l	0.1	0.05	0.01
	Cd	mg l	0.005	0.004	0.018
	Cr_{total}	mg l	0.1	0.05	.
	Co	mg l	0.1	.	0.011
	Cu	mg l	1.0	0.02	1.5
	Ni	mg l	0.1	0.04	0.730
	Hg	mg l	0.005	0.001	0.00063
	Se	mg l	0.05	0.01	0.18
	Ag	mg l	0.1	.	0.18
	V	mg l	0.2	.	0.18
	Zn	mg l	3.0	0.04	11

[1)] In case of arsenic in the Czech Republic, the concentrations higher than the mentioned above indicators of pollution are common due to geochemical conditions in the rock environment. In such cases, only arsenic concentrations exceeding natural background under site-specific conditions of the evaluated location represent the indication of pollution.

The values of indicators of pollution do not represent rehabilitation limits and should not be used as rehabilitation limits either [8]. However, the found out exceeding of the values of indicators is assessed as an indication of pollution that should be further studied and evaluated. If the filling mixture with its properties could negatively affect the environment, its use may be considered as handling harmful substances, i.e. operating activities pursuant to Act No. 167 2008 Coll., on the prevention and remedying of environmental damage and on amendments to certain Acts.

The risk of using products of coal combustion residues was assessed within the project SP 2f3 118 08 entitled "Investigation of real properties of waste that is considered a suitable source of non-standard raw materials (especially energy by-products) according to current legal requirements for public-health and environmental protection and the evaluation of the gained information for establishing safe procedures and requirements for their use". In the years 2008-2010, in addition to the researcher – Univerza-Středisko odpadů Praha, s.r.o., also the National Institute of Public Health (NIPH) in Prague and the Institute of Chemical Technology in Prague participated as co-organizations in the project. In conclusions, based on the obtained objective information about the properties of energy by-products and based on the processing of model risk analyses, the researchers note that the release of products of coal combustion residues into an unprotected environment can cause damage to the environment [9].

It is clear that both products and wastes can contain toxic materials, and if not properly handled or controlled, they may pose a risk to human health or the environment. Additionally, industrial and mining wastes often have characteristics that in comparison with products may pose a risk to the environment. This relates to the fact that while the composition of products is generally specifically formed and controlled, the composition of wastes may be less apparent. That means that from an environmental point of view, it is extremely important the materials to be correctly classified as waste or non-waste substances [10].

6 REACH REGULATION REQUIREMENTS

Doubts arising from the conclusions of this study can be eliminated by a consistent application of the requirements of Regulation of the European Parliament and Council (EC) No 1907 2006 concerning the Registration, Evaluation, Authorisation and Restriction of Chemicals (REACH) to the certification of consolidated backfill materials. If wastes are used as materials and reusable substances and their mixtures are recovered from them, the newly acquired materials switch from the scope of waste legislation to the scope of REACH. The main responsibility the manufacturer shall have to fulfill under this Regulation is the obligation to register substances produced in quantities of 1 tonne or more per year.

The Technical Guidelines No 09 13 01 include the REACH Regulation in the list of technical regulations that apply to the product „Consolidated backfill material – CBM". With regard the input materials, the Technical Guidelines No. 9.13.02 for UBM does not include any reference to the REACH Regulation.

The consolidated filling mixture consists of several substances. Every single substance, representing a component of the mixture, must be identified and registered in accordance with the REACH Regulation. The Substance Identity Profile (SIP) defines the requirements for the substance origin, its production process and chemical composition (a concentration range for each component). This is particularly important in the case of UVCB substances which also include ashes which are components of CBM. Only the substance, meeting the parameters defined in the SIP, can be registered.

A manufacturer of filling mixtures is a so-called downstream user, thus a user of chemicals who is not directly required to register substances, but the registration process concerns even him, especially in the case of cooperation in setting up so-called exposure scenarios for substances, defining the allowed use of a substance.

Part of the documentation of a registered substance is a chemical safety report. The chemical safety report shall be completed for all substances that subject to the registration under the REACH Regulation in quantities of 10 tonnes or more per year per registrant. A downstream user of the very substance or the substance included in a preparation shall prepare a chemical safety report for any use outside the conditions described in the exposure scenario or in the category of use and exposure submitted to him in a safety data sheet or for any use not recommended by its supplier, except the cases referred to in Article 37 (4) of the REACH Regulation [11].

The chemical safety report contains the information on risk management measures, information on substance properties, the assessment of hazard to human health and the environment, assessment of exposure, and risk characterization [12]. The certificate issued in accordance with the requirements of the valid Technical Guidelines No. 09.13.01, accompanied by the documentation of the registered substance, is a guarantee for the organization that the use of such certified filling mixture is in accordance with a responsible approach to the environment.

7 CONCLUSIONS

The REACH Regulation, which came into effect five years ago, supports the objective of protecting health and the environment. Provisions of the regulation are underpinned by the precautionary principle. Its implementation in the area of certification of backfill materials is gradually applied, although expert opinions on the scope of the REACH Regulation for specified construction products are not uniform.

So far the legislation for integrated pollution prevention and control (IPPC) in the Czech Republic allows to operate plants for recovery of hazardous wastes and produce filling mixtures from them, although it is excluded by the relevant technical guidelines.

The legislation concerning products and other legal regulations, such as the REACH Regulation, are intended for the protection of human health and the environment from potential environmental impacts of products and other materials not considered wastes. If this tool is consistently applied in the legislation on reserved construction products, the certification system will ensure the filling mixtures to be produced and used in a safe manner. Besides the Czech Trade Inspection (CTI), controlling whether the properties of the specified products placed on the market comply with the relevant technical requirements or not, the Czech Environmental Inspectorate (CEI) will control the compliance with the obligation of registration, because without the registration number, chemicals cannot be marketed.

A state-owned company that will buy filling mixtures will be confident when strictly applying this legislation that the performed disposal of mine working will not jeopardize the quality of groundwater and does not become a danger for the future.

The principles and legal regulations for the disposal of mine workings and remediation of mining operations, of course, applies also to ore districts and private companies dealing with this issue in the Czech Republic.

REFERENCES

[1] SOCHOR, Zbyšek: Útlum hornictví v procesu dokončení jeho restrukturalizace, HPVT 2003 http: slon.diamo.cz hpvt 2003 sekce 1 L06.htm [online] 2003 [cit. 05 03 2013].

[2] LAPČÍK, Vladimír: Oceňování antropogenních vlivů na životní prostředí (monografie), VŠB-TU Ostrava, HGF, IEI, 2011. ISBN 978-80-248-2440-6.

[3] Czech Republic. Sbírka zákonů. Zákon č. 100/2001 Sb., o posuzování vlivů na životní prostředí a o změně některých souvisejících zákonů.

[4] Czech Republic. Sbírka zákonů. Vyhláška ČBÚ č. 52/1997 Sb., kterou se stanoví požadavky k zajištění bezpečnosti a ochrany zdraví při práci a bezpečnosti provozu při likvidaci hlavních důlních děl.

[5] SLIVKA, Vladimír a kol.: Využití vedlejších produktů z těžby a zpracování energetických surovin, VŠB-TU Ostrava, HGF, 2010. ISBN 978-80-7225-316-6.

[6] Czech Republic. Sbírka zákonů. Nařízení vlády č. 163/2002 Sb., ve znění 312/2005 Sb., které stanoví technické požadavky na vybrané stavební výrobky.

[7] http: www.tzus.cz technicke-navody 27 [online] 2013 [cit. 2013-03-05] Technické návody.

[8] http: www.mzp.cz cz metodiky ekologicke zateze [online] 2013 [cit. 2013-03-05] Metodický pokyn MŽP Indikátory znečištění

[9] Univerza-Středisko odpadů Praha, s.r.o.: Extrakt ze závěrečné souhrnné zprávy o řešení projektu VaV SP 2f3 118 08, http: www.univerza.cz files Extrakt-z-projektu-VEP-08-11.pdf [online] 2010 [cit. 05 03 2013].

[10] KOM(2007) 59 SDĚLENÍ KOMISE RADĚ A EVROPSKÉMU PARLAMENTU o Interpretační sdělení o odpadech a vedlejších produktech.

[11] Nařízení Evropského parlamentu a Rady (ES) č. 1907/2006 o registraci, hodnocení, povolování a omezování chemických látek (nařízení REACH).

[12] http: www1.cenia.cz www reach o-reach [online] 2012 [cit. 2013-03-05] Zpráva o chemické bezpečnosti.

RESUMÉ

Pozůstatky hlubinné těžby uhlí jsou závažným problémem. Zahlazování následků bývalé hornické činnosti je hrazeno ze státních prostředků státním podnikům, které nakupují potřebný materiál a služby v režimu veřejných zakázek. Tato náprava realizací hornické činnosti nepodléhá posuzování vlivů záměru na životní prostředí. Likvidace dolu je povolována podle báňských předpisů. Za správnost zvoleného řešení odpovídá organizace, která prokazuje pro hornickou činnost odbornou způsobilost. Při likvidaci dolu používané základkové směsi jsou tedy posouzeny z hlediska jejich možného vlivu na horninové prostředí a podzemní vody pouze postupem certifikace stanoveného stavebního výrobku. Podmínky certifikace se komplikují při využití základkové směsi, vyrobené na bázi vedlejších energetických produktů. Zatím umožňuje legislativa pro integrovanou prevenci a omezování znečištění i využívání ostatních a nebezpečných odpadů. Výrobce i uživatele se dostanou do oblasti střetu odpadové a výrobkové legislativy. Článek je zaměřen na hodnocení přínosu implementace evropské legislativy chemických látek – nařízení REACH do technických návodů, dojde tak z hlediska působnosti k rozšíření kontroly vlastností stanovených výrobků kromě ČOI i ze strany ČIŽP.

11

INSTITUTE OF CLEAN TECHNOLOGIES FOR MINING AND UTILIZATION OF RAW MATERIALS FOR ENERGY USE – A NEW POTENTIAL OF RESEARCH IN OSTRAVA

Václav DOMBEK

RNDr.,CSc., Institute of Clean Technologies for Mining and Utilization of Raw Materials for Energy Use, Faculty of Mining and Geology, VŠB – Technical University of Ostrava, 17.listopadu 15/2172, Ostrava-Poruba, 708 33

vaclav.dombek@vsb.cz

Abstract

A year ago, in Ostrava, one of the major RDI (Research and Development for Innovation) projects was initiated in the Moravian-Silesian Region, called the *Institute of Clean Technologies for Mining and Utilization of Raw Materials for Energy Use*. During the first year, many of the top and often unique research laboratories and workplaces were built with a budget of over CZK200m, such as "Workplace of Electron Microprobe", „Workplace of Tomographic Methods", "Workplace of Hydrochemistry and Hydrobiology", "Workplace of Thermal, Hydraulic and Mechanical (THM) Processes in Rocks", "Workplace of Water Jet", "Isotope and GC-TOF Laboratory" and many others. This laid the basis for various research programmes with truly extraordinary extent and impact not only on the Czech industry but also economy of other EU member states.

Abstract

V Ostravě byl před rokem zahájen jeden z významných projektů VaVpI (Věda a výzkum pro inovace) v Moravskoslezském kraji s názvem *Institut čistých technologií těžby a užití energetických surovin.* V průběhu prvního roku byla s rozpočtem více než 200 miliónů korun vybudována řada špičkových a mnohdy i unikátních výzkumných laboratoří a pracovišť, jakými jsou „Pracoviště elektronové mikrosondy", „Pracoviště tomografických metod (RTG-CT)", „Pracoviště hydrochemie a hydrobiologie", „Pracoviště tepelných, hydraulických a mechanických (THM) procesů v horninách", „Pracoviště vodního paprsku", „Izotopová a GC-TOF laboratoř a řada dalších. Tím byl položen základ k řešení výzkumných programů se skutečně mimořádným rozsahem a dopadem nejen na český průmysl, ale i ekonomiku ostatních států EU.

Key words: energy resources, clean technologies, RDI

1 INTRODUCTION

A year ago, in Ostrava, one of the major RDI (Research and Development for Innovation) projects was initiated in the Moravian-Silesian Region, called the *Institute of Clean Technologies for Mining and Utilization of Raw Materials for Energy Use* (hereinafter referred to as ICT).

This project is financed by the EU Structural Funds –Research and Development for Innovations Operational Programme with a total budget of nearly CZK 300 million (excluding VAT).

2 BUILDING UP ICT WORKPLACES

A bearer of the project and simultaneously its leader is the Faculty of Mining and Geology of the VSB-Technical University of Ostrava whose newly built workplaces represent two-thirds of the whole ICT budget. However, colleagues from other faculties of the VSB-Technical University of Ostrava were invited, namely from the Faculty of Mechanical Engineering, Faculty of Civil Engineering and the Faculty of Safety Engineering to address research programmes.

A project of this importance and scope neither can do, of course, without close cooperation with other leading scientific institutions outside the university. Such key partner of the HGF VSB-TU Ostrava in the Institute of Clean Technologies for Mining and Utilization of Raw Materials for Energy Use is the Institute of Geonics of the Academy of Sciences of the Czech Republic v.v.i. with a high-quality professional base and a number of reputable research teams. The intention is also to address and attract experts from abroad to contribute to the excellence of expected results.

As for its implementation, the project is divided into two stages. The first stage (so-called Start-up) started on 1st January 2011 and shall be finished on 31st December 2014. In this period, most of the costs associated with the construction, start and overall operation of workplaces and laboratories are financed by the EU Structural Funds. Further viability of the entire project will critically depend on the results of this stage. The second project stage shall be the sustainability phase, which will continue in another five years - i.e. until the year 2019. During this period, the project will not be financed by the EU any longer and its activities must be covered by scientific projects and contractual cooperation with industry.

The first essential part of implementing this type of research projects is to build suitable facilities of laboratories and workplaces and to provide them with scientific instruments and equipment. Today, after the first year of ICT operation, we can say that this stage of the ICT project has been successfully completed. More than 380 m^2 of areas of laboratories have been substantially renovated and modernized and more than 60 public contracts for the acquisition of high-end devices (Fig. 1) and systems have been arranged.

The Institute of Clean Technologies, as results from its whole name, should help the Czech Republic with finding new ways for obtaining energy resources that would have minimal impact on the environment. Such exploitation must be, on one hand, the most efficient from an economic point of view, while environmentally friendly on the other hand. It should help as far as possible to ensure the maximum possible raw material self-sufficiency of the Czech Republic. The project consists not only in the work on untapped raw materials, but also finding new approaches even to so-called industrial waste. In the Ostrava Region, for example, these are heaps of coal-mining waste. The materials, from which they are formed, should in future be regarded as secondary raw materials with other potential use.

2.1 ICT research programme

The project consists of two basic research programmes. The task of the research programme "Poly-phase geological environment" is to gain knowledge about the physical, chemical, isotopic, structural and mechanical properties of environmental components using the most modern devices to significantly increase the level of knowledge and the possibility of its generalization for given geological conditions by means of mathematical modelling. Partial scientific objectives of the research programme then include such research spheres as "Definition of violations of geomaterials based on their internal structure, mode of loading and physical conditions", "Use of raw materials and wastes to produce modified clays and geopolymers with applications in construction and BAT technologies for the human living environment" and "Identification of gas origin in rock mass and its usefulness for increasing exploitability of deposits". This information should then be an essential prerequisite and cornerstone for the design of environmentally friendly technologies in exploiting mineral resources.

The research programme "Environmentally friendly technologies" shall solve the problems of utilizing by-products for the introduction of non-waste technology in extracting mineral resources and creating the conditions to minimize safety risks based on the knowledge of causal processes. While minimizing safety risks, the environmentally friendly mining of mineral resources represents under actual economic conditions a complex process that requires addressing the entire spectrum of specific issues starting from the application of new knowledge of the properties and behaviour of the rock environment through the development of new mining technologies and treatment of energy resources to their safety aspects and environmental impacts. The following are then defined as partial research objectives: "Ensuring technology for treatment of mine water in compliance with the Water Framework Directive 2000/60/EC", "Development of technology for the preparation of mineral precursors and carriers of nanoparticles via high-speed water jet disintegration", "Technology of mining protective pillars in coal mines with minimum effects on surface deformation and determination of operating parameters of mining techniques" and "Security aspects of environmentally friendly technologies associated with the mining of mineral resources in terms of explosiveness, flammability, spontaneous combustion and air conditioning of mines".

The results of the research activities, as for the technical as well as personnel aspects, of such top teams should be in practice utilized not only by important companies in the region, such as OKD, a.s.; Diamo, state enterprise; Severočeské doly, a.s.; Litvínovská uhelná a.s.; RWE; MND a.s., but by many other energy mining companies as well.

2.2 ICT laboratories and workplaces

The success of implementation of the ICT project is heavily dependent on two factors, both excellently equipped workplaces and individual teams of experts. Over CZK 200 million has been invested into the technical equipment of the laboratories, most of which are now ready for intensive research work. A personal factor, thus the creating of truly productive research teams based on both already mature real scientists (so-called senior researchers) and younger and less experienced junior researchers and "scientific rookies" (doctoral students), was to the greatest extent covered from own sources of the VSB-TU Ostrava and the Institute of Geonics AS CR, partly then by recruiting highly skilled workers from nonschool workplaces.

From the truly key laboratories we can mention in particular "Workplace of Electron Microprobe", "Workplace of Tomographic Methods", "Workplace of Hydrochemistry and Hydrobiology", "Workplace of Thermal, Hydraulic and Mechanical (THM) Processes in Rocks", "Workplace of Water Jet", "Isotope and GC-TOF Laboratory", "Laboratory of Instrumental Neutron Activation Analysis (NAA)", "Workplace of X-ray Diffraction", "Laboratory of Raman Spectroscopy", "Laboratory of TGA-MS", "Laboratory of Physical Methods", "Laboratory of Microcalorimetry and Electrochemistry", "Laboratory of Thermal Analysis and Rheology of Building Materials", "Laboratory of Stimulation of Boreholes and Hydrocarbon Deposits" and "Workplace of Safety Analysis".

The Workplace of Electron Microprobe is equipped with the FEI Quanta 650 FEG cutting-edge scanning electron microscope (FEI, USA) provided with WDX, EDX and EBSD analysers (Fig. 1). It is a system fitted with analytical tools for detailed chemical, crystallographic and morphological characteristics of various materials. The microscope alone allows working in an extremely large range of zoom (6x to 2 000 000x) even at very low and variable vacuum (up to 4000 Pa in the ESEM mode) and without surface finish of samples by plating. The microscope as one of these analytical tools is equipped with energy- and wave-dispersive and cathodoluminiscence analyser including a spectrometer and an apparatus for diffraction of backscattered electrons [1]. An integral accessory is a cooling table for the analysis of biological samples and an apparatus for ion polishing and cutting.

Fig. 1 Laboratory of Electron Microprobe

The Workplace of Tomographic Methods can boast of the first system of this type and performance in our country. It consists of two X-ray tomography systems – XTH 450 2D/3D and XTH 225 ST (NIKON Metrology NV, Japan). The first one has a maximum acceleration voltage and power output of X-ray source 450kV/640W with a focal spot size of 80μm/300μm (for 200W/600W). The maximum weight, diameter and height of scanned objects can reach up to 100 kg/approx. 0.6 m/0.8 m with a theoretically radiable thickness of analysed materials of 395 kg/m2. The performance of the other device is slightly smaller, but it can analyze samples with a weight of up to 50 kg.

The Isotope (Fig. 2) and GC-TOF Laboratory is equipped with systems for determining isotope ratios and the study of products of pyrolytic cleavage. The new and unique sector Delta V Advantage Spectrometer (IRMS) (Thermo Fisher Scientific, Inc., USA) in conjunction with the Trace GC Ultra Gas Chromatograph (Thermo Fisher Scientific, Inc., USA) is characterized by excellent parameters for the measurement of isotope ratios $^{13}C/^{12}C$, $^{15}N/^{14}N$, $^{18}O/^{16}O$, $^{34}S/^{32}S$. The H/D ratio measurement in a dynamic mode excels in high sensitivity, monolithic analyser with fixed settings of all components of ion optics [2, 3]. The pyrolytic system Trace GC

Ultra Gas Chromatograph (Thermo Fisher, USA) with the PyroProbe 5000 device (CDS Analytical, USA) and with TOF Mass Spectrometer (Time-Of-Flight) (ALMSCO International, United Kingdom) will allow both monitoring the behaviour of chemical compounds in a selected temperature regime, and their targeted degradation before the very analysis.

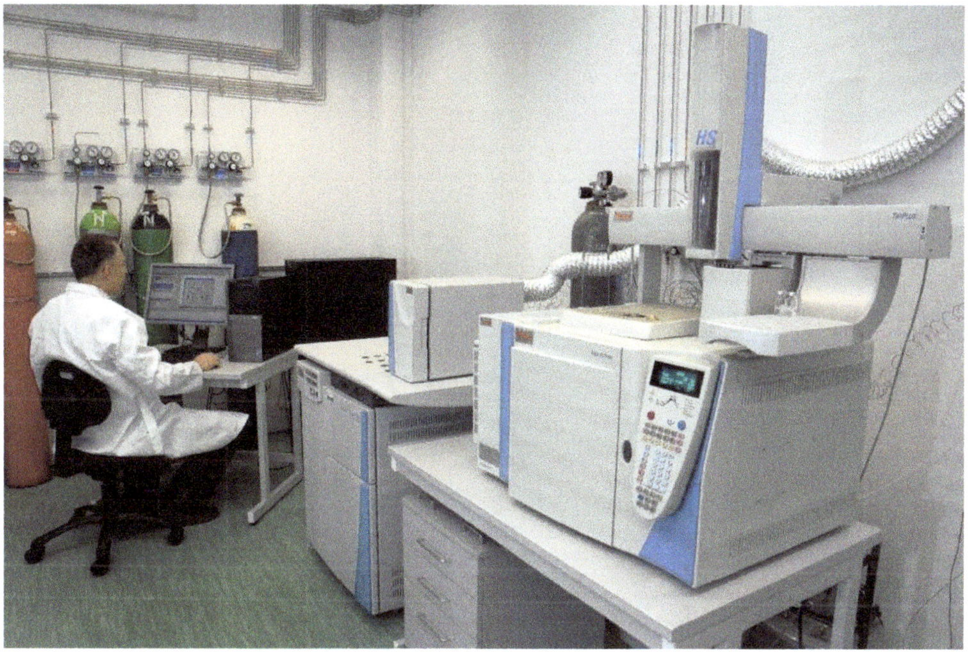

Fig. 2 Isotope Laboratory

The Workplace of Thermal, Hydraulic and Mechanical (THM) Processes in Rocks boasts of a large complex of instrumentation and software, such as MTS Servohydraulic Test System and 656.06 Triaxial Cell for tests of strength and deformation properties of rocks, CSM Instruments Micro-Hardness Tester (USA), SETSYS TG-DTA/DSC 24 Thermal Analyser with a mass spectrometer (Setaram Instrumentation, Canada), NICOLET 6700 FT-IR Spectrometer with the NICOLET NXR FT-Raman module (Thermo Scientific, USA), NICOLET iN10 FT-IR Microscope (Thermo Scientific, USA), NIKON Eclipse 80i (Nikon, Japan) and OLYMPUS BX 50 (Olympus, Japan) Fluorescence Microscopes, OLYMPUS LEXT OLS 3100 Confocal Laser Scanning Microscope (Olympus, Japan) including the image processing and analysis software – NIS Elements and MATLAB Image Processing Toolbox.

The Workplace of Water Jet involves the ABB IRB 6640-180//2.55 Master Robot (Fig. 3) for handling of a water jet cutting head with high pressure pumps – Hammelmann GDP 253 (maximum working pressure of 160 MPa, maximum flow rate of 67 l/min) and PTV75-60 with two pressure multipliers (a working pressure of 40 to 415 MPa, maximum flow rate of 7.8 l/min at 415 MPa). This machine is completed with a system for visualization and measurement of flow velocity fields (2x PIV camera, Imager Pro X 2M CCD with a double pulse laser system with optics to create a light sheet NL 135-15 PIV and HighSpeedStar 3G CMOS high-speed camera). Part of the workplace is a computing system for flow modelling equipped with the ANSYS CFD software.

With regard to its mission in the area of research of explosion risk analysis, the Workplace of Safety Analyses is equipped with a device (Hartmann Tube) for determining the minimum ignition energy (OZM Research, CR), CLASIC 5506 SZ ovens for self-heating (Classic CZ, CR) for determining conditions of the self-heating of solids and GX 51 Inverted Metallographic Microscope (Olympus, Japan) for the identification of changes in materials during combustion and explosion with an automatic analysis of records, and highly prized FLACS advanced modelling software (GexCon AS, Norway) [4,5].

The Laboratory of thermogravimetric and mass analysis (TGA-MS) boasts of a unique system, which is the first of its kind in Europe and the twelfth in the world, of a high-pressure thermal analyser with the TGA-MS-HP50U Mass Spectrometer (TA Instruments, USA). The mass spectrometer with a range of 1-300 amu at the output analyzes gaseous products leaving the reactor.

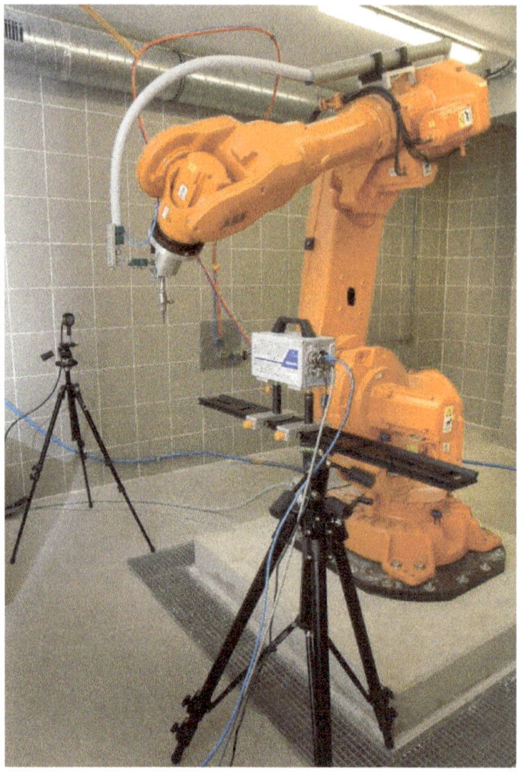

Fig. 3 Robot for Water Jet Cutting

By its concept, the Workplace of X-ray Diffraction is focused on qualitative and quantitative analysis of mineral-phase composition of samples of various kinds. Using the Bruker Advance D8 Powder Diffractometer (Bruker, USA) (Fig. 4) equipped with the LynxEye linear semiconductor detector and energy dispersive SOL-XE Detector, it is possible to analyse the composition and to identify almost any material (rock samples, waste, chemicals, pharmaceutical ingredients, etc.).

The Laboratory of Raman Spectroscopy is equipped with the multifunctional 3D Scanning Confocal Microscope with the Nanofinder S Spectrometer (Tokyo Instruments, Japan). The system uses simultaneously the 2D optical microscopy with high resolution, 3D laser, 3D luminescent and 3D Raman microscopy, which enables a multi-functional micro-structural three-dimensional analysis with a high spatial and spectral resolution.

The instrumentation of the Laboratory of Microcalorimetry and Electrochemistry consists of a high-end system to measure the Zeta-potential, dependence of the Zeta-potential on the environment stability and determination of isoelectric points of colloidal suspensions Zetasizer NANO ZS (Malvern Instruments Ltd., United Kingdom), ED-Z electro-membrane separation unit (MEGA, CR) using electro-dialyse units fitted with replaceable polymer membranes and the TAM III adiabatic simultaneous micro-calorimeter of a new generation offering maximum sensitivity, flexibility of measurement and performance (TA Instruments, USA).

The Laboratory of thermal analysis and rheology of building materials is equipped with the AR G2 Rheometer (rotational viscometer), TAM Air eight-channel isothermal micro-calorimeter (TA Instruments, USA) with the SDT Q600 high thermal module (TA Instruments, USA), allowing simultaneous measurement of mass changes and heat flow up to 1500°C. An essential part of the laboratory is the C-40/1000-S Climatic Chamber used to study the properties of building materials.

Fig. 4 X-ray Diffractometer

3 CONCLUSIONS

Within one year, 15 key workplaces of the Institute of Clean Technologies for Mining and Utilization of Raw Materials for Energy Use were built. These are formed not only by research devices and systems at the global level, but also by highly erudite and experienced personnel having all the prerequisites to achieve and meet individual objectives of the two research programmes.

LIST OF ABBREVIATIONS

BAT	- Best Available Technologies
CFD	- Computational Fluid Dynamics
EBSD	- Electron Backscatter Diffraction
EDX	- Energy-Dispersive X-ray spectroscopy
FT-IR	- Fourier Transform – Infrared
GC-TOF	- Gas Chromatography with Time-off-Flight
H/D	- Hydrogen/Deuterium
IRMS	- Isotopic Ratio Mass Spectrometry
NAA	- Neutron-Activation Analysis
RTG	- X-ray
TGA-MS	- Thermogravimetric Analysis – Mass Spectrometry
THM	- Thermal, hydraulic and mechanical
RDI	- Research and Development for Innovations
WDX	- Wavelength Dispersive X-ray Spectroscopy

REFERENCES

[1] VOORDOUW, R., GUTZMER, J., BEUKES, N.J. Zoning of platinum group mineral assemblages in the UG2 chromitite determined through in situ SEM-EDS-based image analysis. *Mineralium Deposita*. 2010. Vol. 45, pp. 147-159. DOI 10.1007/s00126-009-0265-z. ISSN: 0026-4598.

[2] HILKERT, E. *irm*-GC/MS: 13C/12C Isotope Ratio Monitoring Acyclic Biomarkers. *Thermo Fisher Scientific, Bremen, Germany*. 1995. First published as Finnigan MAT Application Flash Report No. G12 .

[3] KRACHT, O. Simultaneous N, C and S Isotope Ratio Determination on a DELTA V Isotope Ratio MS using a Flash Elemental Analyzer. *Thermo Fisher Scientific, Bremen, Germany.*

[4] DAVIS, S.G., HINZE, P.C., HANSEN, O.R., VAN WINGERDEN K. Does your facility have a dust problem: Methods for evaluating dust explosion hazards. *Journal of Loss Prevention in the Process Industries*. 2011. Vol. 24. Issue6. pp. 837-846. ISSN 1873-3352.

[5] GAVELLI, F., DAVIS, S.G., HANSEN, O.R. Evaluating the potential for overpressures from the ignition of an LNG vapor cloud during offloading. *Journal of Loss Prevention in the Process Industries*. 2011. Vol. 24, Issue 6. pp. 908-915. ISSN 1873-3352

RESUMÉ

V Ostravě byl před rokem zahájen jeden z významných projektů VaVpI (Věda a výzkum pro inovace) v Moravskoslezském kraji s názvem Institut čistých technologií těžby a užití energetických surovin. Klíčovým partnerem Hornicko-geologické fakulty VŠB-TU Ostrava je v ICT ostravský Ústav geoniky Akademie věd České republiky se svou skutečně kvalitní odbornou základnou a řadou uznávaných vědeckých týmů. Na realizaci výzkumných programů se však budou podílet i výzkumní pracovníci z dalších fakult VŠB-TUO, konkrétně z Fakulty strojní, Fakulty stavební a Fakulty bezpečnostního inženýrství. Záměrem je rovněž oslovit i a přitáhnout i odborníky ze zahraničí, kteří by mohli přispět ke špičkové kvalitě očekávaných výsledků.

Realizačně je projekt členěn do dvou bloků. V prvním (do roku 2014) je většina nákladů spojená s vybudováním, rozjezdem a celkovým provozem pracovišť a laboratoří hrazena ze Strukturálních fondů EU. Druhou časovou etapu projektu tvoří tzv. fáze udržitelnosti, která potrvá po dobu dalších pěti let – tedy až do roku 2019. Během tohoto období již projekt ICT nebude financován z prostředků EU a na svoji činnost si musí zajistit finanční prostředky komerčně poskytovanými službami.

První podstatnou částí realizace bylo vybudovat vhodné prostory laboratoří a pracovišť a vybavit je vědeckými přístroji a technikou. Dnes, po roce velice intenzívní a někdy vyčerpávající práce, lze konstatovat, že tato etapa projektu ICT byla úspěšně naplněna. Bylo zrekonstruováno a podstatně zmodernizováno více než 380 m^2 ploch laboratoří a proběhlo více než 60 veřejných zakázek na pořízení špičkových (a v řadě případů i unikátních) přístrojů (obrázek 2) a systémů, z nichž dokonce některé byly instalovány jako vůbec první v Evropě.

Institut čistých technologií, jak už ostatně vyplývá i z jeho celého názvu, by měl České republice pomoci s hledáním nových způsobů získávání energetických surovin a zdrojů, které by měly minimální dopady na životní prostředí. Úkolem projektu je nejen práce na dosud nevytěžených surovinách, ale i nalézání nových přístupů a to i k takzvaným průmyslovým odpadům

Projekt je tvořen dvěma základními výzkumnými programy. Zadáním výzkumného programu „Vícefázové horninové prostředí" je získat poznatky o fyzikálních, chemických, izotopových, strukturních a mechanických vlastnostech složek prostředí pomocí moderní instrumentální techniky, která zásadním způsobem zvyšuje úroveň poznání a možnosti jejich zobecnění pro dané geologické podmínky pomocí matematického modelování. Výzkumný program „Environmentálně šetrné technologie" se bude zabývat problematikou využití vedlejších produktů pro zavedení bezodpadových technologií při těžbě nerostných surovin a vytvoření podmínek pro minimalizaci bezpečnostních rizik na základě poznání příčinných procesů.

Úspěch realizace celého projektu ICT je silně závislý na dvou faktorech – unikátních a špičkově vybavených pracovištích a skvěle kvalifikovaných odborných týmech. Do technického vybavení laboratoří bylo investováno přes více než 200 miliónů korun a většina z nich je již nyní připravena k náročné vědecké práci. Bylo vybudováno 15 klíčových laboratoří a pracovišť: „Pracoviště elektronové mikrosondy", „Pracoviště tomografických metod (RTG-CT)", „Pracoviště hydrochemie a hydrobiologie", „Pracoviště tepelných, hydraulických a mechanických (THM) procesů v horninách", „Pracoviště vodního paprsku", „Izotopová a GC-TOF laboratoř" (obrázek 3), „Laboratoř instrumentální neutronové aktivační analýzy (INA)", „Pracoviště RTG difrakce", „Laboratoř Ramanovské spektroskopie", „Laboratoř TGA-MS", „Laboratoř fyzikálních metod", „Laboratoř mikrokalorimetrie a elektrochemie", „Laboratoř termické analýzy a reologie stavebních hmot", „Laboratoř stimulace vrtů a ložisek uhlovodíků" a „Pracoviště bezpečnostních analýz".

12

SILTING OF FOUM EL GHERZA RESERVOIR

Remini BOUALEM, Bensafia DJILLALI, Missoum MOHAMED

Département des sciences de l'eau et de l'environnement, Faculté de Technologie, Université Blida 1
BP 270, Blida, Algéri
E-mail: reminib@yahoo.fr, bensafiadjillali@yahoo.fr, missoum.maazouz@gmail.com

Abstract

The Foum El Gherza Reservoir fails to satisfy the irrigation of 300,000 palms of Sidi Okba and Seriana due to advanced siltation. Based on data from bathymetric surveys conducted by the National Agency for Dams, the siltation rate was estimated at 0.9 million m^3/year. With a total capacity of 47 million of m^3, the Foum El Gherza dam is silted up with more than 65 million m^3. In this case, the life of the dam will not exceed 11 years of operation. The total silting of the reservoir will occur in 2023. Desilting the dam proves an emergency to save the palms of Seriana and Sidi Okba.

Keywords: Dam, Foum El Gherza, Siltation, Current density, Desilting.

1 INTRODUCTION

The phenomenon of erosion in Algeria is spectacular. The specific erosion in 30 studied catchments varies between 30 and 3350 t/km^2/year (Mekerta, 1993). According to Demmak (1982), the erosion rate can reach 4000 t/km^2/year in the chain of coastal Dahra. By against, it reaches 62 t/km^2/year in the Seine River basin, and 1500 t/ km^2/year in the basin of the Durance River (France). It exceeds 2500 t/km^2/year in several basins in China and the island of Java in India (Mechin, 1980). Values of the specific erosion rates exceeding 4500 t/km^2/year were evaluated in the catchment area of Wadi Agrioum (Remini et al, 2009a, 2009b). About 180 million tons of soil is eroded annually in the catchment areas of northern Algeria (Demmak, 1980). This number has never been updated; it can be revised upwards given the degradation of our catchments in recent years. This situation has a negative effect on dam reservoirs. In fact, each year, silt in a volume of 45 million of m^3 is deposited in 70 major Algerian dams (Remini, 2011). In 2030, the loss of storage capacity of the studied reservoirs may reach 43% of their initial capacities (Ben Mammou and Louati, 2007). The Foum El Gherza dam does not escape this phenomenon. It is classified among the most silted dams in Algeria. It's a barrage of great economic importance for the region of Zibans. Reaching the age of 63 years, the Foum El Gherza dam is intended to irrigate palm groves of the downstream area of the dam. However, given its accelerated siltation, the dam cannot satisfy irrigation needs in the short and medium terms. The study of the evolution of siltation and evaluation of the life of the Foum El Gherza dam proves a necessity.

2 SITUATION AND CHARACTERISTICS OF STUDY AREA

2.1 Catchment of Foum El Gherza dam

The catchment area of El Abiod Wadi is located in the southern slope of the eastern part of Saharan Atlas. With a length of 85 km, El Abiod Wadi raises in the Massif des Aures (the mountains of Chélia). Covering an area of 1300 km^2 in a perimeter of 205 km, the catchment has an elongated shape (Fig. 1). The catchment area of El Abiod Wadi is a semi-arid region (northern part). It is composed of a naked part without vegetation, and then forests with dominance of Aleppo pine occupy much of it.

There are also reserved cereal and tree parts. El Abiod River floods are sudden and dangerous. For example, in October 10, 1966 a flood occurred on the river whose peak discharge reached the value of 1980 m^3/s. During the month of March 2004, the peak discharge was estimated at 1000 m^3/s. At the Foum El Gherza Reservoir the worst flooding took place in March 30, 2004, the water level has exceeded the rating of 202, which corresponds to a nape of 3.65 m, a maximum evacuated discharge of 570 m^3/s.

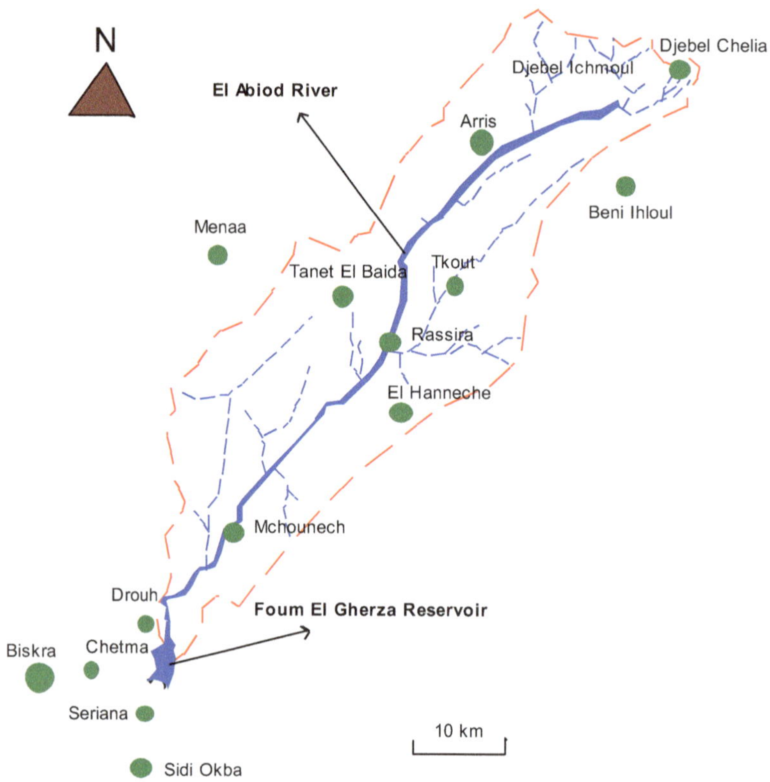

Fig. 1 Catchment boundaries of El Abiod River

2.2 Situation of Foum El Gherza Reservoir

The Foum El Gherza Reservoir is located 18 km east of the city of Biskra and about 600 km southeast of Algiers (Fig. 2). It allows the irrigation of palms of Sidi Okba in the southwest of Seriana in the north and Thoudra. This is an arch dam with an initial capacity of 47 million of m^3 (Fig. 3).

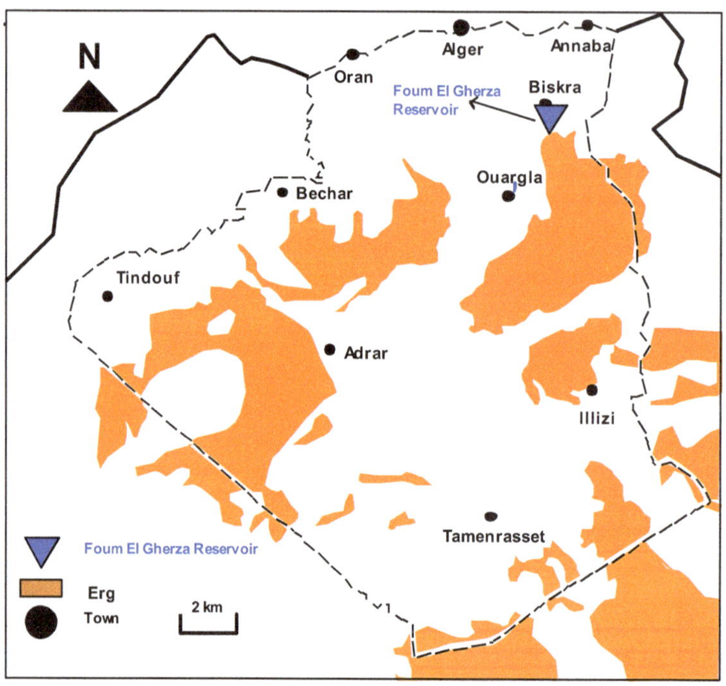

Fig. 2 Location of Foum El Gherza dam

Fig. 3 Overview of Foum El Gherza dam (Remini, 2004)

2.3 Field investigation and data used

The Foum El Gherza Reservoir is intended to irrigate over 300,000 palms of Sidi Okba, Seriana and Thoudra oases. However, the dam is threatened by rapid siltation. The floods of the El Abiod River bring sediments in concentrations exceeding 100g/l. This causes the formation of density currents that drain sediment up to the base of the dam. In order to dredge the dam during the year 2015, we wanted to study the evolution of silting in the dam. For this purpose, nine historical bathymetric surveys collected at the National Agency of Dams (NAD) were used. These are surveys conducted in 1950, 1952, 1957, 1967, 1975, 1986, 1993, 2001 and 2004.

3 RESULTS AND DISCUSSIONS

3.1 Erosion and undercutting of banks

The erosion of the catchment of El Abiod Wadi recorded high values. The annual evaluation of the specific erosion rates exceeded 1500 t/km^2/year during the year 1976. The average specific erosion rate approaches the value of 300 t/km^2/year. Undermining the banks of El Abiod Wadi is very important, except that we have not been able to quantify this phenomenon. However, we have located more than 20 outbreaks of bank erosion along the River of El Abiod (Fig. 4).

Fig. 4 Sapping of banks of El Abiod Wadi upstream the dam (Remini, 2004)

3.2 Sediment transport in El Abiod River

Sediment transport in El Abiod Wadi is very important, especially in times of flood. The amount of soil eroded in the watershed and the banks of the river are drained by the river to the dam. Concentrations of fine particles have been registered at the entrance of the reservoir. With a slope of the wadi bed greater than 1 per thousand, the bed load is very significant in the river wadi of El Abiod.

3.3 Density currents in Foum El Gherza Reservoir

The arrival of flood waters brings to the Foum El Gherza Reservoir charged fine particles causing the formation of density current that propagates down to the foot of the dam (Remini, 2008). Generally, it happens during spring (March and April) when density currents emerge at the entrance of the reservoir (Fig. 5 and 6).

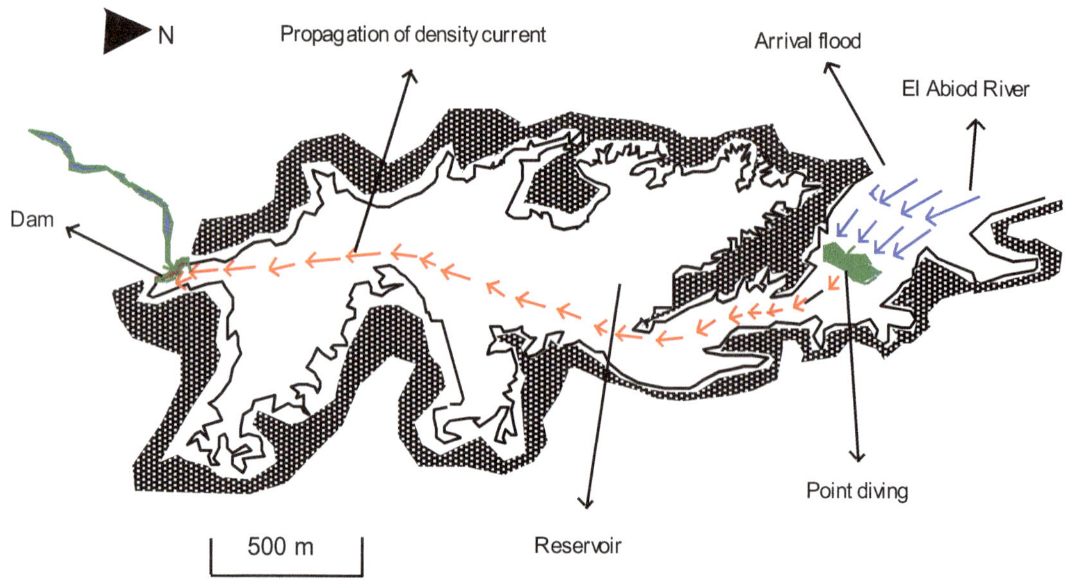

Fig.5 Path of current density of Foum El Gherza Reservoir

Fig. 6 Propagation of density current in channel

3.4 Silting of Foum El Gherza Reservoir

The dam is located in a Saharan region. Streams are characterized by torrents, speed and violence of the floods. The rivers carry a large amount of solid materials from bare slopes as density currents. Arriving at the foot of the dam, the density currents settle those solid particles to the bottom of the dam. Muddy deposits accumulate annually in the reservoir. Bathymetric surveys were established in 1950, 1952, 1957, 1967, 1975, 1986, 1993, 2001 and 2004 in order to monitor regularly the roof of the vase and discount curves of height/capacity. From 1950 to 2004, silt in a quantity of 32 million of m^3 was deposited in the reservoir, causing the siltation of 68% which represents an annual average siltation rate of 0.6 million m^3/s. Fig. 7 shows the temporal evolution of the silting of Foum El Gherza Reservoir. The sedimentation rate increased from 0.3 million of m^3/s in the 1986-1993 period to 0.9 million m^3/year in the 1993-2004 period.

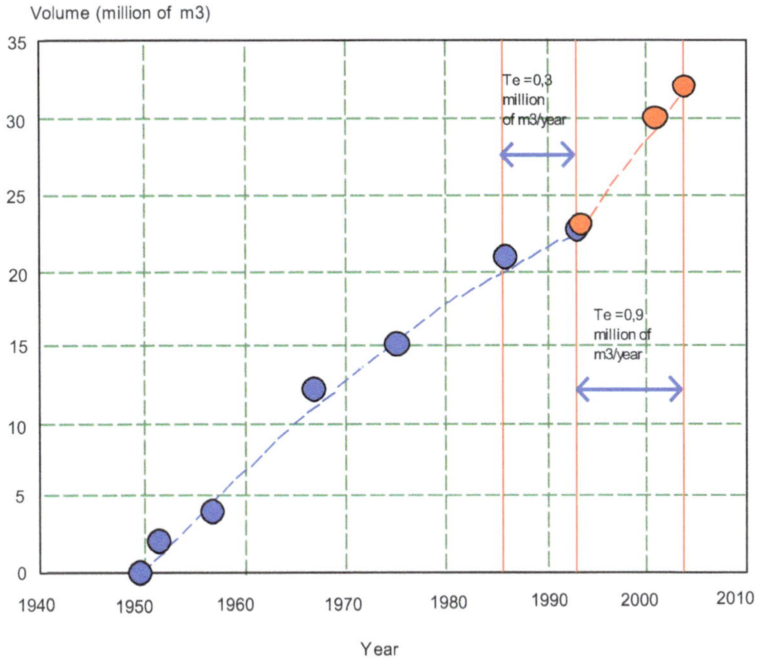

Fig. 7 Evolution of silting of Foum El Gherza Reservoir

Taking into account the temporal evolution of the silting of the Foum El Gherza Reservoir, we represent in Fig. 8 the evolution over time of the capacity of the dam and the lake area of the dam. It is noted that in 2004 the capacity of the dam is 15 million of m^3.

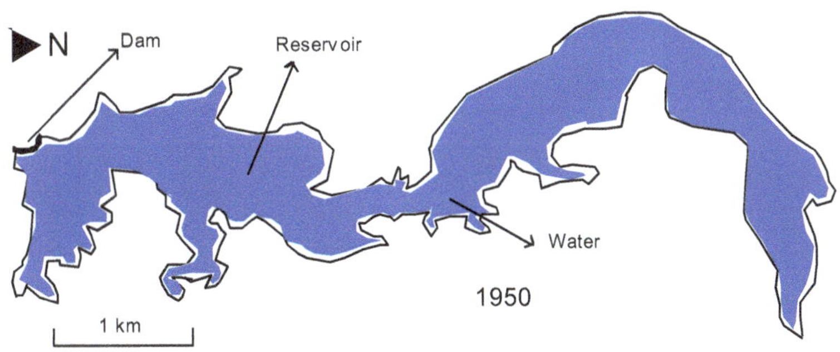

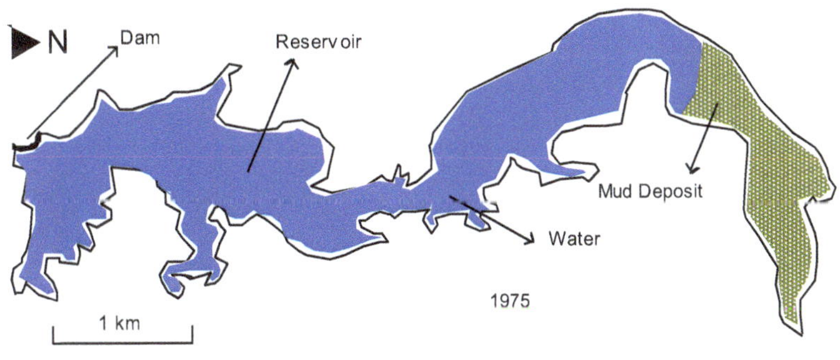

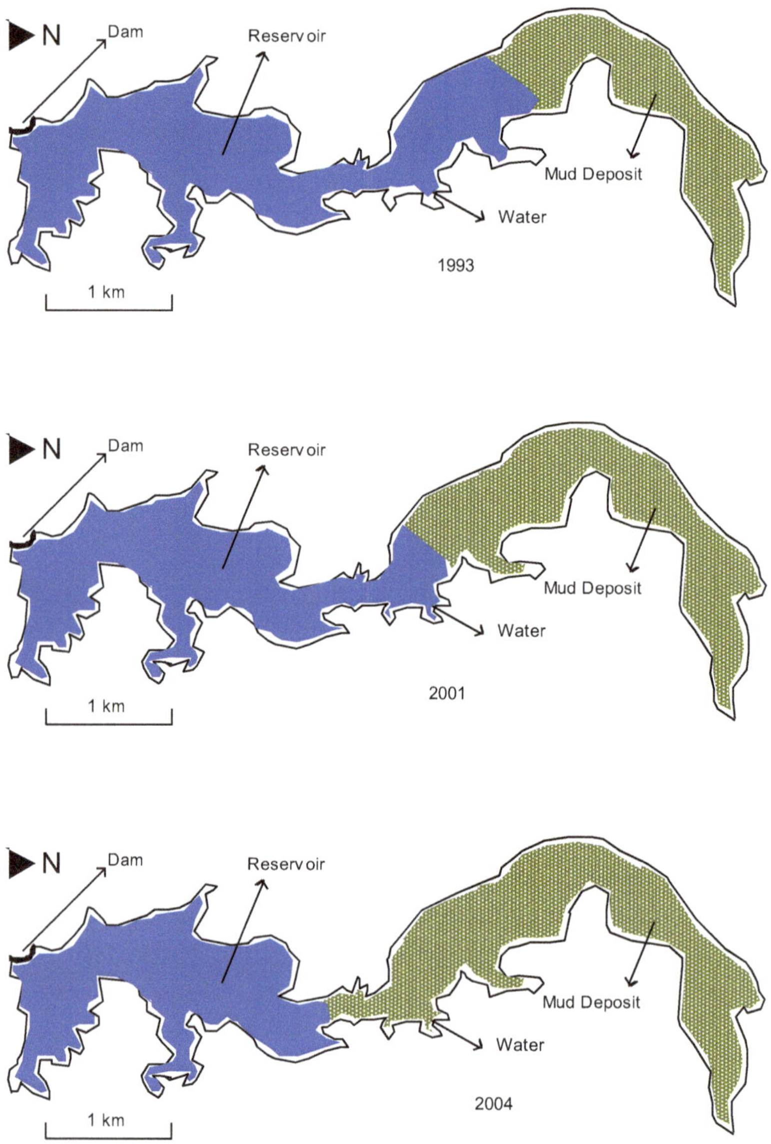

Fig. 8 Temporal evolution of Foum El Gherza Reservoir capacity

3.5 Technical means to fight against silting of Foum El Gherza Reservoir

The most widely used medium to fight against this phenomenon consists in the evacuation of current density by opening the bottom valve. This technique helped evacuate about 0.5 million m^3 in the year 1989/1990. But due to the rapid siltation, this valve was blocked after 7 years (1982-1989) of operation. From 1990 until 1993, an amount of 0.1 million of m^3 of silt was evacuated. In addition to this process, several techniques were used at the catchment to minimize the rate of erosion. For example, there may be mentioned the reforestation of catchment and the treatment of ravines by the construction of weirs. Despite those technical means of control, the state of dam siltation deteriorated, which obliged the hydraulic services to carry out the desilting of the dam by dredging (Fig. 9). A volume of 4 million m^3 was evacuated during 21 months (Fig. 10 and Fig. 11). The operation was launched in March 2005.

Fig. 9 Dredging Foum El Gherza Reservoir (National Agency for Dams)

Fig. 10 Rejection of vase (Remini, 2006)

Fig. 11 Vase of Foum El Gherza dam (Remini, 2006)

3.6 Calculation of dam lifetime

In 2006, the volume of silt deposited at the bottom of the reservoir is around 35 million m^3. In 2008, the volume of mud was 31 million of m^3, since 4 million m^3 of silt have been removed. From 2006 to 2012, a vase in an estimated volume of 3.6 million of m^3 was introduced in the reservoir by various floods. So in 2004, the volume of mud is around 35 million of m3. In regard to the sedimentation rate of 0.9 million m^3/year, the lifetime is 11 years. The dam will cut off in 2023 if the provisions of desilting will not be taken as soon as possible.

4 CONCLUSION

As mentioned at the beginning of this study, the erosion in the catchment of the El Abiod Wadi is very worrying due to its very poor shape. In 1993 its values exceeded 600 t/km^2/year as evaluated. Floods recorded on the El Abiod River, which occur during the months of March, April, May and June, are extremely violent and sudden draining high concentrations of fine particles. Values exceeding 100 g/l were measured at the entrance to the Foum El Gherza Reservoir during the years 2004 and 2006. A significant quantity of sediment comes from watershed erosion and from the undermining of banks of the River of El Abiod. Such concentration automatically generates the formation of density currents at the entrance of the reservoir which propagate over a distance of 6.5 miles below the clear waters of the reservoir to reach the foot of the dam. Closing the bottom valve during floods caused its closure in 1986. Successive deposits of sediment at the bottom of the reservoir caused accelerated silting the dam. Thus during the period 1993-2004, the sedimentation rate increased from 0.3 to 0.9 millions of m^3/year for unknown reasons. The desilting of the reservoir of more than 4 million m^3 of silt during the period of 21 months has not solved the problem. We estimated the lifetime of the Foum El Gherza Reservoir to 11 years, and that is to say the dam will have shut off by 2023 if the technical measures are not taken promptly.

REFERENCES

[1] Ben Mamou A. and Louati M.H., 2007. Temporal evolution of the silting of dams in Tunisia. Journal of Water Science, vol. 20, n° 2, 2007, p. 201-210.

[2] Mekerta B., 1993. Study of the sedimentation of reservoirs, mechanical properties of fine-grained soils. Communication First National Congress of Large Dams. Algiers, 24 and 25 May, pp. 1-9.

[3] Demmak A., 1982. Contribution to the study of erosion and sediment transport in Algeria Septentrionnale. These Doctor Engineer, University of Pierre and Marie Curie, Paris XI.

[4] Mechin Y., 1980. General introductory report on desilting reservoirs. International seminar on desilting reservoirs. Tunis, 1, 2, 3 and 4 July, 5p.

[5] Remini B. et Bensafia D., 2012. The impact of climate changes on the acceleration of reservoir siltation in arid regions the Algerian case study. Elixir International journal Bio. Diver. 44 . 7074-7076.

[6] Remini B., Hallouche W., et Achour B., 2009. Algeria: more than a century of desilting dams. Book entitled State of water resources in the Maghreb in 2009, published by UNESCO Office in Rabat, pp. 123-142.

[7] Remini B., Leduc C. et Hallouche W., 2009. Evolution of large dams in arid regions: some Algerian examples. "Sécheresse" journal, n°20, vol.1, pp.1-8.

[8] Remini B., 2011. Racking turbidity currents: a way to reduce siltation. Case of the dam Wadi Fodda. National Seminar hydraulic. Chlef, 23 and 24 November.

13

CORROSION TESTS AT THE WATER TREATMENT PLANT IN JAKUBANY

MATÚŠ GALÍK [1], JOZEF KRIŠ [2], JÁN ILAVSKÝ [3]

[1] *Ing. Matúš Galík, Faculty of Civil Engineering, Slovak University of Technology, Radlinskeho 11, 813 68 Bratislava, Slovak Republic*
matus.galik@stuba.sk

[2] *prof. Ing. Jozef Kriš, PhD., Faculty of Civil Engineering, Slovak University of Technology, Radlinskeho 11, 813 68 Bratislava, Slovak Republic*
jozef.kris@stuba.sk

[3] *doc. Ing. Ján Ilavský, PhD., Faculty of Civil Engineering, Slovak University of Technology, Radlinskeho 11, 813 68 Bratislava, Slovak Republic*
jan.ilavsky@stuba.sk

Abstract

External corrosion, which depends on environmental and operating conditions, is the main cause of structural deterioration of all metallic mains. Internal corrosion, on the other hand, can cause significant functional (hydraulic, water quality) deterioration within a distribution system. This work deals with the corrosion of water pipes which greatly affects the durability and failure rate of water systems. The test is evaluated in accordance with STN 75 7151 and ASTM D2688-11. The corrosion tests were carried out on raw and treated water at the water treatment plant in Jakubany.

Abstrakt

Hlavnou príčinou poškodenia konštrukcie všetkých kovových sietí je vonkajšia korózia, ktorá je závislá od prostredie a prevádzkových podmienok. Vnútorná korózia môže spôsobiť významné funkčné zhoršenie (hydraulické, kvalita vody) v rámci distribučného systému. Táto práca sa zaoberá koróziou vodovodného potrubia, ktorá výrazne ovplyvňuje odolnosť vodovodných systémov. Skúška je hodnotená v súlade s STN 75 7151 a ASTM D2688-11. Korózne testy boli vykonané na surovej a upravenej vode na úpravní vody Jakubany.

Key words: The corrosion, water treatment, incrustations, change in water quality

1 INTRODUCTION

A water supply system represents a complex mechanism which is affected by many factors. In terms of economy, it is also an exacting system whose service life is limited by the durability of used materials. The durability of materials has a significant impact on reliability of water supply systems and on the quality of supplied water for human consumption.

The majority of water supply systems were built in past when increasing water consumption was assumed. Therefore, the water supply systems are overdesigned. By decreasing the water consumption in overdesigned water supply systems, the extension of water retention period occurs and it may negatively affect the sensorial character of supplied water as well as result in microbial contamination.

The interaction between water and pipeline materials occurs during the transport of water in water supply network, which may lead to a decrease in water quality (Vreeburg 2007) [1]. Water supply operators (companies) often deal with this problem in water supply systems which consist of metal pipelines. In water supply systems, corrosion processes occur which have a negative impact on water quality (Munka 2005) [2]. The corrosion is mainly caused by electrochemical processes which are formed at the water and pipe material interface. The basis of all these reactions consists in oxidation-reduction (redox) reactions which lead to the degradation and depreciation of material (Slavíčková, 2006) [3].

Lytle et al. [4] found that metal levels rapidly increased with respect to stagnation time. High flow velocities can assist the formation of protective coatings by diffusing effectively protective ingredients to the metal surface. However, high flow velocities can mechanically remove the protective wall coating and pipe material, resulting in erosion corrosion or impingement attack. High flow velocities increase the rate, at which dissolved oxygen comes into contact with pipe surfaces, which can affect corrosion rates due to the involvement of oxygen in the corrosion chemical reactions. Volk et al. [5] and Fang et al. [6] studied the effect of temperature on the corrosion of iron in water and clearly demonstrated that the corrosion increased as a function of temperature.

Many other authors who monitored the course of processes corrosive to water pipes made of iron came to the conclusion that corrosion is a major cause of colour changes in water piping [7],[8].
The negative effects of corrosion are mainly expressed by:

- Decreasing the water quality (by its secondary ferrization, decreasing the content of solved oxygen, leaching of lead, zinc, copper and other substances from armature into water),
- Deterioration of hydraulic conditions of water flow
- Reduction of pipeline service life
- Increase in number of disorders and water decrements of pipeline
- Increasing costs of water pumping (Singley et al, 1984) [9].

The corrosion of pipeline has a significant impact on water supply system reliability. Specifically, it is just a reliable and sustained drinking water supply system which represents the basic requirement for social and economic development of a certain area.

Corrosion costs encountered in the water distribution division included expenditure for replacing aging infrastructure, lost water from unaccounted-for leaks, corrosion inhibitors, internal mortar linings, external coatings, and cathodic protection (Devietti 2011) [10].

This study aimed at quantifying key water quality parameters, such as flow velocity, pH, biofilm growth, temperature, and evaluating the influence of these parameters on corrosion and iron release in a drinking water distribution system. The test is evaluated in accordance with STN 75 7151 and ASTM D2688-11. The corrosion tests were carried out on raw and treated waters at the water treatment plant in Jakubany.

2 WATER AGRESIVITY DETERMINATION

It is possible to determine whether the water has corrosive effects or not by using chemical water analyses through a direct $CaCO_3$ test, by means of various calculations or corrosion tests. Determining the water aggressivity based on chemical analyses represents some advantages, such as a relatively quick way of obtaining results, possibility of periodical testing, which means regular control of water quality. The changes in water quality and the results of the tests can be compared. On the other hand, the water aggressivity calculations count only the water agsressivity that is caused by aggressive CO_2, which represents a disadvantage described by many authors. No calculation takes into account the amount of dissolved oxygen in water or the velocity of flowing water which may represent two dominating factors in influencing the corrosion progress significantly.

The method for the corrosion test is mentioned in the standard STN 75 7151 "Requirements for quality of water in piping systems" [11] and is based on measuring differences in mass decrease of tested samples on the 30th and 60th day after their exposition to the effects of flowing water. For the tests, 42x42 mm coupons with a thickness of 1 mm are used.

From the obtained results of corrosion decrement, corrosion velocities are calculated. The velocities show the reduction in pipe wall thickness. If it is necessary to determine the corrosion type in addition to the corrosion decrement, the time period of the test is prolonged to one year. The advantage of this test lies in the effect of water on testing samples which is as complex as the effect of water on the pipeline. The impact of dissolved oxygen in water may occur and be denoted as the dominating factor causing the corrosion or as the positive factor causing the passivation of metal by forming a layer with a good protective properties which separates the transported water from the pipeline surface (Dubová et al, 2010) [12].

From the decrement of testing coupons, the corrosion velocity was determined according to the formulas of STN 75 7151:

Average corrosion decrement ($g.m^{-2}$) is calculated as an arithmetic average of 5 testing coupons fixed in one coupon holder according to the following formula:

$$K' = \frac{1}{n}\sum_{i=1}^{n} K_i \tag{1}$$

The calculation of corrosion decrements of particular samples in $g \cdot m^{-2}$:

$$K = \frac{m_1 - m_2}{s} \tag{2}$$

Where:

m_1 – weight before exposition $\left[\frac{g}{m^2}\right]$,

m_2 – sample weight after exposition $\left[\frac{g}{m^2}\right]$,

S – sample surface [m²].

The calculation of corrosion decrements, U_t, in µm during exposition:

$$U_t \; \frac{1}{7.86}.\left(\bar{K} - K\right) \qquad (3)$$

Where:

$\bar{K}$ - average of corrosion decrements of 5 samples $\left[\frac{g}{m^2}\right]$,

K' - decrement of sample during the blank sample test, which is performed for controlling the possibility of testing samples dissolution in removal process of corrosion waste products in HCL, $\left[\frac{g}{m^2}\right]$,

7.86 - specific weight of steel $\left[\frac{g}{cm^3}\right]$,

The calculation of corrosion velocity, v_u, in µm in 1 year time period:

$$v_u \; \frac{365.(U_{t2} - U_{t1})}{t_2 - t_1} \qquad (4)$$

Where:

t_1 – shorter exposition time of sample [d],

t_2 – longer exposition time of sample [d],

Ut_1 – corrosion decrement of sample in shorter exposition time [µm],

Ut_2 - corrosion decrement of sample in longer exposition time [µm].

Based on the corrosion test results obtained between the 30th and 60th day, the water is classified by the aggressivity level depending on the corrosion velocity in µm for 1 year as shown in Tab. 1.

Tab. 1 Water aggressivity level classification according to corrosion velocity [11].

Aggressivity level	Corrosion velocity v_t [µm.year^{-1}]	Classification
I	‥ 50	moderately aggressive
II	50 until 150	medium aggressive
III	¨ 150	strongly aggressive

The water is classified according to the aggressivity levels into three following categories:

I. No anti-corrosion precaution required
II. Considered individually; the decision about anti-corrosion actions is made with regard to required pipeline service life and based on the results of technical and economic analyses
III. Anti-corrosion actions must be provided with regard to required pipeline service life

As the next equipment for classification of aggressive properties of transported water, an ASTM D2688-11 [13] device was used, in which the testing coupons were installed. The coupon size was 74x9 mm and the thickness 1 mm. The corrosion velocity was calculated according to the following relations.

The calculation of corrosion decrements of particular samples in mg:

$$W \; W_i - W_f \qquad (5)$$

Where:

W_i – sample weight before exposition [mg],

W_f – sample weight after exposition [mg].

The obtained weight values are put into the calculation of corrosion velocity:

$$KR = \frac{\Delta m}{\Delta t} .factor \tag{6}$$

Where:

Δm – determined decrement of weight [mg],

Δt – onset period of coupon [day],

factor – 0.025 (for steel of class 11).

3 CORROSION TEST AT THE WTP JAKUBANY

The corrosion tests were performed in cooperation with the company PVPS, a.s. (Podtatranská Water Operating Company) at the water treatment plant (WTP) in Jakubany, a village in the Stará Ľubovňa District (Fig. 1). Raw water for treatment is sampled directly from the Jakubianka brook and is transported by gravity through a DN 500 pipeline to the water treatment plant. The treatment plant is capable to produce 150 liters per second. Currently, the production is about 60 liters per second. The water treatment technology at the Jakubany treatment plant varies according to season. Over the winter, the brook water quality is at a very high level and the detritus tank is weaned because of ice-cover formation and the only treatment of water by using pressure filters is provided.

Fig. 1 WTP in Jakubany.

The treatment process at the WTP in Jakubany is chosen with regard to the quality of water input. Therefore, the processes used at the WTP in Jakubany include mechanical precleaning, simple sedimentation and consequent 1-step coagulative filtration which is used only in case of impaired raw water quality. In case of need, aluminium sulphate, a coagulation reagent, is added to pressure mixing tanks and the consequently formed flocks are removed by means of pressure sand filters.

Health safety of filtered water is ensured by gaseous chlorine, or by chloramination, which is accumulated in a tank with a capacity of 2500 m^3. The water is transported from the tank through gravity pipes to the consumption area.

This work consists in long-time monitoring of corrosion in the water supply pipeline. Measurements were performed from 11th May 2013 to 12th May 2014. Two devices for monitoring the corrosion velocity were used in order to make the operational classification of aggressive character of water. Testing coupons (according to STN) were used for tests performed on one of the devices and coupons according to ASTM were used for the tests on the other one. These devices were placed into a raw water intake location of the WTP, downstream the pressure filters and upstream the disinfection (Fig. 2).

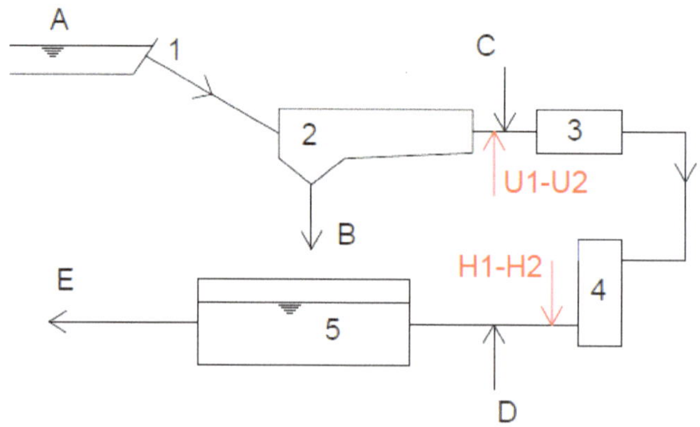

Fig. 2 Technological scheme of WTP in Jakubany

A – Jakubianka water source, B – draining of settled sludge, C – aluminium sulphate dosing, D – water disinfection, E – conduit to water storage, U1,H1 – corrosion equipment according to STN, U2, H2 – corrosion equipment according to ASTM, 1 – sampling object, 2 – sedimentation tank, 3 – pressure mixing tank, 4 – pressure sand filter, 5 – accumulation tank

The water quality was monitored also during the change of testing coupons in corrosion devices. In terms of chemical changes, the quality of water at the WTP in Jakubany did not change significantly. The quality values were: pH 7.22-8.32, temperate 0.2 – 17 °C, $KNK_{4,5}$ 1.4-2.2 $mmol.l^{-1}$, ZNK4,5 below 0.06 $mmol.l^{-1}$, Fe 0.008-0.067 mg. l^{-1}, Mn below 0.01 mg. l^{-1}, Ca^{2} 28.1-40.5 $mg.l^{-1}$. Turbidness of raw water was observed in a range of 0.00-1.60 ZF. Fig. 3 shows the temperature progress during the measurement. The average year temperature was 7.6 °C.

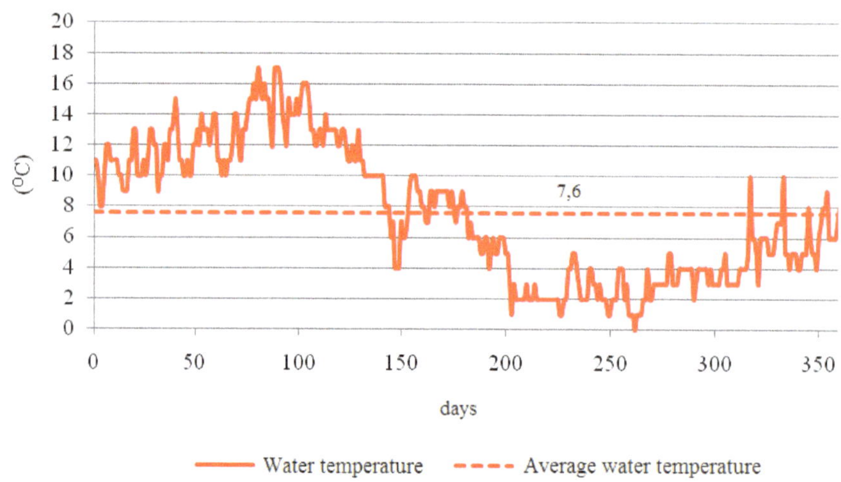

Fig. 3 Water temperature progress in a period of 11.5.2013 - 12.5.2014

4 RESULTS

4.1 Results according to STN

From the performed measurements, it is obvious that the progress of corrosion velocities of raw water and treated water is a bit different. Fig. 4 shows the consideration of corrosion velocities between 30 and 60 days of measuring. Due to these standardized results, the water is classified into I and II aggressivity levels. During the first two tests (5 13 and 6 13), the corrosion velocity of raw water showed a higher value as it was in case of using the water that passed the treatment process. This was probably induced by a water flow decrease (down to stopping) through the corrosion equipment, which was caused by strong storm-related turbidity. During the 5th and 9th tests, the corrosion velocity also increased.

In case of seven tests, the corrosion velocity of water that passed the treatment process was higher than the velocity of raw water which might be caused by a higher flowing velocity, disinfection reagent (gaseous chlorine), using which the pre-disinfection before the filter is provided, or which might be caused by coagulant (aluminium sulphate).

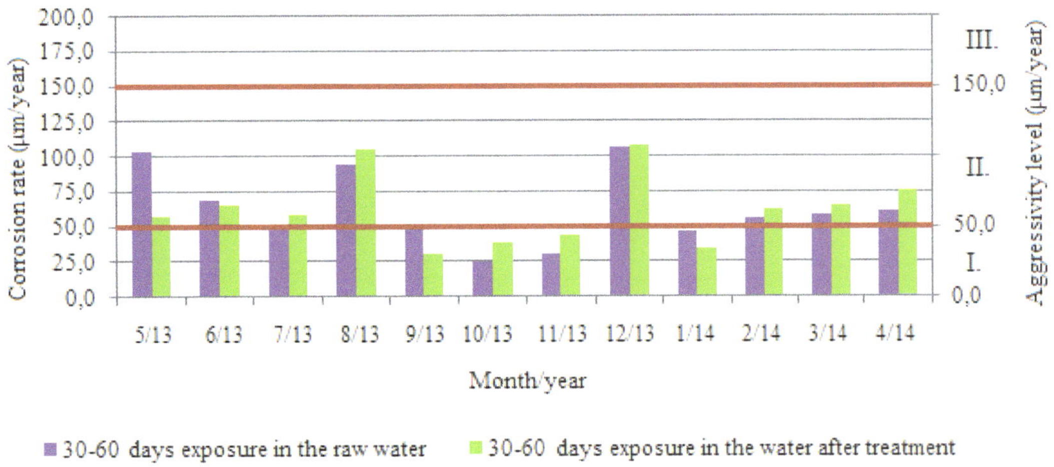

Fig. 4 Consideration of corrosion velocities of raw water and water which passed treatment after 30-60 days of measurements

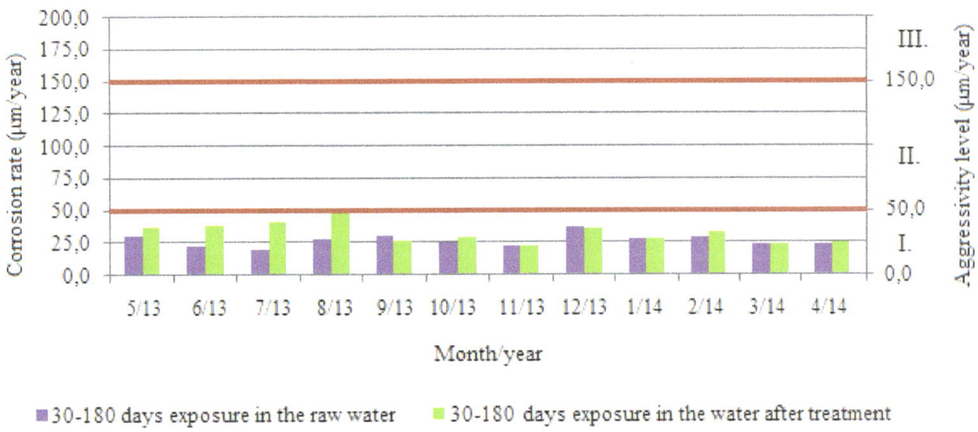

Fig. 5 Comparison of corrosion velocities of raw water and water after treatment during 30-80 days of measurements

Fig. 5 shows the comparison of corrosion velocities when using the raw water and the water after treatment during 30-180 days of measurements. The results refer to the linear progress of corrosion velocity, just in case of water after treatment tests numbered 5 13-8 13 when a moderate increase is presented. In comparison with standardized 30-60 days of measurement, the lower corrosion velocity of water classifies the aggressivity to level I.

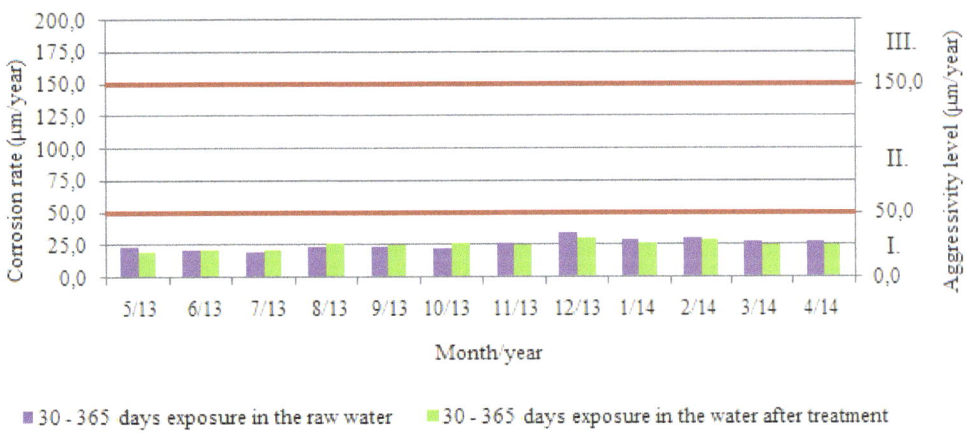

Fig. 6 Comparison of corrosion velocities of raw water and water after treatment during the 30-365 days of measurements

The progress of corrosion velocities of raw water and water after treatment during 30-365 days of experiment measurements is shown in Fig. 6. With regard to the annual results of corrosion velocities, the water is classified into the aggressivity level I which represents the lowest value in comparison with the measurements during 30-60 and 30-180 days. The comparison of corrosion velocities for 30-60 days of measurements of raw water and water after treatment after filtration with a change in water temperature during the monitoring period is shown in Fig. 7.

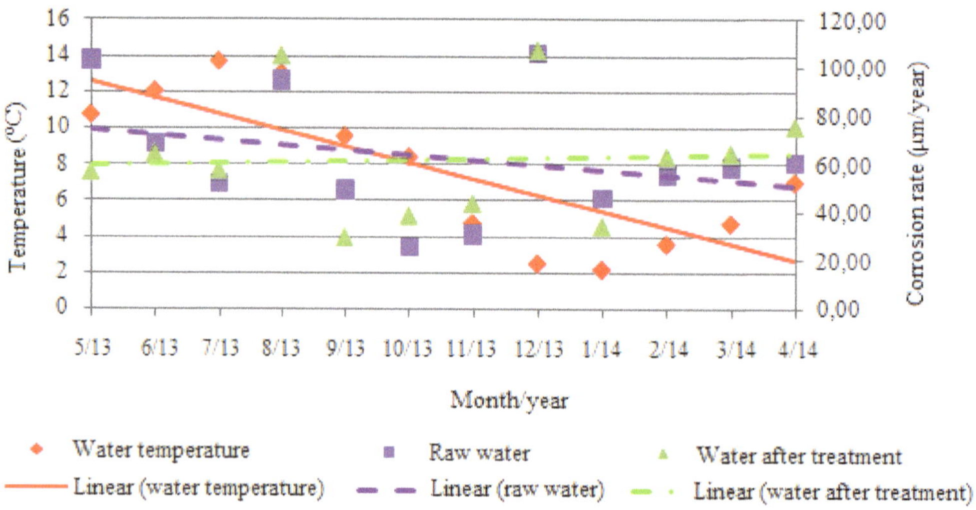

Fig. 7 Corrosion velocities of raw water and water after treatment after filtration during 30-60 days of measurements in comparison with temperature progress

The estimations of service life of pipeline were calculated from the obtained results (annual exposition) for raw water and water after treatment. The service life estimations are listed in Tab. 2. In the tests performed with raw water, the value of estimated service life fell within a range of 14.57-19.50 $\mu m.year^{-1}$ (annual average of 16.57 $\mu m.year^{-1}$), and with water after treatment, after filtration 14.54-23.52 $\mu m.year^{-1}$ (annual average of 19.19 $\mu m.year^{-1}$).

Tab. 2 Estimation of service life of pipeline

Period of corrosion tests	Estimation of service life of pipeline $v_{u(ž)}$ (μm/year)	
	Raw water	Water after treatment
11.5.-10.7.2013	14,74	14,54
10.7.-8.9.2013	14,57	16,18
8.9.-7.11.2013	19,50	22,26
7.11.2013-6.1.2014	17,41	20,59
6.1.-7.3.2014	18,39	23,52
7.3.-12.5.2014	14,81	18,03

According to our results, it is obvious that the aggressivity of water was changing over the testing period. A significant variation of velocities mainly occurred during 30-60 days of measurement. Short-time measurements showed much higher aggressivity of water than it was in case of long-time measurements. Therefore, it is necessary to perform measurements for a longer time in order to obtain more exact results or support the short-time measurement results. The corrosion velocities within 30-180 days of measurements of raw water had almost linear progress, while the measurements of water after treatment decreased moderately. In case of 30 – 365 days of measurement, both water types reflected linear progress of their corrosion velocities. Therefore, it may be asserted that the short-time (30-60 days) measurements may overvaluate the water aggressivity.

The impact of a change in water temperature was found mainly within short-time standardized measurements (60 days), while during long-time (30-180, 30-365 days) measurements, the variation of corrosion velocities was not significant.

4.2 Results according to ASTM

The corrosion test assemblies and corrosion specimens employed in the study were prepared according to the ASTM D2688-11 standard "Test Method for Corrosivity of Water in the Absence of Heat Transfer" (Weight Loss Method). This test method covers the determination of water corrosivity by evaluating pitting and by measuring a weight loss of metal specimens. Pitting is a form of localized corrosion: the weight loss is a measure of the average corrosion rate. The rate of corrosion of a metal immersed in water is a function of the tendency of the metal to corrode and is also a function of the tendency of water and materials contained to promote (or inhibit) corrosion. The test method employs flat, rectangular-shaped metal coupons which are mounted on pipe plugs and exposed to water flowing in metal piping in municipal, building, and industrial water systems using a side stream corrosion specimen rack [13].

The comparison of corrosion velocities in testing raw water and water after treatment during the 30-day experiment is shown in Fig. 8. The corrosion velocity is higher in the raw water experiment. In comparison with the 60-day experiment, the corrosion velocities were lower in the short-time 30-day experiment (Fig. 9).

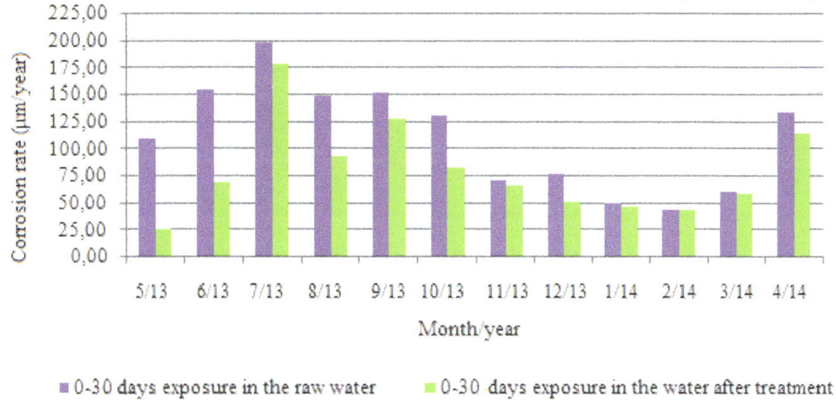

Fig. 8 Comparison of corrosion velocities for raw water and water after treatment during 30-day experiment/measurements

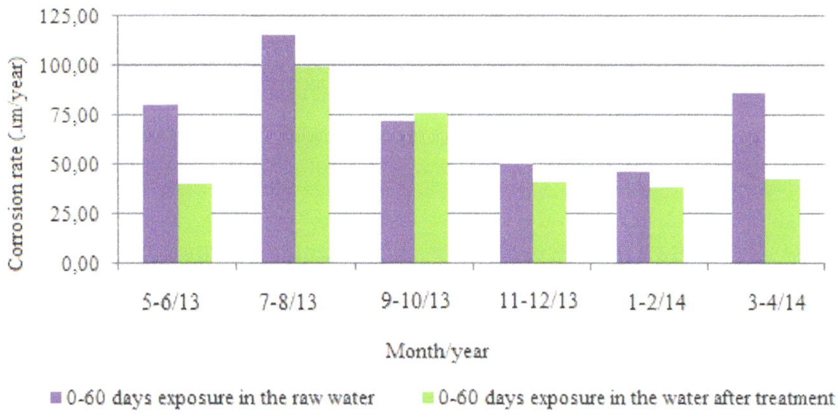

Fig. 9 Comparison of corrosion velocities for raw water and water after treatment during 60-day measurements

The comparison of corrosion velocities of 30-day measurement (Fig. 10) and 60-day measurement for both type of water with water temperature pointed out the impact of temperature on water aggressivity. Following the comparison of corrosion velocities, it may be asserted that the raw water seems to be more aggressive than the water after treatment. In the case of 60-day measurements, the corrosion velocities decreased twice compared to normal (Fig. 11).

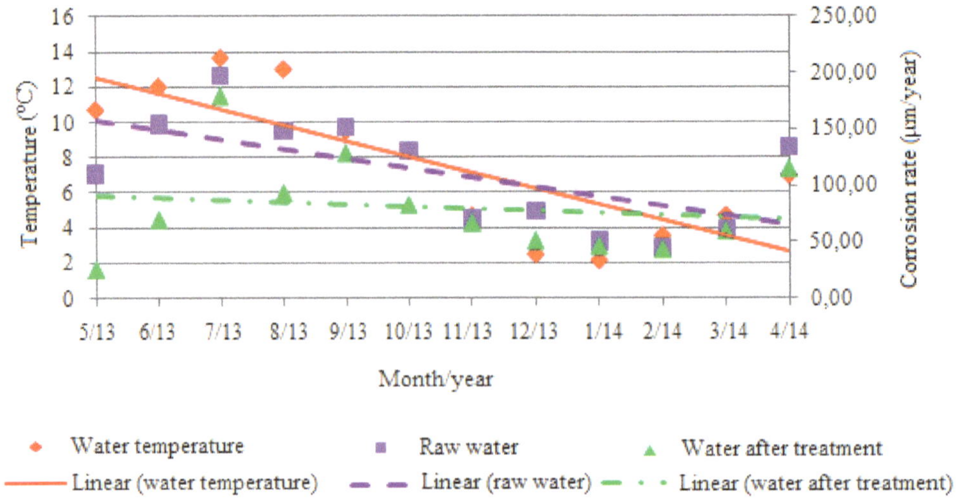

Fig. 10 Corrosion velocities after 30-days of measurements for raw water and water after treatment after filtration in comparison with water temperature progress

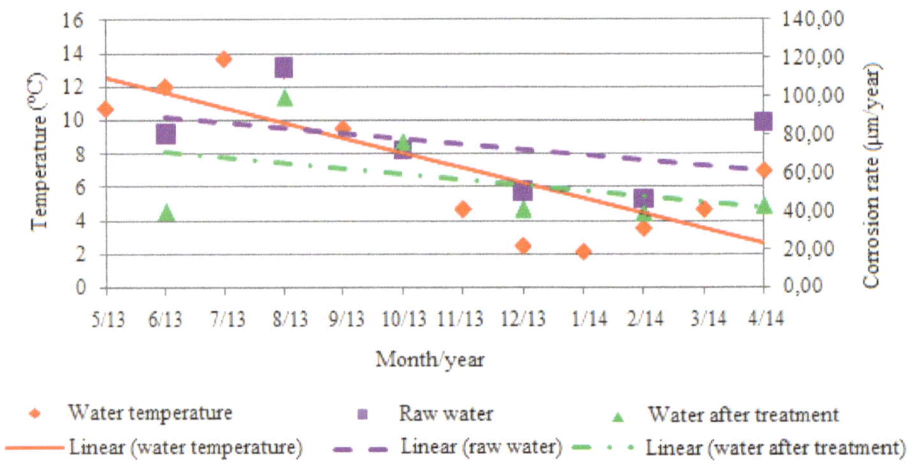

Fig. 11 Corrosion velocities after 60-days of measurements for raw water and water after treatment after filtration in comparison with water temperature progress

5 DISCUSSION

According to the STN 75 7151 and ASTM D2688-11 results, the difference between short-time measurements is obvious. But long-time (annual) measurements show almost similar corrosion velocities. A higher average corrosion velocity in the raw water tests than in the tests for water after treatment (water after filtration) was observed in both types of measurements. From the temperature change progress over the seasons, it is obvious that the change in temperature of flowing water through the testing corrosion devices influenced the corrosion progress of tested samples. The most significant impact of water on corrosion was observed within short-time measurements (30-60 days), while in half-year and annual measurements the impact was minimal.

Annual measurement results refer to the possibility to classify the water (both types, raw water and water after treatment after filtration) as the water of the first aggressivity level (moderately aggressive water – corrosion velocities to 50 µm.year^{-1}). The type of corrosion was determined based on annual testing coupons. In case of raw water, surface area corrosion and sporadically point corrosion prevailed, while in case of water after treatment, more significant point corrosion in combination with fewer area corrosion prevailed. The estimation of service life of pipeline system was calculated by using the annual test results. The annual average of service life estimations was 16.57 µm.year^{-1} for raw water and 19.18 µm.year^{-1} for water after treatment (after filtration).

6 CONCLUSIONS

Iron corrosion is an extremely complex process. Because of the large variability in distribution system conditions, a particular factor may by critical in one system but relatively unimportant in another system. Iron pipe corrosion is extremely complicated and is affected by practically every physical, chemical, and biological parameter in water distribution systems. Because distribution system pipes are in place for long periods of time (50 years), corrosion control is critical to maintain microbial, water quality and pipe integrity.

This work consists in long-time monitoring of corrosion in the water supply pipeline. It showed that corrosion rates were strongly related to water temperature (and or other seasonal factors). According to the obtained results of corrosion velocities, it is obvious that water aggressivity varied over the monitored period of time. The results obtained from annual monitoring of water aggressivity point out a decrease in corrosion velocity in long-time tests in comparison with short-time tests for both types of water. This implies that further arrangements for decreasing the corrosion formation of pipeline system are not required.

Acknowledgement

The article has been written with the support of the VEGA 1 1243 12 project solved at the Department of Sanitary and Environmental Engineering of the Faculty of Civil Engineering of the Slovak University of Technology Bratislava.

REFERENCES

[1] VREEBURG, J.,H.,G., *Discolouration in Drinking Water Systems: A Particular Approach*. Department of Water Management. Delft, Delft University of Technology, The Netherlands, 2007, ISBN 978-90-74741-91-0

[2] MUNKA, K., KARÁCSONYOVÁ, M., GAJDOŠ, Ľ., *Poloprevádzkové hodnotenie vplyvu kvality vody na vybrané materiály prichádzajúce do styku s pitnou vodou*. In: Voda Zlín 2005. Sborník příspěvků IX. mezinárodní vodohospodářské konference. Zlín, 7.-8.4.2005. Zlín: Zlínská vodárenská, a.s., 2005. s. 85-90. ISBN 80-239-4453-3

[3] SLAVÍČKOVÁ, K., *Meření a hodnocení rychlosti koroze při procesu úpravy vody*, In: Zborník príspevkov X. Medzinárodnej vodohospodárskej konferencie, Voda Zlín 2006, s. 143-148. ISBN 80-239-6523-9

[4] LYTLE, D. A., SCHOCK, M. R., *Impact of Stagnation Time on Metal Dissolution from Plumbing Materials in Drinking Water*. Journal of water supply research and technology, 2000, AQUA 49(5):243-257

[5] VOLK, CH., DUNDORE, E., SCHIERMANN, J., LECHEVALLIER, M., *Practical evaluation of iron corrosion control in a drinking water distribution system*, Water Research, Volume 34, No.6., Elsevier Science Ltd, 2000, s. 1967-1974, ISSN 0043-1354(99)00342-5

[6] FANG, C. S., LIU, B., (2003). *Hydrodynamic and temperature effects on the flow induced local corrosion rate in pipelines*, Chemical Engineering Communications, Vol. 190, s. 1249-1266, ISSN 0098-6445

[7] SMITH, S.E., BISSET, A., COLBOURNE, J.S., HOLT, D.M., LLOYD, B.J., 1997. *The occurrence and significance of particles and deposits in a drinking water distribution system*. Journal of the New England Water Works Association, 1997, Vol. 111 (2), 135-150

[8] MCNEILL, L.S., EDWARDS, M., 2001. *Iron pipe corrosion in distribution system*, Journal – American Water Works Association, July 2001, s 88-100,

[9] SINGLEY, J.,E., BEAUDET, B.,A., MARKEY, P.,H., *Corrosion Manual for Internal Corrosion of Water Distribution System*, Environmental Science and engineering, INC., Florida, 1984, ESE No. 81-227-260

[10] DEVIETTI, J.,L., *Modelling water quality in tropical water distribution systems*. PhD thesis, James Cook University, 2011

[11] STN 75 7151 Kvalita vody. Požiadavky na kvalitu vody dopravovanej potrubím

[12] DUBOVÁ, V., ILAVSKÝ, J., BARLOKOVÁ, D., *Sledovanie agresivity vody na VZ Pernek*, Konferencia vodohospodárov v priemysle, Liptovský Ján, 2010, s. 217-224, ISBN 978-80-227-3393-6

[13] ASTM D2688-11 „Standard Test Method for Corrosivity of Water in the Absence of Heat Transfer“ (Weight Loss Method)

14

MODIFIED ZEOLITES IN GROUND WATER TREATMENT

Danka BARLOKOVÁ [1], *Ján ILAVSKÝ* [2], *Marek SOKÁČ* [3]

[1] *Doc. Ing. PhD., Department of Sanitary and Environmental Engineering, Faculty of Civil Engineering, Slovak University of Technology*
Radlinského 11, 81368 Bratislava
e-mail danka.barlokova@stuba.sk

[2] *Doc. Ing. PhD., Department of Sanitary and Environmental Engineering, Faculty of Civil Engineering, Slovak University of Technology*
Radlinského 11, 81368 Bratislava
e-mail jan.ilavsky@stuba.sk

[3] *Doc. Ing., PhD., Department of Sanitary and Environmental Engineering, Faculty of Civil Engineering, Slovak University of Technology*
Radlinského 11, 81368 Bratislava
e-mail marek.sokac@stuba.sk

Abstract

There are presented results of technological experiments carried out in Water Treatment Plant Kúty. The goal of this study was to compare modified zeolite known as clinoptilolite (rich deposits of clinoptilolite were found in the region of East Slovakia in the 1980s) with the imported modified zeolite from deposit situated in Hungary. Klinopur-Mn and Klinomangan were used for removal of iron and manganese from ground water to meet the requirements of the Regulation of the Government of the Slovak Republic No. 496/2010 on Drinking Water. The materials observed exhibit different efficiencies of manganese removal from water, since the quality of the treated water play a major role (oxygen content and pH value). In the case of the removal of the iron from the water, the quality of the raw water is not a limiting factor; both materials removed Fe from the water to below the limit value (0.20 $mg.l^{-1}$).

Abstrakt

Článok prezentuje výsledky technologických skúšok vykonaných v UV Kúty. Cieľom tejto práce bolo porovnať modifikované (povrchovo upravené) zeolity známe ako klinoptilolit (veľké nálezisko klinoptilolitu bolo objavené na Východnom Slovensku v 1980-tych rokoch) s dovážaným povrchovo upraveným zeolitom z náleziska v Maďarsku. Klinopur-Mn a Klinomangan boli použité pre odstraňovanie železa a mangánu z podzemnej vody na dosiahnutie limitných hodnôt pre pitnú vodu podľa Nariadenia vlády č. 496/2010 Z.z. Sledované materiály vykazovali rôznu účinnosť odstraňovania mangánu z vody, na účinnosť odstraňovania mala významný vplyv kvalita upravovanej vody (obsah kyslíka, hodnota pH). V prípade odstraňovania železa z vody kvalita surovej vody nie je limitujúcim faktorom, obidva materiály odstraňovali železo z vody pod limitnú hodnotu (0,2 $mg.l^{-1}$).

Keywords: treatment of groundwater, removal of iron and manganese, filtration, modified clinoptilolite, Klinopur- Mn, Klinomangan, drinking water

1 INTRODUCTION

In Slovakia there are a number of groundwater resources. Larger resources are unevenly distributed throughout the territory; therefore, the water intended for drinking purposes is supplied from smaller ones. In an evaluation of the quality of water in small resources, more than 300 resources with higher concentrations of iron, manganese, nitrates, ammonium ions, arsenic and antimony have been identified.

According to the 2010 Reports on the Environment in Slovakia, the concentration of iron exceeded the 0.2 mg/L limit in more than 18.9 % of the samples, and the concentration of manganese exceeded the 0.05 mg/L limit in more than 19.4 % of 396 groundwater samples (which represents 175 objects of basic monitoring). The concentration of iron exceeded the 0.2 mg/L limit in more than 37.2 % of the samples, and the concentration of manganese exceeded the 0.05 mg/L limit in more than 40.1% of 698 groundwater samples (which represents 211 objects of operational monitoring). The limit values are defined under the Regulation of the Government of the Slovak Republic No. 496/2010 on Drinking Water [1].

In searching for suitable water treatment technology, an emphasis is placed on new, more efficient and cost-effective methods and materials compared to the technology currently used.

The natural Zeolites are known for more than 200 years. There were Zeolites as mineral substances discovered by the Swedish mineralogist Cronstedt in 1756. In the course of this period were more than 35 types of Zeolites discovered and described, but its practical importance has remained unused for a long time. Only the modern identification methods have contributed to the fact that in the last 30 years have been more than thousands important occurrences of more than 50 types of Zeolites recorded in more than 40 countries.

In the eighties the deposit of natural Zeolite, with high content of Zeolitic mineral – Clinoptilolite in Nižný Hrabovec in the area of the East – Slovakian neovolcanites, was discovered. Besides this deposit, the Zeolite accumulations were discovered in the complex of the Central Slovakian neovolcanites, in the South – Western border of Kremnické vrchy and in Zemplínske pohorie on the territory of the Slovak Republic.

A primary structure unit of Zeolite is an octahedron – the even octahedron formed by atoms of silicium and aluminium with tetrahedrite-coordinated atoms of oxygen. The representation of the silicium and aluminium to the oxygen is usually 1:2. In this way, relatively big cavities mutually interconnected by small channels, the effective diameters of which differ from 0.2 to 0.7 nm and the total volume of vacant places forms from 20 % up to 50 % of mineral volume, are formed in the crystalline structure. There are the mobile mono-valent and bivalent cations of alkaline metals and alkaline earth metals (Ca, K, Na) encircled by molecules of water, which can be substituted by other ones from the surrounding solution, placed in these cavities.

In the last decades the Zeolitic mineral – Clinoptilolite has started to be also used in drinking water treatment. Zeolites are not to be found in the grained condition in the natural deposits, and before their use, Zeolites require crushing and sorting at the required fractions in the waterworks engineering. Sufficient mechanical strength, chemical stability and abrasion values, by which, though, they are classified among the soft filtration materials, and these properties made the natural Zeolites use as a filtration material.

One of the methods for elimination of dissolved iron and manganese is the elimination by means of the oxidized coat on the grains of filter medium. By addition of potassium permanganate (not only $KMnO_4$, but also other strong oxidizing agents) there is the coat form on the surface of filter medium, and this coat serves as a catalyst of oxidation, and the grain of filter medium are covered by higher metal oxides. In this case it can be said about the special filtration, so-called contact filtration, filtration on the manganese filters. These two ways of treatment are very closely connected and they can be hardly separated unambiguously. The oxidation condition of the medium coat MnO_x(s) plays the obvious role in elimination of dissolved manganese, and the efficiency of manganese elimination is the immediate function of MnO_x(s) concentration and oxidative condition. There is the formation of coats with different abilities to eliminate dissolved manganese from water on the various filter medium [1-11]. At present in addition to the traditional modified filtration sand is using a wide variety of materials with oxidation film on the surface of the grains of filter media (Klinopur-Mn, Greensand, Birm, Cullsorb M, Everzit Mn, MTM and Pyrolox).

The objective of the technological experiments in the locality of Kúty (a water treatment plant) was to compare the efficiency of manganese and iron removal in water treatment using a filtration medium based on a chemically modified natural zeolites (Klinopur-Mn, Klinomangan) and monitoring the influence of water quality (pH, concentration of oxygen, iron and manganese) on efficiency removal iron and manganese from water.

2 MATERIALS AND METHODS

Klinopur-Mn - activated zeolite – clinoptilolite $(Na,K)^{6+}(Al_6Si_{30}O_{72})\cdot 20H_2O$ is produced in Slovakia. On clinoptilolite grains are industrially created a thin layer of manganese oxide, which allows the material used for contact filtration. Based on previous experiments (pilot tests) conducted by the Department of Sanitary and Environmental Engineering at the Faculty of Civil Engineering STU in Bratislava can be concluded that the surface of activated clinoptilolite is comparable with foreign materials and can be used for removing Fe and Mn in water [12,13].

Klinomangan - activated zeolite – clinoptilolite $(K,Na,Mn)^{6+}[(AlO_2)_6(SiO_2)_{30}].24\ H_2O$ from the bearing Rátka in Hungary. Superficial layers of manganese oxide as in the case of Mn-Klinopuru allows to use this material in contact filtration for removal of iron and manganese from water. Depending on the quality of the treated water is required after a certain time, the filter cartridge to regenerate the solution of potassium or sodium permanganate [14].

Table 1 shows the content of essential minerals forming clinoptilolite deposit, Nižný Hrabovec and clinoptilolite from the deposit Rátka in Hungary. **Table 2** is the chemical composition of clinoptilolite bearing Nižný Hrabovec and clinoptilolite from the deposit Rátka before activation and **Table 3** shows the chemical composition zeolites after activation with $KMnO_4$ and the content of MnO_2 on the surface of filter media. The selected parameters of the filtration materials used in the experiments are compared in **Table 4**.

Tab. 1 Mineralogical composition of the zeolite deposits from Nižný Hrabovec and Rátka

Mineral	Nižný Hrabovec (Slovakia)	Rátka (Hungary)
	Content [%]	
Clinoptilolite	84	55
Cristobalite	8	15
Feldspar	3-4	10
Illite	4	-
Montmorillonite	-	10

Tab. 2 Chemical composition of the nature zeolites from Nižný Hrabovec and Rátka

Zeolite	Content [%]							
	SiO_2	Al_2O_3	K_2O	CaO	Fe_2O_3	MgO	Na_2O	TiO_2
Nižný Hrabovec	66.40	12.20	3.33	3.04	1.45	0.56	0.29	0.15
Rátka (Hungary)	72.15	12.86	3.72	1.84	1.22	0.53	0.26	0.10

Tab. 3 Chemical composition of the Klinopur-Mn and Klinomangan

Material	Content [%]									
	SiO_2	Al_2O_3	K_2O	CaO	Fe_2O_3	P_2O_3	MgO	Na_2O	TiO_2	MnO_2
Klinopur-Mn	69.56	8.19	5.58	3.79	3.32	1.08	0.81	-	0.74	**6.92**
Klinomangan	61.68	8.29	4.77	3.51	2.87	0.81	0.27	2.56	-	**15.16**

Tab. 4 Grain size and bulk density (in a dry state) of the selected filtration materials

Material	Klinopur-Mn	Klinomangan	Silica sand	Greensand
Grain size [mm]	0.6 – 1.6	0.5 – 1.2	0.5-1.0	0.25-0.8
Bulk density [$g.cm^{-3}$]	0.84	1.04	1.55	1.36

2.1 Water treatment plant Kúty

The water treatment plant in Kúty is a part of the Senica group of water supply systems. The water from two wells with a yield of 80 $l.s^{-1}$ does not meet the requirements of Regulation No. 496/2010 on Drinking Water for iron, manganese, sulfate ions and aggressive carbon dioxide. The technological water treatment process consists of aeration, a dosage of calcium hydrate, rapid mixing, contact filtration and disinfection. The technological scheme of the WTP Kúty is shown in **Fig. 1**. The figure also indicates the location of the filter columns (sampling points) used in experiments.

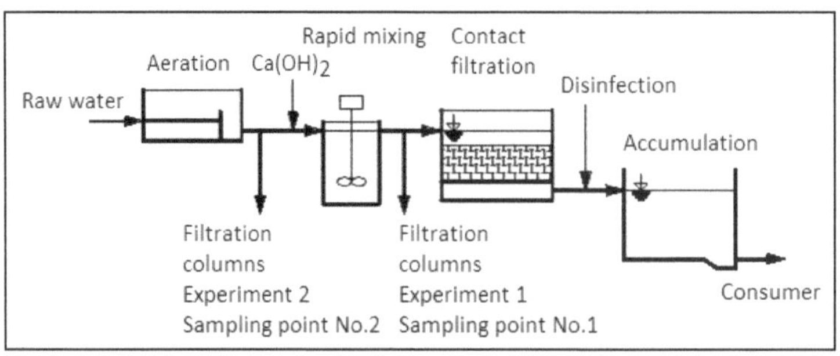

Fig. 1 Scheme of the technology of WTP Kúty and the location of the filter columns

The methodology for the verification of suitable filtration materials for iron and manganese removal is based on their properties and possible technological applications in the water treatment process. The following technological water treatment method was proposed:

raw water → filtration and oxidation (backwashing and regeneration)

Raw water is passed through the filtration equipment, and the removal of the Fe^{2+} and Mn^{2+} ions is carried out directly in the filtration column beds. The following were used as filtration materials:

- activated zeolite from Nižný Hrabovec with MnO_2 on the surface (Klinopur-Mn)
- activated zeolite from Rátka with MnO_2 on the surface (Klinomangan)

The experiments were designed to optimize the filtration rate (contact time of the raw water with the filter media) and washing and regenerating the filter materials (filter length of cycles). The quality of the raw water (Fe and Mn content) and treated water at the outlet from the separate filtration columns was monitored during the experiments. At the same time the amount of water (filtration rate) at the outlet from the columns was measured.

2.2 Filtration Model

To verify the efficiency of iron and manganese removal from the water resources in the locality of WTP Kúty, two filtration columns containing Klinopur-Mn and Klinomangan were used. The parameters of each adsorption column are as follows: diameter = 5.0 cm; height = 2 m; surface = 19.635 cm^2; filtration medium height 110 cm; and volume of filtration medium 2160 [cm^3].

The water was supplied to filtration columns from two different sites for the technological water treatment process. The water for Experiment 1 (sampling point No. 1) was taken after aeration and lime dosing, where the optimal conditions for the removal of the iron and manganese (increased oxygen content and a pH of more than 8) were achieved. The water for Experiment 2 (sampling point No. 2) was taken after aeration of the water, where the content of the oxygen in the water had increased. **Table 5** shows the values of the basic parameters during the experiments.

Tab. 5 The basic parameters during the pilot test

Parameter	Sampling point	
	No.1	No.2
Fe [$mg.l^{-1}$]	2,28 – 5,92	1,18 – 5,96
Mn [$mg.l^{-1}$]	0,210 – 1,154	0,969 – 1,412
pH	8,26 – 8,68	6,81 – 7,28
O_2 [%]	56 – 58	62 – 65

3 RESULTS AND DISCUSSION

The model tests and the results of the experiments are divided into two parts:

a) raw water after aeration and the addition of lime (sampling point No.1)

b) raw water after aeration (sampling point No.2).

3.1 Experiment 1

During the model tests, an average concentration of iron and manganese in raw water were 0.506 $mg.l^{-1}$ for manganese or 3.92 $mg.l^{-1}$ for iron and the filtration rates were 5.16 $m.h^{-1}$ for Klinopur-Mn or 4.74 $m.h^{-1}$ for Klinomangan. Filtration conditions are shown in a **Table 6**.

Tab. 6 Filtration conditions

Parameter	Klinopur-Mn	Klinomangan
Grain size [mm]	0.6-1.6	0.5-1.2
Medium height [cm]	110	110
Average flow through column [$l.h^{-1}$]	168.95	155.07
Average filtration rate [$m.h^{-1}$]	5.16	4.74
Filtration total time [h]	1644	1644
Total volume of water flowing through [m^3]	16.665	15.296
Average residence time in column [min]	12.78	13.93

The results removal of the iron and manganese from *the raw water after aeration and the addition of lime* (sampling point No. 1) are documented by **Fig. 2**, in which the concentration of manganese and iron in the raw water and the values measured after passing through the monitored filter materials are shown. The figure also shows the manganese limit value (0.05 mg.l^{-1}), and respectively the iron limit value (0.2 mg.l^{-1}) in the drinking water as defined under Regulation No. 496/2010 on Drinking Water. The arrow presents the regeneration of the filter media.

Fig. 2 left shows that in optimal conditions for contact filtration (pH 8.2 to 8.6; an oxygen content of 56 to 58% saturation) materials Klinopur-Mn and Klinomangan achieved a high efficiency for removing of manganese from the water. After regeneration materials at the beginning of the experiment (after 60.5 hours of operation model devices) they did not exceed the value of the manganese in the treated water limit value of 0.05 mg.l^{-1} even after 1600 hours of operation of the filter columns. The filter materials were backwashed continuously (approximately every 3 to 4 days). There was no need to regenerate these materials.

Fig. 2 right shows the progress made in removing of the iron from the water for sampling point No.1 (after the water aeration and the addition of lime). The value of the iron in the raw water is quite changed, depending on the production of precipitated $Fe(OH)_3$, which gradually clogged the system. In general both materials removed the iron effectively and, during the operation of the filtration columns, did not exceed the limit value of 0.20 mg.l^{-1} as defined under Regulation No. 496/2010 on Drinking Water.

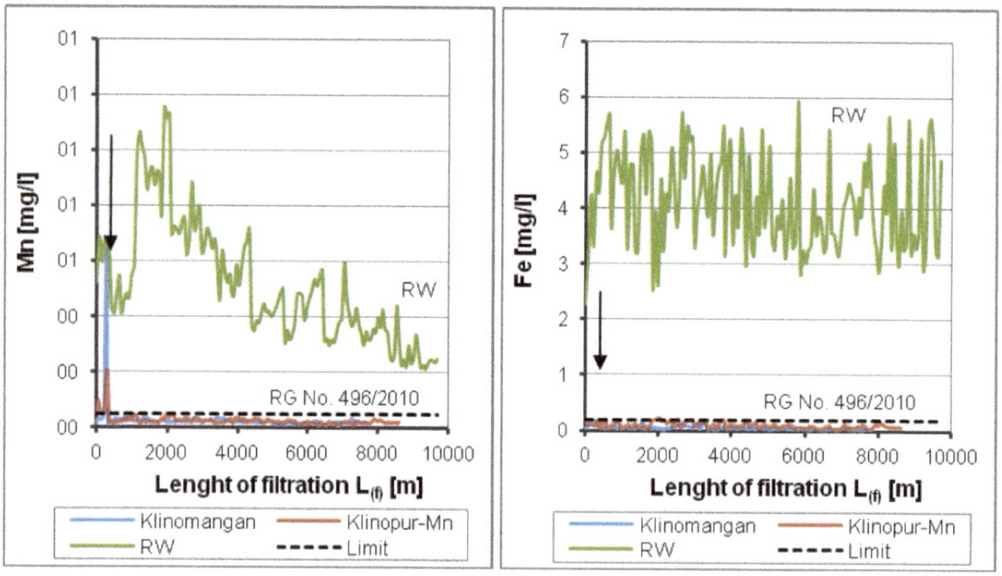

Fig. 2 Course of the removal of the manganese (left) and iron (right) from the water.

3.2 Experiment 2

During the model tests, an average concentration of iron and manganese in raw water were 1.124 mg.l^{-1} for manganese or 3.28 mg.l^{-1} for iron and the filtration rates were 4.74 m.h^{-1} for Klinopur-Mn or 4.60 m.h^{-1} for Klinomangan. Filtration conditions are shown in a **Table 7**.

Tab. 7 Filtration conditions

Parameter	Klinopur-Mn	Klinomangan
Grain size [mm]	0.6-1.6	0.5-1.2
Medium height [cm]	110	110
Average flow through column [l.h^{-1}]	155.94	150.64
Average filtration rate [m.h^{-1}]	4.74	4.60
Filtration total time [h]	1408	1408
Total volume of water flown through [m^3]	13.174	12.726
Average residence time in column [min]	13.85	14.34

Fig. 3 shows the results of removal of the iron and manganese from *the raw water after aeration* (sampling point No.2). The concentration of manganese and iron in the raw water and the values measured after they passed through the monitored filter materials, the manganese limit value (0.05 mg.l^{-1}), and respectively the iron limit value (0.2 mg.l^{-1}) in the drinking water defined under Regulation of the Government of the Slovak

Republic No. 496/2010 on Drinking Water and the regeneration for the filter media, are illustrated. The arrow presents the regeneration of the filter media.

Fig. 3 left shows that the influence of changes in the quality of the raw water (pH 6.8 to 7.2, the oxygen content from 56 to 57% saturation) for the efficiency of the manganese removal from the water through the filtration media improved significantly. Klinopur-Mn and Klinomangan are necessary to modified by the gradual backwashing and regeneration with 2,5% solution of $KMnO_4$. In the first filtration step it exceeded the limit value of 0.05 $mg.l^{-1}$ after 194 hours of operation; in the second filter cycle after 216 hours for Klinopur-Mn or 268 hours for Klinomangan, in the third filter cycle after 232 hours for Klinopur-Mn and 280 hours for Klinomangan; in the fourth filter cycle the limit value was exceeded after 338 hours for both materials. The filtration time without regeneration was gradually extended. This means that the industrially activated clinoptilolite (Klinopur-Mn, Klinomangan) should be modified an on-site directly in water treatment plant. The filter cycles will be extended, and after some time, regeneration will not be needed.

The filter media were backwashed continuously (approximately every 3 to 4 days) (given the amount of precipitated ferric hydroxide collected).

Fig. 3 right shows the course of the iron removing from the water for sampling point No. 2. The value of the iron in the raw water was quite changed, depending on which well was used for pumping or the production of precipitated $Fe(OH)_3$, which gradually clogged the system. In general both materials removed the iron effectively and, during the operation of the filtration columns did not exceed the limit value of 0.20 $mg.l^{-1}$ as defined under Regulation No. 496/2010 on Drinking Water.

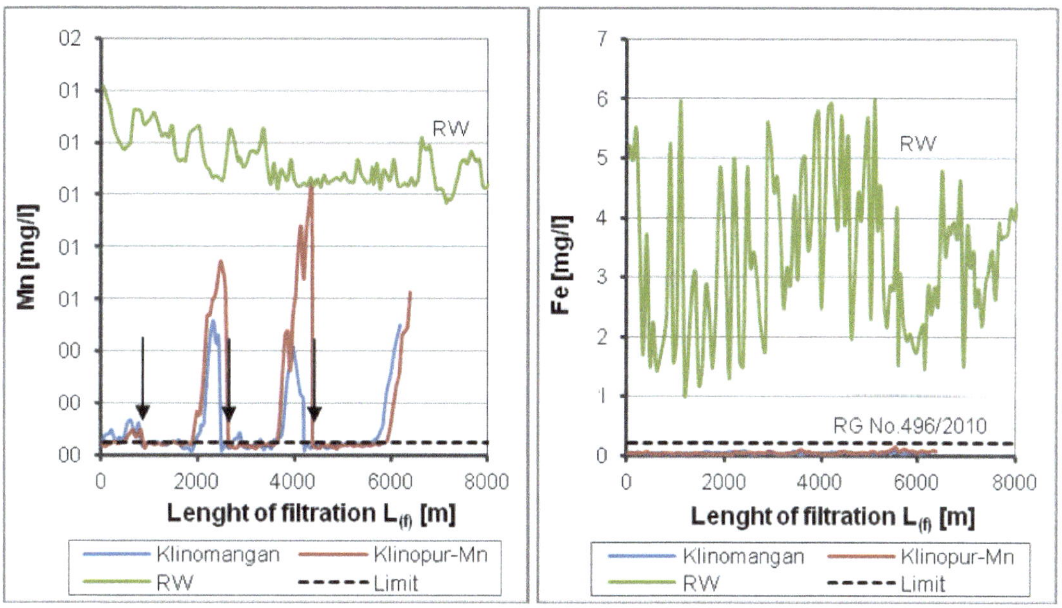

Fig. 3 Course of the removal of the manganese (left) and iron (right) from the water.

4 CONCLUSION

Obtained results prove the possibility to use Klinopur-Mn for removal of iron and manganese in water treatment process. Klinopur-Mn is comparable with imported filtration material Klinomangan. The results also show that the repeated use of Klinopur-Mn leads to extended filtration cycles and based on this fact it is comparable with prepared filtration sand. In comparison with filtration sand, the advantage of Klinopur-Mn is that it requires lower amount of water for backwashing which results in more cost-effective operation.

The content and form of occurrence of iron and manganese in water as well as water pH and dissolved oxygen concentration (min. 15 % of Fe + Mn content) are among the important criteria for Fe and Mn removal efficiency (lower efficiency in removal of iron bound to humic acids). The efficiency of Mn removal is influenced by the contact time between water and filter material (height of filtration layer, filtration time), properties of preparation active layer, i.e. layer thickness and its chemical composition, regeneration method, $KMnO_4$ concentration as well as the time of backwashing and regeneration.

ACKNOWLEDGEMENT

Experimental measurements were carried out with the financial support VEGA 1/1243/12 and VEGA 1/0691/13 research projects from the Scientific Grant Agency of the Slovak Ministry of Education, Science, Research and Sport.

REFFERENCES

[1] Správa o stave životného prostredia Slovenskej republiky v roku 2009, Ministerstvo životného prostredia Slovenskej republiky, Slovenská agentúra životného prostredia, Report on State of the Environment of the Slovak Republic in 2009, the Ministry of the Environment of the Slovak Republic, Slovak Environmental Agency, ISBN 978-80-88833-53-6, p. 308

[2] Doula, M.K. Removal of Mn2+ Ions from Drinking Water by Using Clinoptilolite and a Clinoptilolite–Fe Oxide System. Water Research. 2006. XL. Nr. 17, pp. 3167-3176. ISSN 0043-1354.

[3] Knocke, W.R., et al.: Kinetics of Manganese (II) and Iron (II) Oxidation by Potassium Permanganate and Chlorine Dioxide. Journal of AWWA. 1991. LXXXIII. Nr.6, pp. 80-87. ISSN 0003-150X.

[4] Knocke, W.R., Hungate, R., Occiano, S. Removal of Soluble Manganese by Oxide-Coated Filter Media: Sorption Rate and Removal Mechanism Issues. Journal of AWWA. 1991. LXXXIII. Nr.6, pp. 64-69. ISSN 0003-150X.

[5] Knocke, W.R., Hamon, J.R., Thompson, C.P. Soluble Manganese Removal on Oxide-Coated Filter Media. Journal of AWWA. LXXX. Nr.12, pp. 65-70. ISSN 0003-150X.

[6] Merkle P.B., Knocke W.R., Gallagher D.L., Solberg T Characterizing Filter Media Mineral Coatings, Journal of AWWA. 1996. LXXXVIII. Nr.12, pp. 62-73. ISSN 0003-150X.

[7] DIMIRKOU, A., Doula, M.K. Use of clinoptilolite and an Fe-overexchanged clinoptilolite in Zn2+ and Mn2+ removal from drinking water. Desalination 2008. CCXXIV. Nr.3, pp. 280–292 ISSN 0011-9164.

[8] Ates, A. Role of modification of natural zeolite in removal of manganese from aqueous solutions. Powder Technology. 2014. CCLXIV, pp. 86–95. ISSN 0032-5910.

[9] Taffarel, S.R., Rubio, J. Removal of Mn2+ from aqueous solution by manganese oxide coated zeolite. Minerals Engineering 2010. XXIII. pp. 1131–1138 ISSN 0892-6875.

[10] Bruins, J.H., Dirk Vries, D., Petrusevski, B., Slokar, Y.M., Kennedy, M.D. Assessment of manganese removal from over 100 groundwater treatment plants. Journal of Water Supply: Research and Technology—AQUA 2014. LXIII. Nr.4, pp 268–280. ISSN 0003-7214.

[11] Shalini, Ch., Pragnesh N.D. Removal of iron for safe drinking water. Desalination 2012. CCCIII. Nr.3, pp. 1–11 ISSN 0011-9164.

[12] Barloková, D., Ilavský, J. Removal of Iron and Manganese from Water Using Filtration by Natural Materials. Polish Journal of Environmental Studies 2010. XIX, Nr.6, pp.1117-1122. ISSN 1230-1485.

[13] Barloková D., Ilavský J. Modified Clinoptilolite in the Removal of Iron and Manganese from water. Journal of Civil Engineering. 2012. XX. Nr.3, pp.1-8. ISSN 1210-3896.

[14] http://www.zeoclay.eu/domain/zeoclay/files/klinomangan-1000-data-sheet.pdf, 2014.

[15] Rajic, N., Stojakovic, D., Sanja Jevtic, S., Zabukovec Logar, N., Kovac, J., Venceslav Kaucic, V. Removal of aqueous manganese using the natural zeolitic tuff from the Vranjska Banja deposit in Serbia. Journal of Hazardous Materials 2009. CLXXII. pp. 1450–1457. ISSN 0304-3894.

RESUMÉ

V článku sú uvedené výsledky práce, ktorej cieľom bolo porovnať povrchovo upravené zeolity známe ako klinoptilolit (veľké nálezisko klinoptilolitu bolo objavené na Východnom Slovensku v 80-tych rokoch 20. stor.) s dovážaným povrchovo upraveným zeolitom z náleziska v Maďarsku. Klinopur-Mn a Klinomangan boli použité pre odstraňovanie železa a mangánu z podzemnej vody.

Dosiahnuté výsledky pri odstraňovaní mangánu a železa z vody poskytujú podklad pre využitie filtračných materiálov (Klinopur-Mn, Klinomangan) v úprave vody.

Sledované materiály vykazovali rôznu účinnosť odstraňovania mangánu z vody, nakoľko veľkú úlohu zohráva kvalita surovej vody (obsah kyslíka a hodnota pH) a obsah oxidačnej vrstvy MnO_2 na povrchu filtračného materiálu.

Z Rtg analýzy materiálov vidieť, že Klinopur Mn obsahuje 6,92% MnO, v porovnaní s Klinomanganom (15,16% MnO), čo môže mať vplyv na ich účinnosti pri odstraňovaní rozpusteného mangánu z vody kontaktnou filtráciou, potrebu regenerácie náplne v priebehu prevádzky a „zapracovanie" filtračných materiálov priamo na mieste úpravy.

V prípade odstraňovania železa z vody kvalita surovej vody nie je limitujúcim faktorom, všetky materiály odstraňovali Fe z vody na hodnotu nižšiu ako je limitná hodnota (0,20 $mg.l^{-1}$).

Pri výbere vhodného materiálu je nutné zvážiť viacero faktorov, ktoré môžu ovplyvniť efektívnosť technologického procesu, napr. chemické vlastnosti a zloženie upravovanej vody, požiadavky na upravenú vodu, rovnako je dôležité posúdenie investičných a predpokladaných prevádzkových nákladov. Kontaktná filtrácia cez vhodný materiál predstavuje ekonomicky prijateľnú a nenáročnú technológiu na odstraňovanie železa a mangánu z vody, ktorá z hľadiska obsluhy je zároveň maximálne bezpečná.

15

THE EVALUATION OF SELECTED AIR POLLUTANTS IN KOŠICE FOR THE PERIOD 2008-2010

Marcela GERGEĽOVÁ [1], Žofia KUZEVIČOVÁ [1], Štefan KUZEVIČ [2]

[1] *MSc., PhD., Institute of Geodesy, Cartography and Geographic Information Systems, Faculty of Mining, Ecology, Process Control and Geotechnologies, Technical University of Košice, Park Komenského 19, 043 84 Košice, Slovak Republic*
e-mail: marcela.gergelova@tuke.sk

[1] *assoc.prof., Institute of Geodesy, Cartography and Geographic Information Systems, Faculty of Mining, Ecology, Process Control and Geotechnologies, Technical University of Košice, Park Komenského 19, 043 84 Košice, Slovak Republic*
e-mail: zofia.kuzevicova@tuke.sk

[2] *assoc.prof., Institute of Business and Management, Technical University of Košice, Faculty of Mining, Ecology, Process Control and Geotechnologies, Park Komenského 19 042 00 Košice, Slovak Republic*
e-mail: stefan.kuzevic@tuke.sk

Abstract

In general, air quality is determined from the concentrations of pollutants in ambient air. Air quality criteria (limit and target values, margin of tolerances, upper and lower assessment thresholds) are based on the current legislative framework. In Slovakia, the air quality criteria are imposed by Decree No 360/2010 Coll, on air quality, of the Ministry of Environment [1]. In relation to the implemented measurements, it is necessary to choose effective tools needed for the pre-processing and post-processing of overall air quality assessment. The article aims at applying suitable GIS tools in the assessment process of air quality in the Košice agglomeration for the selected period 2008-2010. In the overall assessment of the area in question, the processing of information on emissions declared as particulate matter (PM) was considered in terms of a conservative approach to the assessment of air quality for PM_{10}. For the assessment, the yearbooks and reports on air quality in Slovakia in 2008-2010 of the Slovak Hydrometeorological Institute (SHMI) and professional publications were used.

Keywords: air quality, GIS tools, agglomeration of Košice, particulate pollutants

Abstrakt

Vo všeobecnosti kvalita ovzdušia je stanovená z obsahu znečisťujúcich látok vo vonkajšom ovzduší. Kritériá kvality ovzdušia (limitné a cieľové hodnoty, medze tolerancie, horné a dolné medze na hodnotenie) vychádzajú z platného legislatívneho rámca. V podmienkach SR kritéria kvality ovzdušia vyplývajú z vyhlášky MŽP SR č. 360/2010 Z.z. o kvalite ovzdušia [1]. V nadväznosti na realizované merania je potrebné zvoliť efektívne nástroje potrebné pre preprocesing a postprocesing procesu plošného hodnotenia kvality ovzdušia. Príspevok si kladie za cieľ aplikovať vhodné nástroje GIS do hodnotiaceho procesu kvality ovzdušia v aglomerácií Košice za vybrané obdobie rokov 2008 - 2010. Pri celkovom zhodnotení predmetnej oblasti bolo uvažované so spracovaním informácií o emisiách deklarovaných ako tuhé znečisťujúce látky (TZL) v zmysle konzervatívneho prístupu k hodnoteniu kvality ovzdušia za PM_{10}. Pre posudzovanie boli použité ročenky a správy pre kvalitu ovzdušia v SR za rok 2008 - 2010 z SHMÚ a odborné publikácie.

Keywords: kvalita ovzdušia, nástroje GIS, aglomerácia Košice, tuhé častice

1 ASSESSED TERRITORY AND INPUT INFORMATION FOR EVALUATION

Based on current legislation, Act No. 318/2012 Coll., on the atmosphere, as amended, the area of the Košice Region belongs to 1 group, i.e. among the zones and agglomerations in which the levels of one or more pollutants are higher than the limit values, or the limit values plus the margin of tolerance. The pollutant due to which the Košice Region and the town of Košice are included in the first group is PM_{10} (particulate matter in the air which passes through a size-selective inlet with a 50 % efficiency cut-off at 10 μm aerodynamic diameter) [2].

Based on further measurements, the region territory was also included in the second group, i.e. among the zones and agglomerations in which the levels of one or more pollutants are between the limit values and the limit values plus margin of tolerance. The zones and agglomerations in which the measured ozone concentration was higher than the long-term objective for ozone, but less than or equal to the target value set for ozone were also included in this group.

The pollutant due to which the area of the Košice Region and the town of Košice is classified into the group 2 is ozone.

Based on further measurements, the Košice Region was included into the group 3, i.e. the air pollution level is below the limit values, or below the limit values plus the margin of tolerance.

The pollutants due to which the Košice Region and the town of Košice is classified into the group 3 are SO_2, NO_2, Pb, CO and benzene. The overview of the inclusion of the area in question into different groups is shown in Tab. 1 [3], [4].

Tab. 1 The inclusion of the area into the zones by pollutants [5]

Group	Zone or agglomeration	The pollutant for which a given zone, or an agglomeration is included in a respective group
1	Košice Region	PM_{10}
	Košice	PM_{10}
2	Košice Region	Ozone
	Košice	Ozone
3	Košice Region	Sulphur dioxide, nitrogen dioxide, lead, carbon monoxide, benzene
	Košice	Sulphur dioxide, nitrogen dioxide, lead, carbon monoxide, benzene

*PM_{10} – *particulate matter in the air which passes through a size-selective inlet with a 50 % efficiency cut-off at 10 μm aerodynamic diameter*

1.1 Characteristics of the analysed particulate matter PM_{10}

PM_{10} and $PM_{2,5}$ (particulate matter – PM).

In general, particulate pollutants (*PM = particulate matter*) are a mixture of airborne solid and liquid fractions. Several particulate pollutants are emitted directly; others are formed in the atmosphere in reactions of other pollutants. The particulates are in a wide range of sizes. The particulates smaller than 2.5 micrometres (*μm)* in diameter ($PM_{2,5}$) are called "fine" particulate matter. They are so small that they can be captured only by electron microscopy. The particulates between 2.5 and 10 micrometres in diameter (PM_{10}) are referred to as *"coarse"*, but their size is smaller than the thickness of a human hair. The listed particulate pollutants have adverse effects on human health. They can get into the lungs, which can cause serious health problems. It is therefore necessary to monitor the presence of PM in the air [6].

Tab. 2 Geographical coordinates of monitoring stations and the list of the monitored pollutants - 2009 [7]

AGGLOMERATION/ Zone	Municipality, locality	Geographical length	Geographical width	Altitude [m]	PM_{10}	CO	Pb	Cd	Ni	As	BaP
KOŠICE	Košice, Strojárska 1)	21°15'07"	48°43'36"	202	*						
	Košice, Amurská 1)	21°17'11"	48°41'28"	201	*						
Košice Region	Veľká Ida, Letná	21°10'30"	48°35'32"	209	*	*	*	*	*	*	*

Note: 1) *Strojárska by 16 September 2009, Amurská from 5 October 2009*

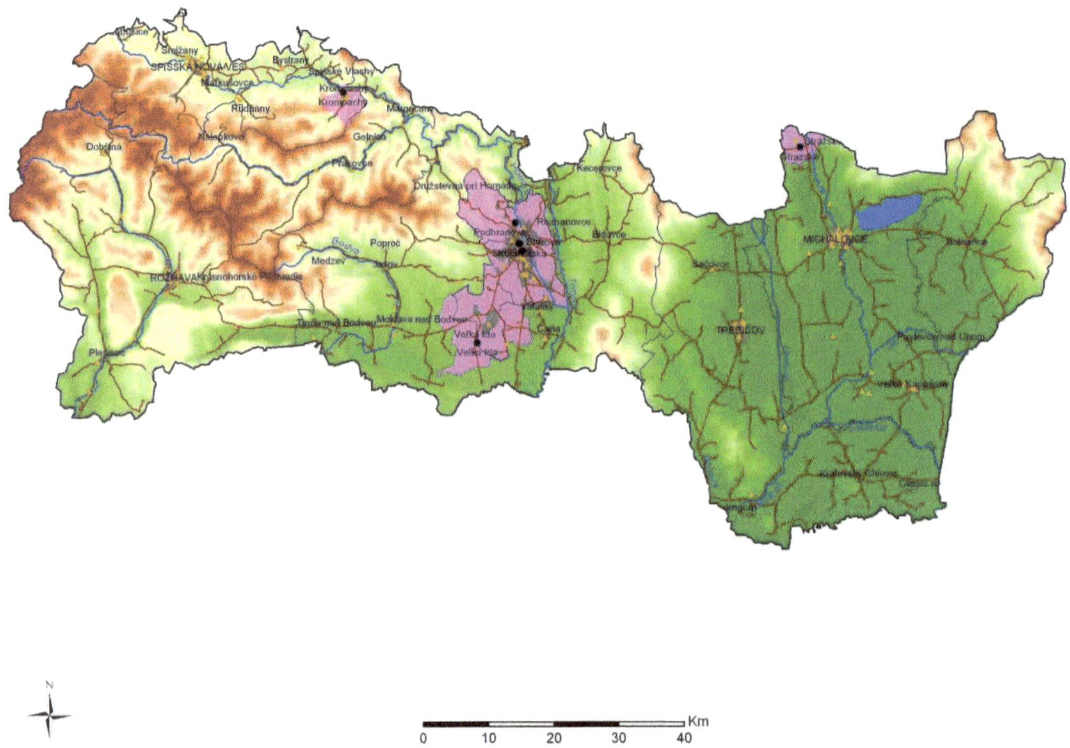

Fig. 1 The spatial localization of the area [8]

Monitoring the levels of air pollution in the territory of the Košice agglomeration is provided by the Slovak Hydrometeorological Institute (SHMI). The measurements are mainly focused on the detection of concentrations of SO_2, NO_x and particulate matter PM_{10}. The air pollution control is provided through a network of automated monitoring stations. Their number for the selected period (2008-2010) changed. In the process of addressing the issues of processing particulate matter by means of tools of the Geographic Information Systems (GIS), the following monitoring stations were used:

- Podhradová,
- Strojárenská,
- Štúrova,
- Amurská,
- Veľká Ida.

The stations were deployed to capture the situation of air pollution in Košice as well as possible. The selection of all sites was conducted in a commission way with the involvement of interested authorities, organizations and institutions. When selecting the sites, wind conditions and dominant town resources were taken into account.

The station ***Košice - Podhradová*** is located in the area of the SHMI workplace in a relatively open air space on the northern side of the Podhradová housing estate and the town itself. Since 2000, the station monitors the level of ground-level ozone pollution only.

The station ***Košice – Strojárenská*** represents the northern part of the historic town. It is located next to the Town Hall in congested urban area about 50 meters from the surrounding buildings and far from the road about 15 meters. When the southern wind flows, the site is exposed to emissions from cars in traffic in the Moyzesova Street.

The station ***Košice – Štúrova*** represents the centre of the town. It is located in an open air area in the middle of the square *Námestie osloboditeľov* between the car park and symbolic cemetery. Far from the station about15 meters north and 50 meters south there are roads of inner circle that lead in the east-west direction [9].

1.2 Climatic characteristics of the assessed area

Climatic conditions of the area in question are affected by the layout of surrounding mountains. From the south-west, the Slovak Karst affects the area, in the north there are the Slovak Ore Mountains, in the east Slanské Hills. Among the mountains, the Košická kotlina basin ranges. In the Kosice Region, several climatic zones exist. In the East Slovak Lowland (in Slovak: Východoslovenská nížina), including the Zemplin Hills (in

Slovak: Zemplínské vrchy), there is a warm area representing the area with an average of 50 or more summer days and annual daily maximum air temperature ≥ 25 °C. From a climatic point of view, also Košická kotlina basin belongs to the same area which is warm, dry, or slightly damp with cold winters. From a climatic point of view, the Slovak Ore Mountains are located in a slightly warm and cold area. In a slightly warm area, there is, on average, less than 50 summer days per year with a daily maximum air temperature ≥ 25 °C and a July average air temperature ≤ 16 °C. In a cold area, there is a July average air temperature < 16 °C with relatively high humidity. The lowest values of average annual rainfall are in southern areas of the region, such as the Eastern Lowland (in Slovak: Východoslovenská nížina) and Bodviany Hills (in Slovak: Bodvianska pahorkatina) (550-600 mm). The highest values of average annual rainfall are in highland areas of the Volov Hills (in Slovak: Volovské vrchy) in the range of 800-900 mm. It is similar with the number of days with snow cover – in southern areas it is 60 days, in highland areas 120 days. [16]

Weather conditions pose a high degree of influence on the spread of pollutants in the air. In the area in question, there is relatively small occurrence of wind calm. The highest incidence of calm is recorded in January; thereupon unfavourable conditions for dispersion of pollutants in the air arise in this month.

The north-south orientation of the basin is the most important factor in shaping the direction of flow, resulting in a significantly narrow wind rose with prevailing wind blow in the north direction and secondary wind blow in the south direction (especially in the cold half of the year). The prevailing wind from the north is characterized by relatively higher speeds, achieving on average 5.7 ms-1. The average wind speed per year in all directions is 3.6 ms-1. The southern part of the Košice basin is open and highly windy. Also, a substantial part of the town of Košice, especially, the Hornád river valley and ridges surrounding the hills are very windy. Fig. 2 below presents the wind rose from the meteorological station spatially localized in the part Košice-Airport for the period of March 2013. The abundance of wind directions is expressed in % [15].

The dispersion conditions in the area are greatly influenced by orography. The subarea of the town of Košice situated in the Hornád river valley shows limited conditions for dispersion of pollutants in the air. Airflow is of a typical north-south nature and the most unfavourable conditions for the dispersion here arise during the heating period at low south flow, or stagnant air.

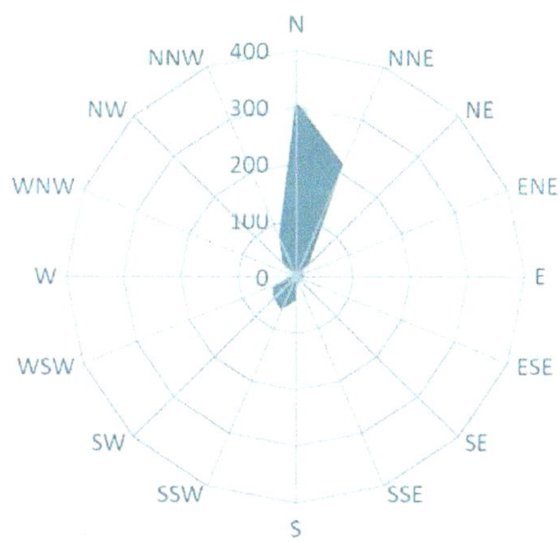

Fig. 2 Wind conditions for selected part

2 ASSESSMENT OF AIR QUALITY USING GIS

In general, the GIS represent an effective tool for the field of processing spatially localized data with versatile use. Their applications can be found in the area of the monitoring and processing of air quality. The "Reports on the state of the environment" for the selected periods were the basis for processing air pollutant values for the selected period 2008-2010 in the GIS environment. The summary reports on air quality under the conditions in the Slovak Republic are compiled by the SHMI. On the basis of available information, the air pollution levels measured at individual stations of the Observatory Network were processed for the Košice agglomeration. Available spatial information given in Tab. 3 and Tab. 4 contributed to a distinct extent to the spatial processing of air pollution. The distribution of the considered automated monitoring stations is presented in Fig. 3. In terms of the deployment of the automated monitoring stations, it is necessary to note a change in the spatial location. By 2009, the air pollution monitoring was carried out through the stations listed in Tab. 3. Since 2009, a change has occurred which necessitated the deployment of new monitoring stations. The list of the new stations which are in force to the present day is characterized in Tab. 4.

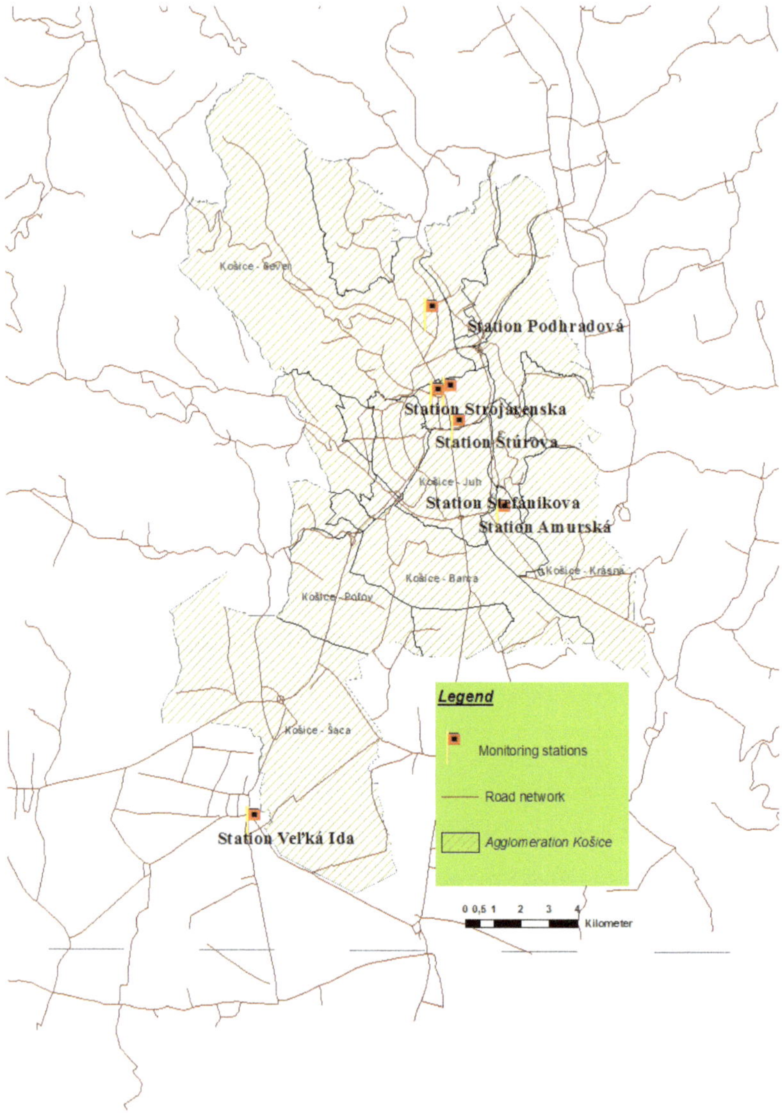

Fig. 3 Spatial distribution of monitoring stations in the agglomeration of Košice

Tab. 3 List of monitoring stations in the year 2008 [9]

Monitoring stations	Longitude	Latitude
Štúrova	E 021°15'39"	N 48°43'02"
Strojárenská	E 021°15'07"	N 48°43'36"
Veľká Ida	E 021°10'30"	N 48°35'32"

Tab. 4 List of monitoring stations since the year 2009 [7]

Monitoring stations	Longitude	Latitude
Amurská	E 021°17'11"	N 48°41'28"
Štúrova	E 021°15'39"	N 48°43'02"
Strojárenská	E 021°15'07"	N 48°43'36"
Veľká Ida	E 021°10'30"	N 48°35'32"

2.1 Processing the results of measurements of air pollution according to air pollutant limits for the period 2008-2010

In 2008, based on the results of the level of air pollution, three different monitor modes are distinguished. Fig. 4 and Tab. 5 specify requirements for the assessment of air quality for each proposed monitoring mode.

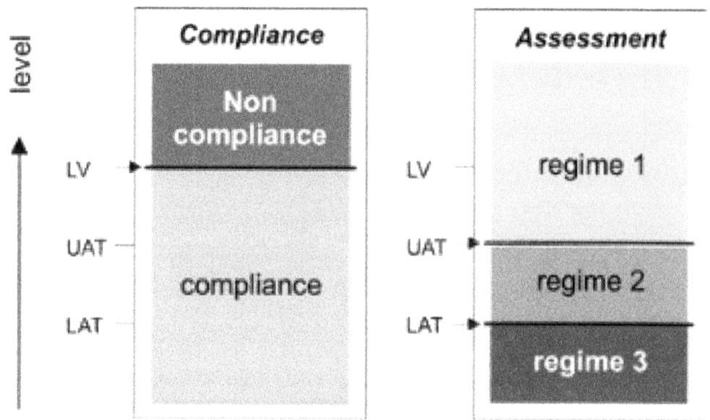

Fig. 4 Regimes of air quality assessment in relation to LV1,UAT2 and LAT3 [9], [7]

Legend:
LV - limit value according to Decree No. 360/2010 Coll., on Air Quality
UAT - Upper Assessment Threshold (Decree 360/2010 Coll.)
LAT- Lower Assessment Threshold (Decree 360/2010 Coll.)

Tab. 5 Requirements for assessment in three different regimes [8]

Maximum level of pollution in agglomerations and zones	**Requirements for assessment**
REGIME 1 Above upper assessment threshold	High quality of measurements is obligatory. Measured data can be supplemented by further information, model computations including.
REGIME 2 Below upper assessment threshold, but above lower assessment threshold	Measurements are obligatory, however to a lesser extent, or to a lesser intensity, under the premise that the data are supplemented by other reliable sources of information.
REGIME 3 Below lower assessment threshold In agglomerations only for pollutants, for which an alert threshold has been set	At least one measurement station is required in each Agglomeration combined with the model computations, expert estimate and indicative measurements. Those are measurements based on simple methods, or operated in limited time. These are less accurate than continuous measurements, but may be used to control relatively low level of pollution and as supplementary measurements in other areas.
In all types of zones, apart from agglomeration zones, for all pollutants for which an alert threshold has been set	Model computations, expert estimates and indicative measurements are sufficient.

Tab. 6 Presentation of the limit values for PM 10 by [1]

Averaged period	Limit value	Threshold of tolerance
	Particulate matter PM_{10}	
1 day	50 μg.m^{-3} may not be exceeded more than 35 times per calendar year	**50%**
Calendar year	40 μg.m^{-3}	**20%**

Tab. 7 Annual average concentrations of PM_{10} [μg.m^{-3}]

Monitoring stations/particulate matter	PM_{10} *2008*	PM_{10} *2008*	PM_{10} *2009*	PM_{10} *2009*	PM_{10} *2010*	PM_{10} *2010*
	(24 hour)	(1 year)	(24 hour)	(1 year)	(24 hour)	(1 year)
Košice - Štúrova	38	**29.5**	*	*	*	*
Košice - Strojárenská	55	**31.6**	*	*	*	*
Košice - Štefániková	*	*	*	*	67	**36.2**
Košice - Amurská	*	*	15	**26.5**	30	**25.2**
Košice - Veľká Ida	157	50	166	**51.3**	132	**46.7**

According to Decree 360/2010, it was necessary to perform further processing to establish limit values for air pollution components. The set upper limits were based on Tab. 5. In accordance with relevant EU directives, the particulate matter is reported as PM_{10}. The graphic processing of the measured values of the amount of particulate matter for the selected period realized by means of a spreadsheet MS Excel is presented in Graph 1. The subject matter of the processing was to compare average annual concentrations of pollutants with corresponding air pollutant limits according to applicable SK legislation. Graph 1 shows the comparison of the average annual concentrations of the pollutant with relevant air pollutant limits. During the monitored period, the air pollutant limit values for particulate matter (PM_{10}) were exceeded at the monitoring stations Veľká Ida in 2009 and Štefániková in 2010.

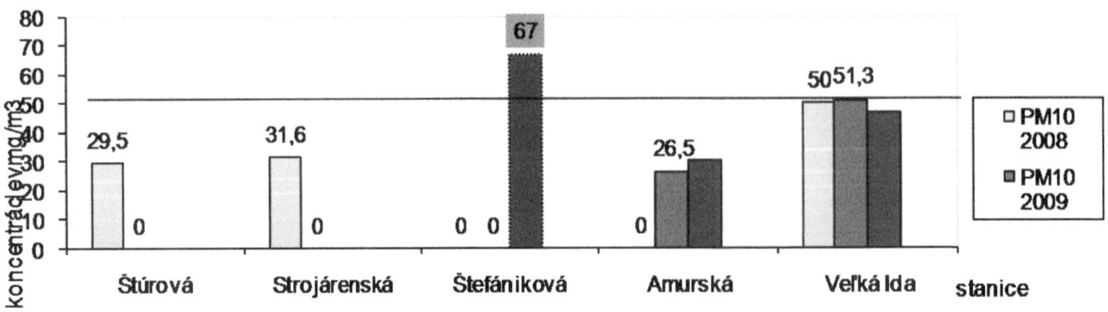

Graph. 1 Graphical data processing concentrations of PM10

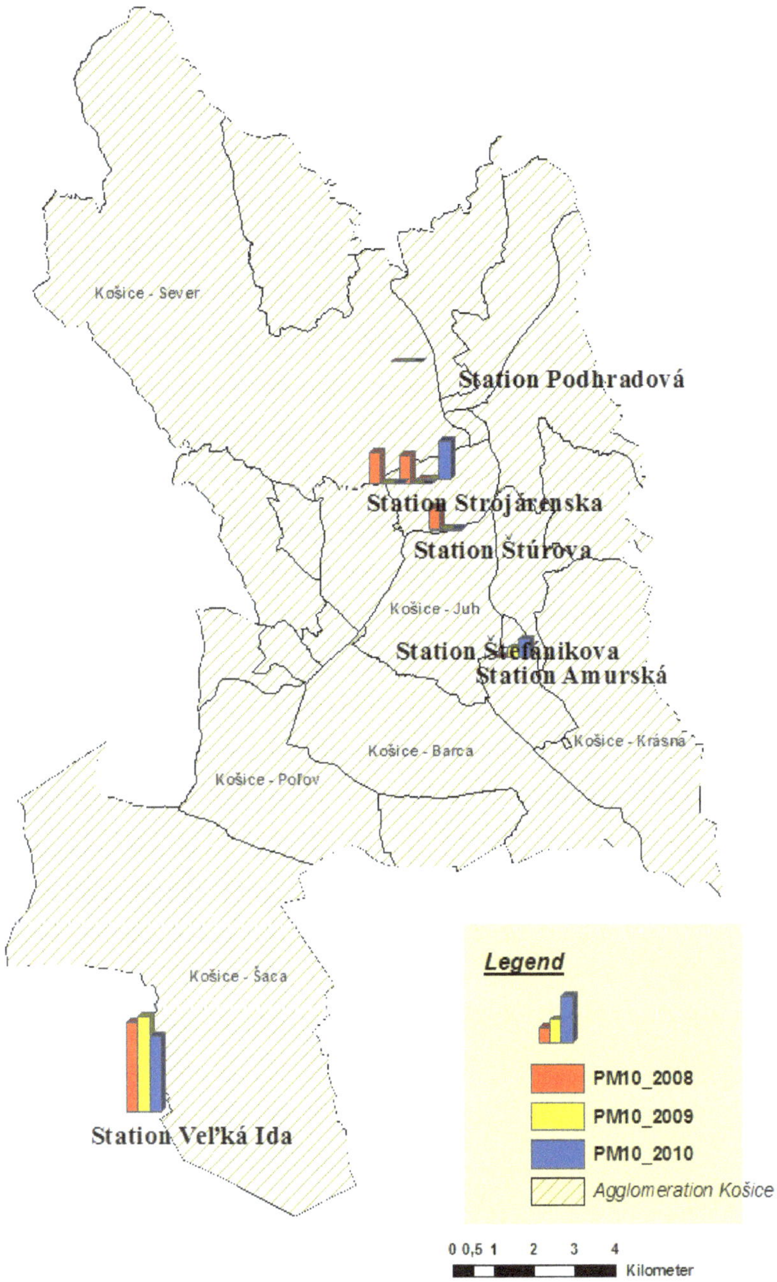

Fig. 5 Spatial processing of measured data in GIS environment

By default, also interpolation methods in addition to statistical methods are applied in processing and evaluating the measured values of concentrations of air pollutants. Together, they form a mathematical basis for the processing of data presenting the level of air pollution in general [10]. For the purpose of determining the extent of air pollution by particulate matter, it is appropriate to apply the following methods:

1. Thiessen polygon method - the nearest neighbour method,
2. Inverse distance weighted (IDW),
3. Spline interpolation
4. Bilinear interpolation,
5. Cubic interpolation. [10] , [11]

In the issues of observations of PM_{10} for a selected period of time interval of one year in the agglomeration of Košice, the IDW method was applied in the GIS environment. The IDW method represents a deterministic method, for the calculation of which the weighted linear average is used. The weight, which is used for calculation, is the reciprocal value of the distance of measurement from a local estimate with the power p. The equation of the simplest form of the IDW method according to [13] looks like this:

$$F(x,y)=\sum_{i=1}^{n} w_i Z(s_i) \tag{1}$$

Where:

n – number of measured points,

$Z(s_i)$ – variable value in measured ith point

w_i – weight

The advantage of using the IDW method in processing the meteorological data is the fact that each cell of a raster is assigned a value according to the following mathematical equation by a weighted linear combination of the values measured at the automated monitoring stations [12]:

$$Z(s_0)=\frac{\sum_{i=1}^{n}\frac{Z(s_i)}{d_{0i}^{\beta}}}{\sum_{i=1}^{n}\frac{1}{d_{0i}^{\beta}}} \tag{2}$$

Where:

$Z(s_0)$ – value of interpolated point

$Z(s_i)$ – value of measurement at ith point

d_{0i} – distance between interpolated point and point of measurement

β – power of weight

At a level of the modelling of spatial interpolation of concentrations of selected pollutants (e.g. PM_{10}, $PM_{2,5}$, ozone and heavy metals), this method is classified in the group of the models which are applied also in the environment of the Slovak Hydrometeorological Institute (SHMI). For the purpose of monitoring the concentrations of the substances present in the air, the IDW is a very successful used method. Under the conditions of SHMI, the interpolation method IDW-A was designed, in which the degree of influence of monitoring stations on the concentration in nodal network points is indirectly dependent on their mutual distances [7]. The interpolation model has a wide field of applicative use. Its options can also be associated for instance for the nationwide assessment of concentration levels of PM_{10}. The mentioned issue is more fully addressed also by the publication [14].

Input data for IDW

The input data should cover the following areas of resources:

- Measured or derived data from monitoring stations
- Attributes depending on an environment nature for each surveying point
    - Presence and significance of resources
    - Weights, geographic integrity
    - Size of built-up area [7]

Results and outcomes

Model calculations for the assessment of concentrations of a selected pollutant (PM_{10}) were performed by applying the above-mentioned method IDW in the GIS environment for the period 2008-2010. In the following part, the presented map outputs identify the areal layout of air pollutants for the selected area. The software support of the ArcView 9.3 interface simplified the solution process of the mentioned issues to a great extent. With the support of analytical tools, the obtained outputs offer the information at the level of air pollution carried over into a map view. The output provides an illustrative overview of the development of air pollution during the selected period 2008-2010 presented in detail in Fig. 6 and in Fig. 7.

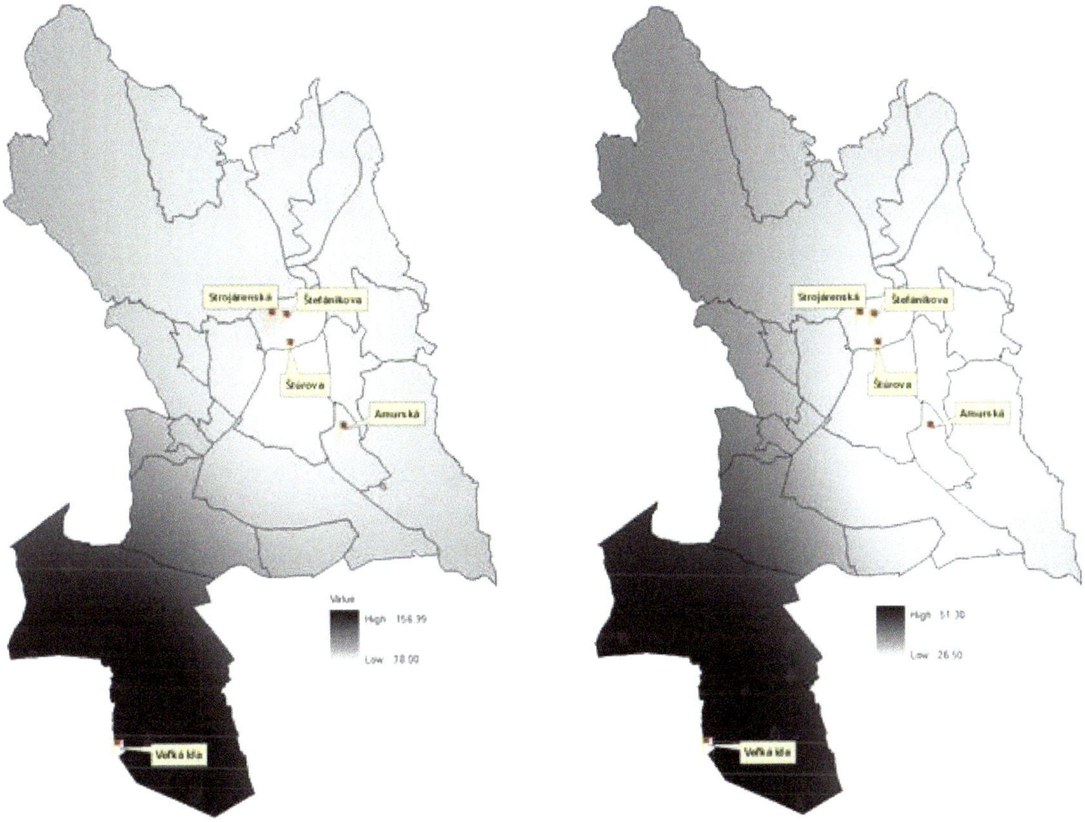

Year 2008 **Year 2009**

Fig. 6 Spatial interpolation of PM_{10} concentrations using IDW for years 2008, 2009

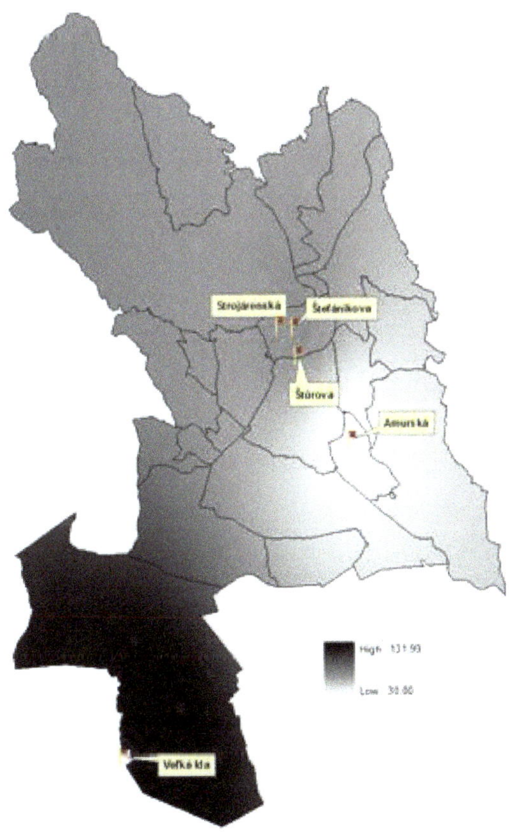

Fig. 7 Spatial interpolation of PM_{10} concentrations using IDW for year 2010

3 DISCUSSION

In general, it can be stated that the main pollutants in the atmosphere are produced by the energy, automotive and transportation industries. Local air pollution of a ground layer caused by local sources is of a great significance to the human environment. Its concentration show a distinct diurnal and seasonal variations depending on the orographic and meteorological factors. In case of weak air flow or calm coupled with a significant temperature inversion, pollutant concentrations can achieve even a value far in excess of the permissible limit value.

In the area of Košice, the most emissions of main pollutants are produced in total within other regions of Slovakia on a long-term basis.

Problems of air pollution in Košice are characterized by:

- High annual amount of produced emissions of pollutants from stationary industrial sources
- Rising trend in air pollution by emissions of exhaust gases from mobile sources - automobile transport
- Air pollution emissions (from high to overlimit concentrations of NO_x and particulate matter)

The overall development of emissions from dominant sources in the area of the town of Košice in terms of air pollution by sulphur dioxide and particulate matter is auspicious. The burden on the town with the harmful substances from the end of the eighties gradually decreases.

The trend of reducing total emissions from stationary sources in Košice is the result of changes in the fuel base, progressive supplying with gas and introduction of heating systems (central heating), the implementation of technical and technological measures at the sources of air pollution.

During the selected period, the emission limit values for particulate matter (PM_{10}) were exceeded at the monitoring stations Veľká Ida in 2009 and Štefániková in 2010.

4 CONCLUSIONS

This contribution presents the analysis and evaluation of the values of particulate matter PM_{10} on the basis of measurements at selected monitoring stations with administrative determination for the area of the town of Košice. For pre-testing, analysis and visualization, a set of measured concentrations of particulate matter – PM_{10} for the period 2008-2010 has been chosen which were taken over from the reports on the state of the environment of the Slovak Hydrometeorological Institute SR. The results of processing are clearly presented through graphical outputs in a form of tables, graphs and thematic maps from the GIS environment. The selected method for the processing of spatial information – IDW is currently an active form of tracking changes in concentrations of particulate pollutants in the air. The introduced method of processing spatial data can serve as a supplement to the estimation of air pollution in the town of Košice.

REFERENCES

[1] VYHLÁŠKA 360/2010 Ministerstva pôdohospodárstva, životného prostredia a regionálneho rozvoja Slovenskej republiky o kvalite ovzdušia. (online), (Citované 2013.04.03) <http://ipkz.enviroportal.sk/nkzp/360_2010.pdf>

[2] ZÁKON Č. 318/2012 Z. Z: o ovzduší v znení neskorších predpisov. (online), (Citované 2013.04.03) <http://www.zbierka.sk/sk/predpisy/318-2012-z-z.p-34874.pdf>

[3] GERHÁTOVÁ, E., JANKOVIČOVÁ, K.: Kvalita ovzdušia na Slovensku z hľadiska prachových častíc. MŽP SR, (online), (Citované 2013.04.03) <http://www.cepta.sk/attachments/article/468/MZP_Kvalita%20ovzdu%C5%A1ia%20na%20Slovensku.pdf>

[4] PROGRAM HOSPODÁRSKEHO A SOCIÁLNEHO ROZVOJA KOŠICKÉHO SAMOSPRÁVNEHO KRAJA – 3. aktualizácia. Košice 2011, 50 s. (online), (Citované 2013.04.03) <http://web.vucke.sk/files/dokumenty/pub/regionalny_rozvoj/phsr/kap13.pdf>

[5] INFORMÁCIE O KVALITE OVZDUŠIA 2009 (online), (Citované 2013.04.03) < http://www.ke.kuzp.sk/>

[6] HUDEC, D.: Zmeny a doplnky č.2 Územného plánu obce Veľká Ida Návrh. Sprievodná správa. Košice 2012, 36 s. (online), (Citované 2013.05.13)

<http://velka-ida-zad.uzemnyplan.sk/pdf-show/velka-ida-zad/t01n.pdf>

[7] SPRÁVA O KVALITE OVZDUŠIA ZA ROK 2009, (online), (Citované 2013.04.03)

<http://www.shmu.sk/sk/?page=996>

[8] SPRÁVA O KVALITE OVZDUŠIA ZA ROK 2010, (online), (Citované 2013.04.03)

<http://www.shmu.sk/sk/?page=996>

[9] SPRÁVA O KVALITE OVZDUŠIA ZA ROK 2008, (online), (Citované 2013.04.03)

<http://www.shmu.sk/sk/?page=996>

[10] PROCHÁZKA, A. et.al: Analýza a zpracování dat reprezentujících znečištení ovzduší získaných pozemním měřením. In: Vyhodnocení připravenosti České republiky splnit požadavky na kvalitu ovzduší podle směrnic EUa Konvence LRTAP. Praha 2000, 50 s. (online), (Citované 2013.05.06)

<http://ebookbrowse.com/du01-2e1-3-pdf-d313422174>

[11] DITTMAN, E., SZALAI, S. et. al. Spatial interpolation of climatological parameters. Deutcher Wetterdienst Forschung und Entwicklung Nr. 57, Offenbach 1999.

[12] HORÁLEK, J., et. al: Vypracování přehledu používaných metod mapování znečišťujících látek za pomoci dostupné literatúry. In: Vyhodnocení připravenosti České republiky splnit požadavky na kvalitu ovzduší podle směrnic EUa Konvence LRTAP. Praha 2000, 50 s. (online), (Cit. 2013.04.08)

<http://ebookbrowse.com/du01-2e1-3-pdf-d313422174>

[13] SHEPARD, D.: A two-dimensional interpolation function for irregularly-spaced data. Proceedings of the 1968 ACM National Conference. pp. 517–524.

[14] SABOLOVÁ, J. et. al.: Modelovanie plôch v 2D skalárnom poli v prostredí GIS. In: Trendy a inovatívne prístupy v podnikových procesoch : 15. medzinárodná vedecká konferencia. Košice 2012, 1-6.s, ISBN 978-80-553-1126-5. (online), (Citované 2013.12.03) <http://www.sjf.tuke.sk/kpiam/TaIPvPP/2012/index.files/Page307.htm>

[15] MESAČNÝ PRIEBEH METEOROLOGICKÝCH PRVKOV KOŠICE - LETISKO, (online), (Citované 2013.12.03)

http://www.shmu.sk/File/ExtraFiles/KMIS/publikacie_let/Marec_2013.pdf

[16] KOŠICKÝ SAMOSPRÁVNY KRAJ: Program hospodárskeho a sociálneho rozvoja Košického samosprávneho kraja v programovacom období 2007 – 2013, (online), (Citované 2013.12.08)

http://web.vucke.sk/files/dokumenty/pub/regionalny_rozvoj/phsr/phsr_kapitola_13_1_2_2.pdf

RESUMÉ

Príspevok je venovaný analýze vybraných tuhých znečisťujúcich látok, ktoré bezprostredne vplývajú na kvalitu ovzdušia vo všeobecnosti. Najväčší problém kvality ovzdušia na Slovensku, ako aj vo väčšine európskych krajín, predstavuje v súčasnosti znečistenie ovzdušia suspendovanými časticami (PM_{10}) [5]. Rozhodujúci podiel produkcií PM_{10} majú regionálne zdroje (prírodné zdroje, poľnohospodárske aktivity). Konkrétne riešenie uvedenej problematiky bolo zamerané na integráciu nástrojov GIS v oblasti posudzovania (hodnotenia) kvality ovzdušia. GIS, ako vhodný informačný systém ponúkol pre riešenú lokalitu širokú škálu analytických nástrojov a funkcií. GIS sa v súčasnosti využíva v širokospektrálnom uplatnení nevynímajúc odbor civilnej obrany preplánovanie možných krízových situácií vrátane oblasti ochrany životného prostredia. GIS na úrovni spracovania príspevku vhodnou kombináciou možnosti analytických nástrojov ponúka silný nástroj pre modelovanie priestorových javov v oblasti ochrany životného prostredia aj v budúcnosti. Komplexné spracovanie informácií týkajúcich sa kvality ovzdušia pre zvolené územie si autori príspevku bez nasadenia GIS nedokážu predstaviť. Ďalšiu neoddeliteľnú výhodu aplikácie GIS je možnosť spracovania a prezentácie dát v jednom systéme, čo ponúka možnosť komfortnej aktualizácie a integrácie aj časových radov pre ľubovoľnú monitorovaciu stanicu.

Štandardne sa v procese spracovania a vyhodnotenia meraných hodnôt koncentrácie znečisťujúcich látok v ovzduší aplikujú rôzne interpolačné metódy. Použitím interpolačnej metódy IDW boli v prostredí GIS spracované priestorové informácie monitorovacích staníc nachádzajúcich sa na území mesta Košice a jeho bezprostrednom okolí. V sledovanom období hodnoty imisných limitov pre tuhé častice (PM_{10}) boli prekročené na monitorovacích staniciach Veľká Ida v roku 2009 a Štefániková v roku 2010. Získané výsledky spracovania koncentrácie tuhých častíc (PM_{10}) za vybrané sledované obdobie rokov 2008 - 2010 sú prezentované formou výstupných zostáv.

POSSIBILITIES FOR SIMULATING THE SMOKE ROLLBACK EFFECT IN UNDERGROUND MINES USING CFD SOFTWARE

Vancho ADJISKI

Ing. Student of Ph.D., Faculty of Natural and Technical Sciences, Mining engineering, "Goce Delchev" University, P.O. Box 201, 2000 Shtip, Republic of Macedonia
e-mail: adziski.v@gmail.com

Abstract

The effect of fire generated smoke rollback in underground mines can be dangerous and a potentially fatal threat to all who are endangered by the fire. Three critical stages in the process of smoke rollback are 3D local phenomena that can be analysed by CFD software simulations. With the help of a 3D-CFD analysis we can observe the critical stages of smoke rollback and their reaction to a ventilation network. The CFD provides the opportunity to expand the range of prediction of smoke spread for a wide spectrum of airflows, fire intensities, mine-section inclinations and layouts. The paper deals with the possibilities for simulating the smoke rollback effect in underground mines using the CFD software. For the purposes of this research paper, the PyroSim software from the company Thunderhead Engineering was used to create three different CFD fire scenarios and then observe the critical stages of smoke rollback effects.

Abstrakt

Zpětné proudění kouře požáru v hlubinných dolech může představovat potencionální hrozbu pro všechny, kteří mohou být jeho dosahem ohroženi. Tři kritické fáze procesu změny proudění kouře jsou 3D místní jevy, které mohou být zkoumány pomocí CFD simulací. Za pomocí 3D-CFD analýzy můžeme sledovat kritické fáze změny proudění kouře a následné ovlivnění větrní sítě dolu. CFD nabízí mnoho příležitostí k rozšíření prognózy proudění kouřů při různých objemových průtocích větrů, intenzitách požárů, v úklonných a horizontálních důlních dílech. Článek se zabývá možnostmi simulace jevu změny proudění kouře v hlubinných dolech za použití CFD softwaru. Pomocí softwaru PyroSim společnosti Thunderhead Engineering byly vytvořeny tři různé CFD scénáře požárů a následně byly pozorovány kritické fáze jevu změny proudění kouře.

Keywords: fighting fires, smoke rollback, CFD analysis, modelling, simulation

1 INTRODUCTION

A fire that occurs in an underground mine could have disastrous consequences for employees if the fire is not put under control in its initial stage. Initially, thermal buoyancy forces generated by a fire will produce a growing plume of flame followed by generating fire gases. With enough air velocity (ventilation), the smoke and fire products will be transported in a direction of positive-pressure ventilation. Once the fire develops into its highest intensity, buoyancy forces generated by the fire can overcome the inert forces of the ventilation, and in this stage, the migration of smoke and fire products spreads versus the positive-pressure ventilation located on the roof of the underground mining facility. In case of underground mine fires under low-velocity motion of air, there is often present the effect of smoke rollback. The effects of smoke rollback and fire products can be dangerous and a potentially fatal threat to miners and mine rescue teams, preventing them to approach close enough and be able to effectively extinguish the fire. This fire effect can occur directly over the heads of the firefighters and miners, enveloping them with hot fire gases, and can cause fatal consequences. The local reversing of the ventilation network which will contaminate the clean air flows and make the process of evacuation more difficult is another important scenario among the many harmful scenarios that can cause the smoke rollback effect.

Risks of explosion of fire gases must be taken at a high alert level when there are effects of smoke rollback. Mine explosions can be the result of ignition of an explosive mixture of flammable gases or explosive dust or a combination of the two. Possible explosions also can occur from ignition of fire gases given off by mine fires. In coal and metal non-metal mines, a flammable gas, usually methane, is ignited by a spark or the flame source which propagates to explosion proportions because of the quantity of gas present or explosive dust picked up by the gas ignition.

The ability to predict the smoke rollback effect in case of fires in underground mines can significantly improve the chances of safely evacuation.

2 SMOKE ROLLBACK EFFECT

A smoke rollback effect usually occurs when the air velocity (ventilation) is too low in dependence of fire intensity, Fig.1.

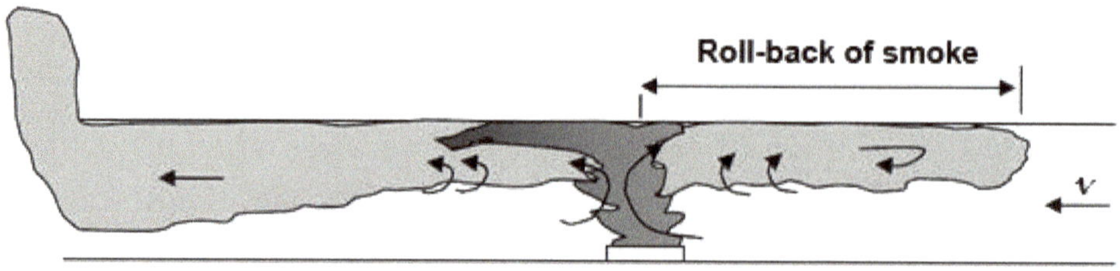

Fig.1 The process of smoke rollback [1]

The critical ventilation air velocity to prevent the effect of smoke rollback is calculated using the following equation [2]:

$$v_{crit} = k\left(\frac{gQ'}{\rho_0 c_p T}\right)^{\frac{1}{3}} \quad (1)$$

Where:

v_{crit} - Critical ventilation air velocity, [m s],
g - Acceleration due to gravity, [m s^2],
k - Constant,
$\dot{Q}'$ - HRR-heat release rate, [kW],
ρ_0 - Ambient air density, [kg m^3],
c_p - Specific heat capacity of air, [J kg.K],
T - Smoke temperature, [K].

The development of smoke rollback in an entrance to a mine or a tunnel occurs in three distinct stages [3]:

1. Critical velocity stage, Fig. 2(a),
2. Partial smoke rollback stage, Fig. 2(b),
3. Complete smoke rollback stage, Fig. 2(c).

The critical ventilation velocity (Fig. 2(a)) is the minimum airflow velocity preventing the smoke from rolling back. The critical velocity has become one of the prime criteria used in the design of underground mining ventilation systems. As the fire grows to thermal intensity, the inertial force of the intake airflow is overcome by the fire-generated buoyancy forces, and smoke begins to migrate along the airway roof counter in the positive-pressure ventilation airflow direction. At this time, two directions of flow co-exist in the airway, with the lower layer of fresh air maintaining its forward direction and the upper layer of hot smoke rolling back, as shown in Fig. 2(b). This scenario is defined as a partial smoke rollback stage. If the fire grows to sufficient rapid intensity, it is possible for the airway to completely fill with hot smoke, causing total airflow reversal, as shown in Fig. 2(c).This phenomenon is known as a complete smoke rollback or smoke reversal. The complete smoke reversal rarely happens in road tunnel fires because the pressure differences between the two ends of the tunnel, which are both open to the atmosphere, is insignificant. However, it can and has occurred during underground mine fires because an underground mine fire exists not only in a single airway but also in a ventilation network with many branches, junctions, various controls, and fans.

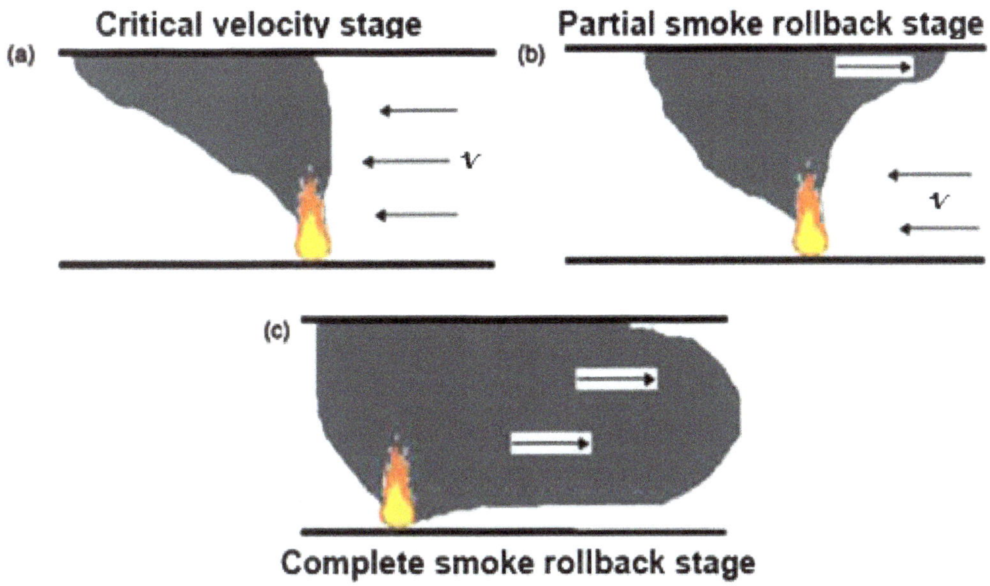

Fig.2 Development phases in the process of smoke rollback or smoke reversal, [3]

In this time, many software packages are available such as [4]: MFIRE, Ventgraph, MineFIRE Pro , VentFIRE™ and many others that can predict the complete smoke rollback stage. However, the partial smoke rollback stage, Fig. 2(b), is a three-dimensional phenomenon which represents a critical point in dealing with this fire effect and is beyond the scope of the above mentioned software packages. The best method for analysing the stage of a partial smoke rollback effect, which represents a key point for dealing with this fire effect, is the method of a CFD (computational fluid dynamics) analysis. It should be noted that the CFD analysis can represent only a small portion of the ventilation network because of the large number of calculations performed by the analysis, it means that we are limited by computer performance so this computing method can be used only for local effects and can't completely replace the above mentioned software packages.

3 POSSIBILITIES FOR SIMULATING THE SMOKE ROLLBACK EFFECT IN UNDERGROUND MINES USING A CFD SOFTWARE

For the purposes of this research paper, the PyroSim software from the company Thunderhead Engineering is used [5]. The PyroSim software is a graphical user interface for the Fire Dynamics Simulator (FDS) which was originally developed by the National Institute of Standards and Technology (NIST) USA. The Fire Dynamics Simulator (FDS) [6] is a computational fluid dynamics (CFD) model of fire-driven fluid flow. The software solves numerically a large eddy simulation form of the Navier-Stokes equations [7] appropriate for low-speed, thermally-driven flow, with an emphasis on smoke and heat transport from fires. This approach is very flexible and can predict and simulate different fire scenarios, including the ventilation in the same process. The FDS and Smokeview are two free computer codes that can simulate the process of fire and smoke development, which was originally developed by the NIST (USA), and are combined to one software interface by the PyroSim software from Thunderhead Engineering.

For the purposes of this research paper, a 3D computer horizontal mining drift with the following dimensions is generated: a width of 4m, a height of 3m and a length of 50 m, Fig. 3.

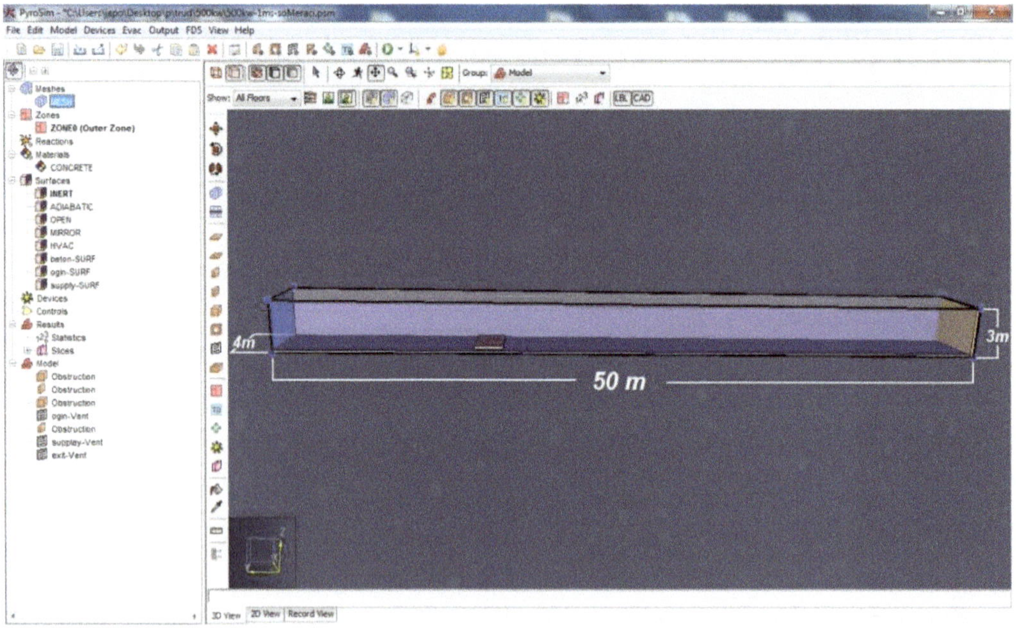

Fig. 3 Process of 3D modelling a horizontal mining drift

The location of the fire is 35 m from the beginning of the 3D horizontal mining drift, and the fire covers an area of 6 m^2 (2 m length, 3m width). To measure the properties of the fire and smoke effects in this simulation, we placed devices and sensors exactly six meters from the fire (we assumed that an optimal place for firefighters) Fig. 4, Tab. 1.

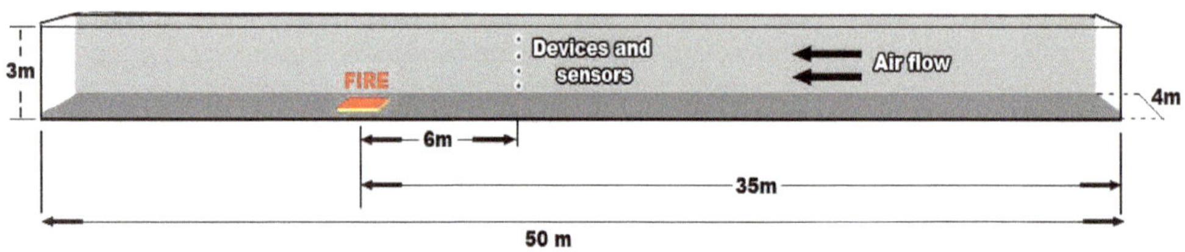

Fig.4 Scheme of horizontal mining drift

In this large eddy simulation (LES), the grid size is an important factor to be considered. A smaller gird size gives more detailed information of the turbulent flow but needs more computation resource and longer computing time. However, the basic methodology of LES is that the accuracy increases as the numerical mesh is refined. The cell size (dx) for the given simulation can be related to the characteristic fire diameter (D*), i.e.; the smaller the characteristic fire diameter, the smaller the cell size should be in order to adequately resolve the fluid flow and fire dynamics. The characteristic fire diameter (D*) is given by the following relationship [8]:

$$D^* = \left(\frac{\dot{Q}}{\rho_\infty c_p T_\infty \sqrt{g}}\right)^{\frac{2}{5}} \tag{2}$$

Where:

$\dot{Q}$ - Expected heat release rate, [kW],

ρ_∞ - Density, [kg m^3],

c_p - Specific heat, [J kg.K],

T_∞ - Ambient temperature, [293K],

g - Acceleration due to gravity, [m s^2].

For this simulation, we take into account a moderate cell size (dx) of 14.88 cm, Fig 5. The mesh line for FDS is as follows:

- actual (dx) size is 0.148(x), 0.139(y) 0.15(z), [m],
- distances are 4(x), 50(y), 3(z), [m],
- total number of cells is 194,400.

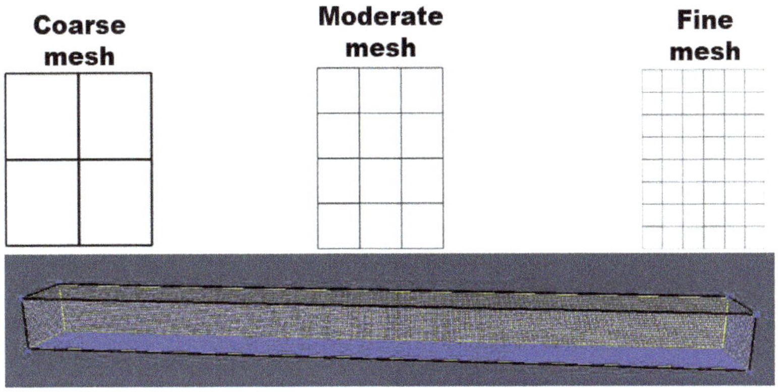

Fig.5 Process of modelling a moderate mesh

Tab. 1 Properties for devices and sensors in horizontal mining drift 6m placed upstream the fire

Thermocouple Name: Thermocouple Quantity: Thermocouple Bead Diameter: 1,0 mm Emissivity: 0,85 Bead Density: 8908,0 kg/m³ Bead Specific Heat: 0,44 kJ/(kg·K)	The temperature of the thermocouple itself, which is usually close to the gas temperature, but not always, since radiation is included in the calculation of the thermocouple temperature, is the output of the thermocouple.
Layer Zoning Device Name: Layer Zoning Device ☑ Measure Layer Height ☐ Enable Setpoint: 0,0 m ☑ Measure Upper Temperature ☐ Enable Setpoint: 0,0 °C ☑ Measure Lower Temperature ☐ Enable Setpoint: 0,0 °C	There is often the need to estimate the location of the interface between the hot, smoke-laden upper layer and the cooler lower layer in a burning compartment. The FDS uses an algorithm based on integration along a line to estimate the layer height and the average upper and lower layers temperatures.
Smoke Detector Description: Model: Cleary Ionization I1 Obscuration Threshold: 3,28 %/m Smoke Species: Default Detector Specification: Cleary Model Alpha C: 0,8 Alpha E: 2,5 Beta C: -0,9 Beta E: -0,7	A smoke detector measures obscuration at a point with two characteristic fill-in or "lag" times. The percent obscuration per meter is the output.

On the same 3D horizontal mining drift, we will perform a total of three simulations in the following order, Tab 2.

Tab. 2 Properties for performing simulations

	Location of fire (m)	Area of fire (m^2)	Heat release rate HRR (kW m^2)	Ventilation air velocity (m s)
1.	35 m	6 m^2	500 kW m^2	1 m s
2.	35 m	6 m^2	500 kW m^2	1,5 m s
3.	35 m	6 m^2	500 kW m^2	2 m s

All three simulations will be analysed from the same time at the 200-th second from the development of the fire. The development of peak intensity of the fire is from 0 to 60 seconds.

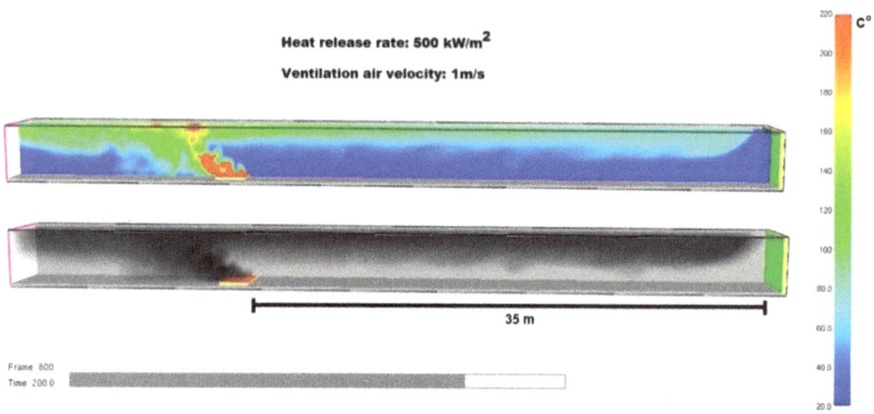

Fig. 6 Simulation 1

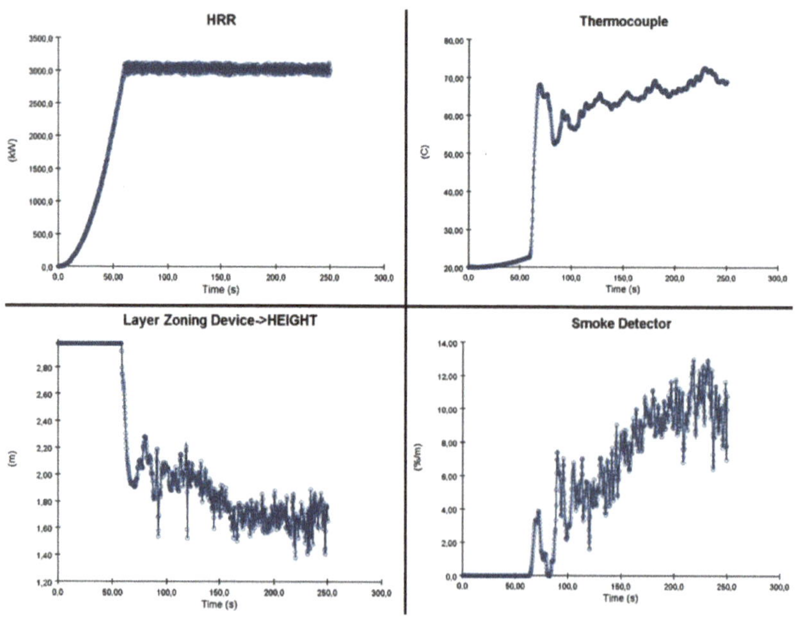

Fig.7 Results of devices and sensors for simulation 1

Based on Simulation 1 (Fig. 6), we can conclude that the fire with a heat release rate of 500 kW m^2 and a ventilation air velocity of 1m s can cause a complete smoke rollback stage with a mean temperature in the first 35m of smoke rollback layer of 70 °C versus the positive-pressure ventilation. With the help of devices and sensors placed 6 m in the upstream position of the fire, different parameters of smoke and fire products can be observed depending on time (Fig. 7). This critical stage of complete smoke rollback can cause difficulties in dealing with the fire because of the inability of firefighters to approach close enough to the fire. The risk of possible explosions that can occur as a result of the ignition of accumulated fire gases given off by the mine fire is another possible critical factor. This complete smoke rollback stage also have detrimental effects of contaminating and putting to risk other underground mining areas with a flow of fresh air.

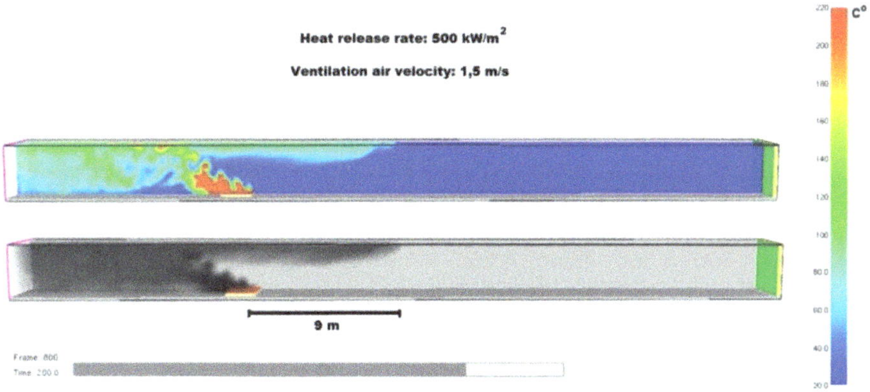

Fig. 8 Simulation 2

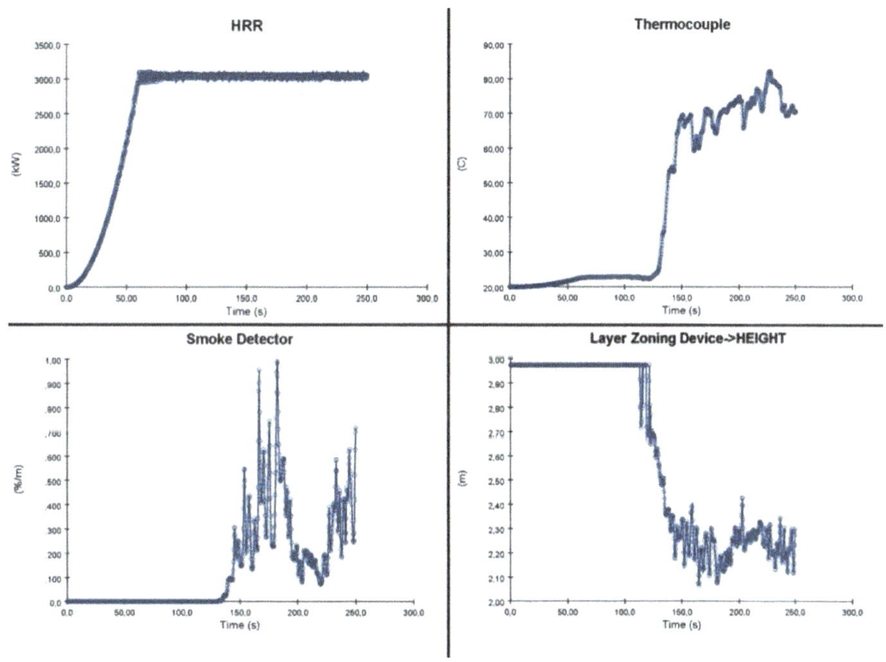

Fig.9 Results of devices and sensors for simulation 2

Based on Simulation 2 (Fig. 8), we can conclude that the fire with a heat release rate of 500 kW m^2 and a ventilation air velocity of 1.5 m s can cause a partial smoke rollback stage versus the positive-pressure ventilation with a length of rollback (reversal) smoke layer of about 9 m and a mean temperature of the same smoke layer of about 75 °C. The information about the length of the rollback (reversal) smoke layer and its properties that can be observed from the placed devices and sensors (Fig. 9) is crucial information for possible methods that can be implemented to handle and control a fire.

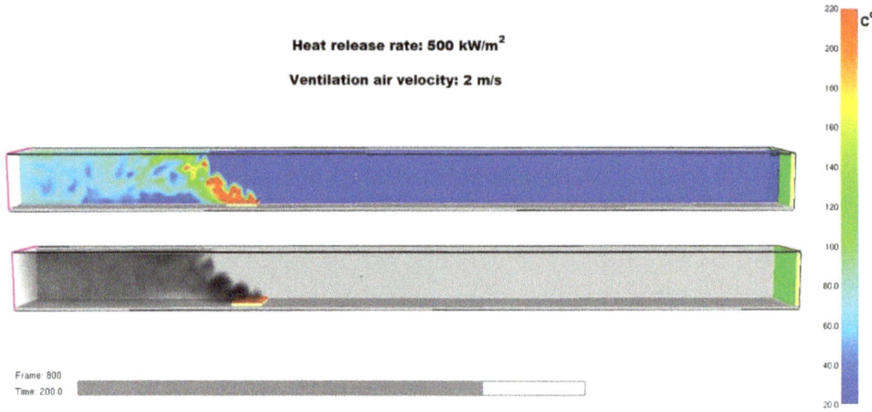

Fig. 10 Simulation 3

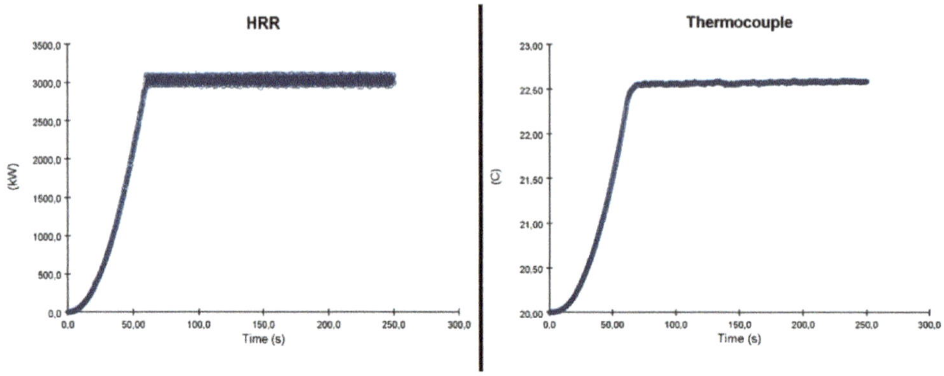

Fig.11 Results of devices and sensors for simulation 3

Based on Simulation 3 (Fig.10), we can conclude that the fire, with a heat release rate of 500 kW/m^2 and a ventilation air velocity of 2 m/s for this scenario and, under these circumstances, completely direct (blows) smoke and fumes in a direction of positive-pressure ventilation and space in front (upstream) of the fire, is free of smoke and fire gases, which is a favourable and conducive environment for firefighters to be able to come close and successfully extinguish the fire.

From these three simulations we have performed we can conclude that the process of smoke rollback mostly depends on the critical ventilation air velocity which is closely related to the power of fire (HRR-heat release rate) which can be calculated from Equation 1. The above three performed simulations give crucial information for possible and safe ways of extinguishing and controlling the fire.

4 CONCLUSIONS

Fires that occur in underground mines pose one of the most serious hazards. At a low air velocity in case of fire in underground mines, the effect of smoke rollback is often present. The process of smoke rollback can be dangerous and a potentially fatal threat to miners and mine rescue teams. In case of underground mining fire, evacuation of people can be extremely difficult. Opportunities for safe evacuation out of the affected area are strongly associated with the development and spread of smoke and fire products through a mining ventilation network. The ability to predict the dangerous smoke rollback effect in case of fire in underground mines can help to a successful response of mine rescue teams. The partial smoke rollback stage represents a critical moment for getting the fire under control. This critical stage can be observed and analysed using CFD methods, by which we are able to analyse the time of its development and the possibilities to prevent this effect from transforming itself in a complete smoke rollback stage that can bring catastrophic consequences. The above three performed CFD simulations give crucial information for possible and safe ways of extinguishing and controlling the fire. This CFD computing method for predicting the stages of smoke rollback during an underground mining fire can be applied to different fire scenarios and can help us gather crucial information. The CFD modelling is a viable method for analysing potential hazards associated with smoke from an underground mining fire.

REFERENCES

[1] Ingason, H.: Tunnel Fire Safety, *(handbook)*, Chapter 11, Thomas Telford Publ. London, 2005, pp.241-245, ISBN:0-7277-3168-8 .

[2] John, C., Edwards, R. Franks, V, et al.: Experimental and modeling investigation of the effect of ventilation on smoke rollback in a mine entry", National Institute for Occupational Safety and Health Pittsburgh Research Laboratory, Pittsburgh, PA, 2006, p.53-58

[3] Lihong, Z., Smith, A.: "Improvement of a mine fire simulation program — incorporation of smoke rollback into MFIRE 3.0", Office of Mine Safety and Health Research, National Institute for Occupational Safety and Health, Pittsburgh, PA, 2011, p.78-85

[4] Friel, GF., Yuan, L., Edwards JC., et al.: "Fire-generated smoke rollback through crosscut from return to intake-experimental and CFD study", Proceedings of 11th U.S./North American Mine Ventilation Symposium, Mutmansky, 5-7 June 2006, PA, p.142-151

[5] PyroSim User Manual-PyroSim Documentation, *(user manual of computer code)*, Thunderhead Engineering Consultants, Inc.Manhattan, KS 66502-6081, USA, 2012

[6] McGrattan, K.B., Forney, G.P., Prasad, K., at al.: "Fire Dynamics Simulator (Version3)-User's Guide," U.S. Dept. of Commerce, National Institute of Standards and Technology, 2002, p.82-94

[7] Edwards, JC., Hwang, CC.: "CFD analysis of mine fire smoke spread and reverse flow conditions". Proc. of 8th U.S. Mine Ventilation Symposium. University of Missouri-Rolla, Rolla, MO, 1999, p.91-99

[8] McGrattan, K., Hostikka, S., Floyd, J., et al.: "Fire Dynamics Simulator (Version 5) - Technical Reference Guide" , National Institute of Standards and Technology, NIST Publ. 1018-5, 86 pages, October 2007.

RESUMÉ

Důlní požáry představují jedno z nejzávažnějších rizik hlubinného hornictví. Nízké objemové průtoky větrů v důsledku důlních požárů mají často za následek způsobení jevu změny proudění kouře. Proces změny proudění kouře představuje závažnou hrozbu nejen pro pracovníky ale především pro báňské záchranáře. Důlní požár může také značně zkomplikovat evakuaci pracovníků. Evakuace v případě důlního požárů je značně ovlivněna vývinem a prouděním kouřů a produktů hoření obsažených ve větrném proudu. Možnost předpovědi vzniku změny proudění kouře může pomoci báňským záchranářům k úspěšnému zásahu. Částečná změna proudění kouře představuje kritickou fázi bezpečného zásahu při potlačení důlního požáru. Tato fáze může být zkoumána a analyzována pomocí CFD metod, pomocí kterých můžeme analyzovat čas vývoje a možnosti prevence tohoto jevu s cílem dosáhnout změny stavu zpětného postupu kouřů, který by mohl způsobit katastrofální následky. Simulace pomocí CFD poskytuje související informace k možnosti účinného hašení a kontroly při potlačení požárů. Výše uvedená CFD metoda predikce vzniku jevu změny proudění kouře může rovněž poskytnout rozhodující informace také pro odlišné důlní požáry. CFD modelování je jednou z možných použitelných metod analýzy rizik spojených s vývinem kouřů v průběhu důlních požárů.

17

VERMICULITE WITH Ag AND Cu USED AS AN ANTIBACTERIAL NANOFILLER IN POLYETHYLENE

Marianna HUNDÁKOVÁ [1], Marta VALÁŠKOVÁ [2], Magda SAMLÍKOVÁ [3], Erich PAZDZIORA[4]

[1] *Ing., Ph.D., Nanotechnology Centre, IT4 Innovations Centre of Excellence, VŠB-Technical University of Ostrava*
17. listopadu 15, Ostrava
e-mail: marianna.hundakova@vsb.cz

[2] *RNDr., DSc., Nanotechnology Centre, IT4 Innovations Centre of Excellence, VŠB-Technical University of Ostrava*
17. listopadu 15, Ostrava
e-mail: marta.valaskova@vsb.cz

[3] *Ing., Dis., Nanotechnology Centre, IT4 Innovations Centre of Excellence, VŠB-Technical University of Ostrava*
17. listopadu 15, Ostrava
e-mail: magda.samlikova@vsb.cz

[4] *RNDr., CSc., Institute of Public Health in Ostrava*
Partyzánské náměstí 7, Ostrava
e-mail: Erich.Pazdziora@zuova.cz

Abstract

Vermiculite (Ver) enriched with silver and copper was used as nanofiller to the polyethylene (PE) matrix. Specifically, the low density polyethylene (LDPE) was chosen as a matrix. The samples Ver-Ag,Cu were prepared by shaking of Ver with the aqueous solutions of silver and copper nitrate. The mixtures of the Ver nanofillers and PE were homogenized by melt compounding technique and further thin plates were pressed from stiff matter of PE with Ver nanofiller. The exfoliation of the powdery Ver nanofillers in PE matrix was characterized by the X-ray diffraction analysis of thin plates. Distribution of Ver nanofiller in PE matrix was observed by Light microscopy. The reinforcing effect of nanofillers onto PE matrix was studied by creep experiment. Antibacterial activity of powder Ver-Ag,Cu samples and surfaces of PE/Ver-Ag,Cu samples was tested on the Gram-positive bacteria *Enterococcus faecalis*. All tested PE/Ver-Ag,Cu surfaces showed good antibacterial behaviour after 24 h in comparison to pure PE. The number of colonies decreased from the countless number to several hundred colonies.

Abstrakt

Vermikulit (Ver) obohacený stříbrem a mědí byl použit jako nanoplnivo v polyethylenové (PE) matrici. Konkrétně byl jako matrice vybrán nízko hustotní polyethylen (LDPE). Vzorky Ver-Ag,Cu byly připraveny mícháním Ver s vodnými roztoky dusičnanu stříbrného a měďnatého. Směsi Ver nanoplniv a PE byly homogenizovány postupem míchání taveniny a následně byly ze ztuhnuté směsi PE s Ver nanoplnivem vylisovány tenké destičky. Exfoliace práškového Ver nanoplniva v PE matrici byla hodnocena pomocí Rentgenové difrakční analýzy tenkých destiček. Distribuce Ver nanoplniva v PE matrici byla studována Světelnou mikroskopií. Vyztužující efekt nanoplniv na PE matrici byl studován pomocí krípového experimentu. Antibakteriální působení práškových vzorků Ver-Ag,Cu a povrchů vzorků PE/Ver-Ag,Cu bylo testováno na grampozitivní bakterii *Enterococcus faecalis*. Všechny testované povrchy PE/Ver-Ag,Cu vykazovaly dobré antibakteriální působení po 24 h oproti povrchu čistého PE. Počet kolonií poklesl z nespočetného množství na několik stovek kolonií.

Key words: Vermiculite, Polyethylene, X-ray diffraction, Antibacterial activity

1 INTRODUCTION

Vermiculite belongs to the layered silicates with structure consist from type of layer 2:1, thus, one octahedral sheet is sandwiched between two tetrahedral sheets. Central octahedral cations (mainly Si^{4+}) and tetrahedral cations (mainly Al^{3+}, Fe^{3+}) are usually substituted by cations with lower valency (Al^{3+}, Fe^{3+}, Fe^{2+}, Mg^{2+}, etc.) and a negative charge arising on vermiculite layers from these subtitutions. The interlayer space of vermiculite is occupied by exchangeable hydrated cations which compensate a negative layer charge [1]. Due to structural properties the clay minerals are suitable substrates for various inorganic and/or organic constituents. Vermiculite is occasionally used as silver or copper particles support [2-5].

Polymers with layered silicate nanofillers represent the wide group of the clay mineral-polymer nanocomposites (CPN). The most commonly used clay minerals as a nanofiller are montmorillonite, vermiculite, kaolinite and less often halloysite, hectorite etc. The layered silicates are often modified with various inorganic and organic compounds. The main reason to their modification with organic surfactants is to change their hydrophilic nature to hydrophobic for better compatibility with polymer. These clay mineral nanofillers can improve the mechanical and thermal properties [6-7], anticorrosion properties [8], gas permeability resistance [9] and fire retardancy [10] of polymers. Various preparation procedures of the antibacterial CPN were published [8-9,11-13].

Nigmatulin et al. (2008) prepared Nylon-6/clay nanocomposites by melt extrusion. They used an organically modified montmorillonites as clay nanofillers in mass content of 1, 2, 5 or 10 wt.%. Antibacterial activity was studied on bacteria *Escherichia coli* (*E. coli*) and *Staphylococcus aureus* (*S. aureus*). Authors found that antibacterial activity increased with an increase loading of the organoclay nanofiller. They also observed that antibacterial effect is a function not only of the organic components activity, but also of the nanocomposites structure [12].

Xue et al. (2011) prepared nano-Cu/organo-montmorillonite/LLDPE (Linear Low Density Poly-Ethylene) via melt mixing and melt extruding process. Authors observed that antibacterial effect against bacteria *S. aureus* increased with nano-Cu content [8].

Bruna et al. (2012) studied antimicrobial activity of Cu^{2+}-montmorillonite/LDPE prepared by melt mixing LDPE with 1, 2, 3 and 4 wt.% of Cu^{2+}-montmorillonite. The best antimicrobial effects against *E. coli* was found for CPN containing 4 wt.% of the nanofiller in the LDPE [11].

In this work, vermiculite enriched with silver and copper without organic modifiers was used as a nanofiller to the PE matrix. Structural changes, level of filler dispersed in PE and antibacterial effect of prepared materials were studied.

2 EXPERIMENTAL PART

2.1 Materials

As starting material, vermiculite from Brazil – Santa Luzia (material obtained from Grena, a.s., CZ) was used. Its fraction under 40 µm (labelled as Ver) was prepared by milling in a planetary ball mill, sieving and afterwards used for the experiment. The structural formula $(Si_{3.12}Al_{0.88})(Mg_{2.32}Fe^{3+}_{0.48}Ti_{0.05})O_{10}(OH)_2K_{0.23}Ca_{0.15}Na_{0.07}$ was calculated from the elemental analysis.

Silver nitrate ($AgNO_3$) and copper nitrate ($Cu(NO_3)_2.5H_2O$) with analytical purity (Vitrum, CZ) were used for preparation of Ver with Ag and Cu.

Polyethylene (PE) was prepared from three commercial low density PE (LDPE) in the mass ratio: 33 wt.% of Bralen VA 20-60 powder, 42 wt.% of Bralen VA 20-60 granulated and 25 wt.% of Bralen FB2-17 powder (Slovnaft, SK).

2.2 Samples preparation

Preparation of Ver-Ag,Cu nanofillers

Ver was mixed with 0.1 mol. dm^{-3} aqueous solution of silver and copper nitrate (samples were labelled VerAg and VerCu) and with both of these solutions in molar ratio 1:1 (sample was labelled VerAgCu). Detailed preparation procedure was published in our previous work [5]. The elemental analysis revealed 5.6 wt.% of Ag in VerAg, 3.5 wt.% of Cu in VerCu and 2.1 wt.% of Ag and 1.4 wt.% of Cu in VerAgCu.

Preparation of PE/Ver-Ag,Cu samples

PE samples were prepared by melt compounding at 160°C in Brabender kneader. The prepared mixtures of PE with nanofillers were pressed into the thin plates (125 × 125 × 1 mm) according the same conditions as in our previous works [14,15]. The prepared materials were labelled as PE/VerAg, PE/VerCu, PE/VerAgCu and PE/VerAgCu-b. Samples PE/VerAg, PE/VerCu, PE/VerAgCu contained 7 wt.% of the nanofiller VerAg, VerCu and VerAgCu in PE, and sample PE/VerAgCu-b contained only 3.5 wt.% of the nanofiller VerAgCu.

2.3 Methods

The X-ray diffraction (XRD) patterns of powder samples and composite thin plates were measured on the X-ray diffractometer Ultima IV Rigaku (reflection mode, Brag-Brentano arrangement, CuKα radiation) under ambient temperature, and working conditions 40kV, 40mA, range of angle 2 – 50 ° 2θ.

The images of the samples with spot elemental analysis were obtained by Scanning electron microscopy (SEM) on PHILIPS XL-30 with energy dispersive spectrometer.

Nanoparticles of silver on Ver surface were observed by Transmission electron microscope Jeol JEM 2010, operating at 80–200 kV.

Distribution of vermiculite fillers in PE was observed using the Light microscopy (LM). The images were obtained by Olympus BX51 with camera UC30 at transition mode, using bright field of the polarized light.

The reinforcing effect of Ver filler onto PE matrix was studied by creep testing applying constant stress of app. 1 MPa at temperature 25°C.

2.4 Antibacterial test

Testing of powder Ver-Ag,Cu nanofillers

The antibacterial activity of powder Ver-Ag,Cu samples was studied on Gram-positive (G^+) bacteria *Enterococcus faecalis* (*E. faecalis, CCM 4224*) by the broth dilution method [16]. The minimum inhibition concentration (MIC) was determined, as the lowest concentration of the sample that absolutely stops bacterial growth. The bacterial suspension was prepared by twenty hour cultivation in glucose broth. The density of *E. faecalis* glucose suspension was 1.1×10^9 CFU per ml. The tested samples were prepared as 10% (w/v) aqueous suspensions, further diluted to 3.33%, 1.11%, 0.37%, 0.12% and 0.04%. Detailed procedure of antibacterial test was described in [4].

Testing of PE/Ver-Ag,Cu samples

The antibacterial activity of PE/Ver-Ag,Cu samples was studied on G^+ bacteria *E. faecalis* by the fingerprints method. The bacterial suspension (density 1.5×10^8 CFU per ml) was applied on the surface of the thin plates and also on control pure PE plate (25 μl of suspension on plate 50 × 50 mm). The plates were placed into the laminar box. The fingerprints were made on blood agar in Petri dishes after 24 h, 48 h, 72 h and 96 h of exposure. The Petri dishes were placed into the thermostat at 35°C for 24 h. The numbers of colony forming units (CFU) were recorded. The three fingerprints were made for each plate and the countless number of bacterial colonies was expressed as CN. Detailed procedure was described in [14].

3 RESULTS AND DISCUSSION

3.1 Characterization of materials

The XRD patterns of the studied powder samples in Fig. 1 (green patterns) show basal reflections (002) of Ver that changed after saturation with silver and copper nitrate.

The original Ver (Fig. 1b) showed reflections corresponding to the interlayer space value with the cations surrounding with one layer of water molecules (d = 1.28 nm), two layers of water molecules (d = 1.44 nm) and their mixed-layered hydrate domains (d = 2.38 nm) [17,18].

The XRD pattern of VerAg (Fig. 1c) showed no reflection with d = 2.28 nm and reflection with d = 1.28 nm has shifted to d = 1.20 nm. A new reflection with d = 1.00 nm proved formation of mixtures of less hydrated phases. These changes mean that Ver structure was partially disturbed and probably Ag^+ was intercalated into the Ver interlayer. Besides, Ag is anchored in the form of nanoparticles on the Ver substrate as confirmed by TEM images in Fig. 2c. The XRD pattern of VerCu (Fig. 1e) showed only slight widening of reflections profile. Position of basal reflection d(002) was not significantly changed in VerCu. Nevertheless, value d = 1.42 nm is in agreement with data from literature for Cu-vermiculite [19], thus, Cu^{2+} can be intercalated in the Ver interlayer.

The XRD pattern of PE (Fig. 1a) shows the crystalline region with the reflections, corresponding to the values d = 0.42 nm and d = 0.38 nm, over the amorphous region between 15–25 °2θ [20]. The XRD patterns in Fig. 1 show that the vermiculite fillers structure has not changed significantly after mixing with PE. We can say that no intercalation or exfoliation of the fillers in PE occurs and only traditional microcomposite structure is formed in prepared samples. Ver fillers are in the form of individual particles distributed in PE matrix without interaction with PE.

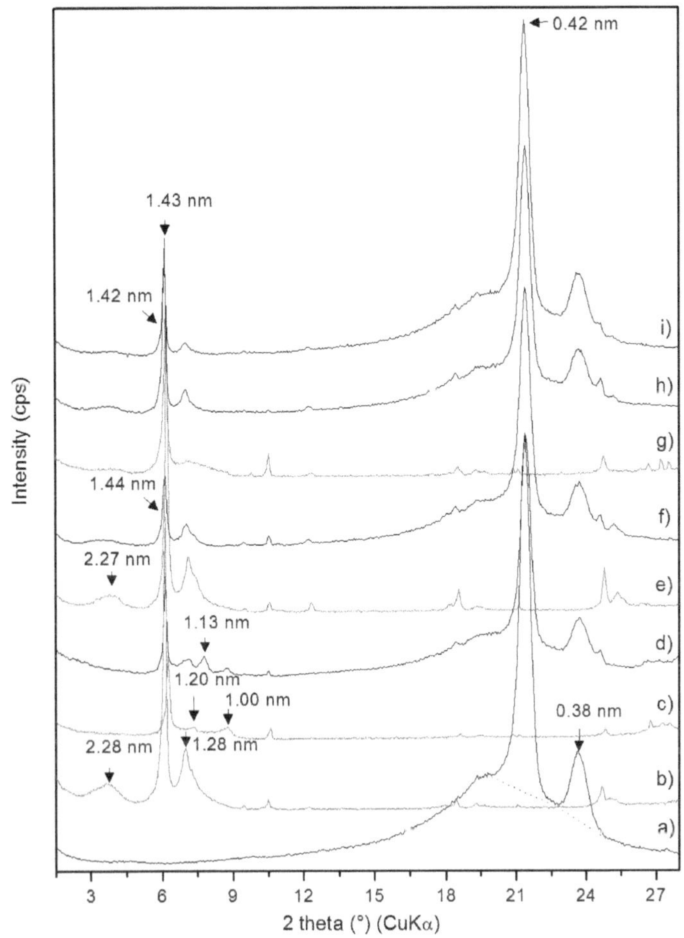

Fig. 1. XRD patterns region with reflections of nanofillers and composites PE (a), Ver (b), VerAg (c), PE/VerAg (d), VerCu (e), PE/VerCu (f), VerAgCu (g), PE/VerAgCu (h) and PE/VerAgCub (i).

The images of the samples VerAg and VerAgCu with spot elemental analysis (Fig. 2) show that characteristic vermiculite lamellar morphology was particularly disturbed mainly in the sample VerAg what confirmed also XRD patterns (Fig. 1). TEM image of sample VerAg (Fig. 2c) shows nanoparticles of silver on Ver surface. The elemental mapping showed that the distribution of Ag and Cu in sample VerAgCu was inhomogeneous (Fig. 3).

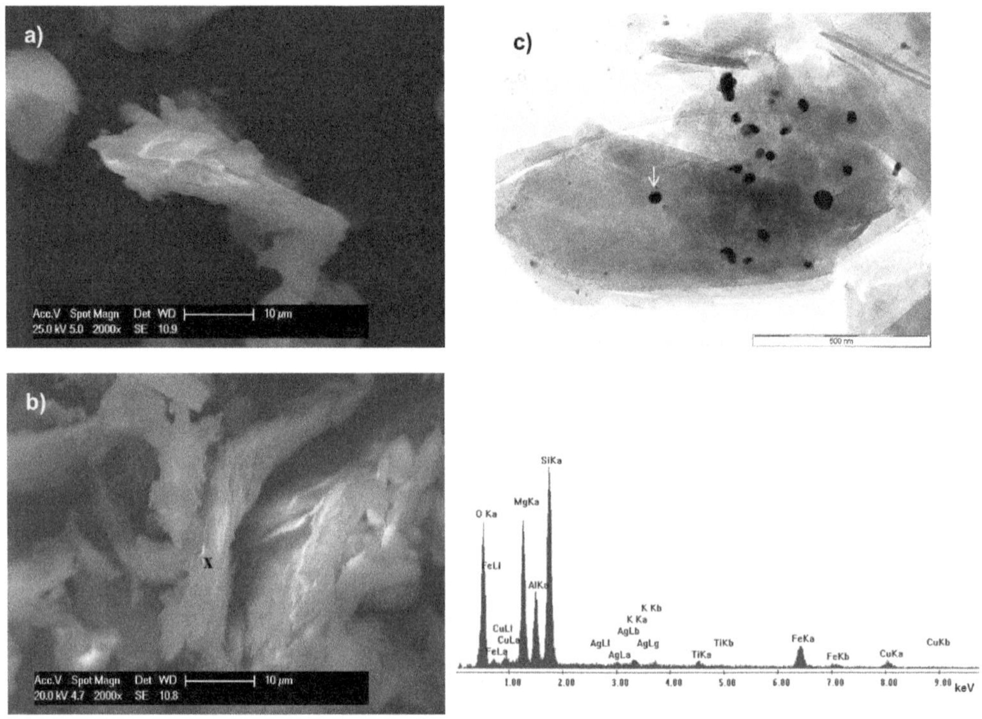

Fig. 2. SEM images of the sample VerAg (a) and VerAgCu with spot elemental analysis (b) and TEM image of sample VerAg with nanoparticles of silver (c).

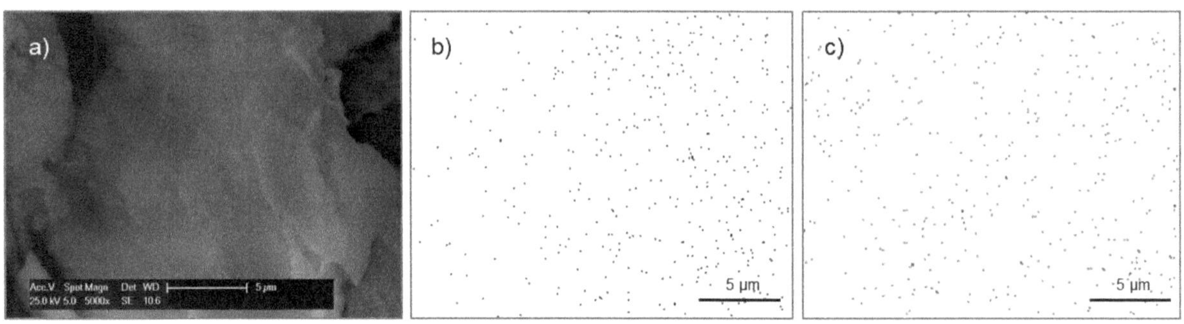

Fig. 3. SEM image of VerAgCu sample (a) with elemental maps of Ag (b) and Cu (c).

The images of samples obtained by LM in Fig. 4 showed no significant differences of the nanofiller particles distribution in PE matrix, except the sample PE/VerAg. The frequency of visible particles was observed with approximate average size in the ranges from 80-60 μm, 60-40 μm, 40-20 μm and under 20 μm (Fig. 5). The clay nanofiller particles in PE formed agglomerates from smaller particles. The largest number of visible particles was in the sample PE/VerAg with higher amount of particles under 20 μm – 75% (Fig. 5a).

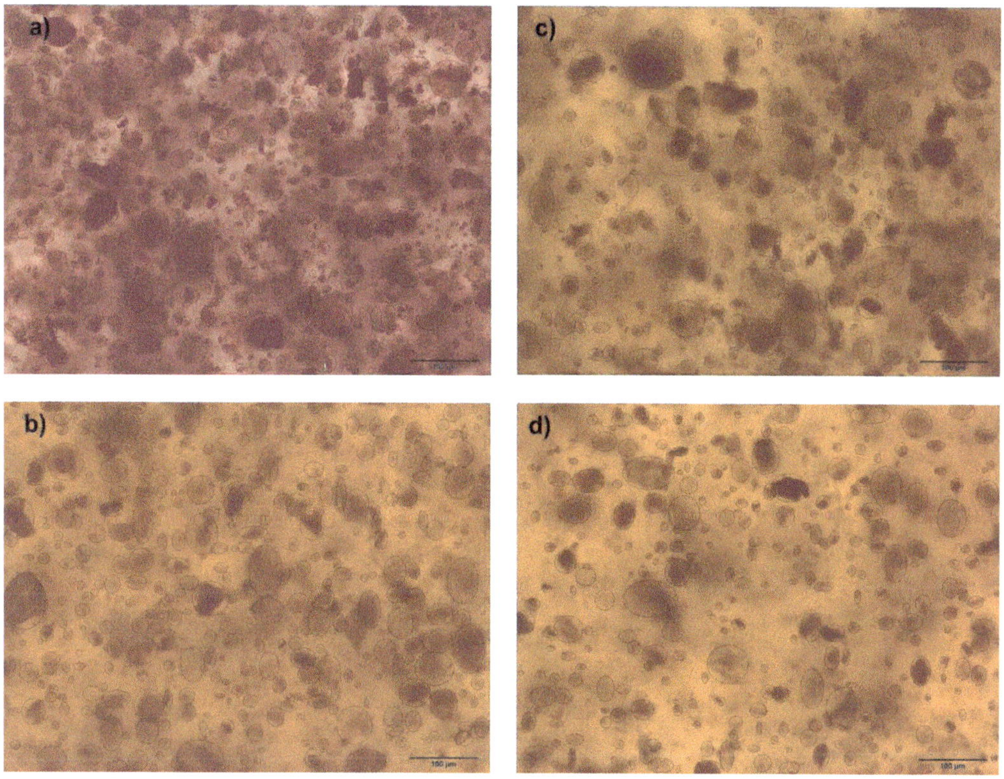

Fig. 4. LM images of composite samples PE/VerAg (a), PE/VerCu (b), PE/VerAgCu (c) and PE/VerAgCu-b (d) with highlight Ver filler particles. The size of the observed area is 710.5 x 527.4 µm.

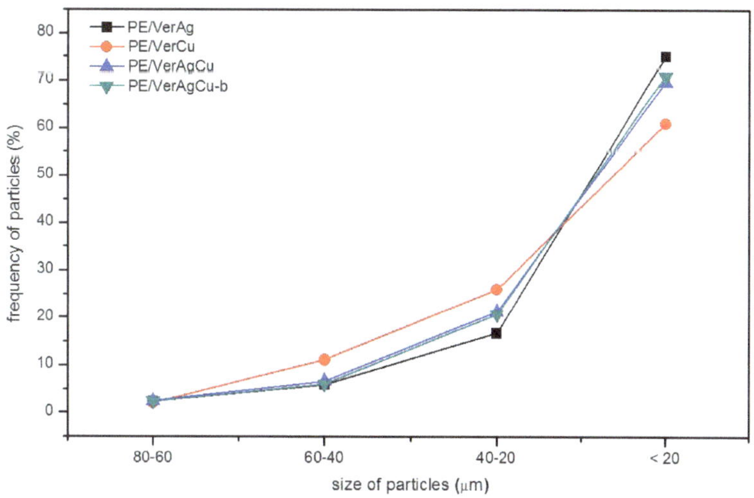

Fig. 5. Frequency of visible particles in the size ranges on the plates observed at the area of 710.5 × 527.4 µm.

3.2 Creep test

Experimental data in Fig. 6 and Tab. 1 compare pure PE matrix with selected PE/Ver-Ag,Cu samples. The results represent time dependence of tensile creep compliance data. These data show significant modification which indicates that the presence of vermiculite based nanofillers increases rigidity of PE matrix. These differences can be quantified also by initial tensile creep modulus representing elastic part of deformation. Values of initial tensile creep modulus in Tab. 1 represent increase about 43% for PE/VerAg, 25% for PE/VerCu and 57% for PE/VerAgCu compared to pure PE. The highest reinforcement effect showed sample PE/VerAgCu.

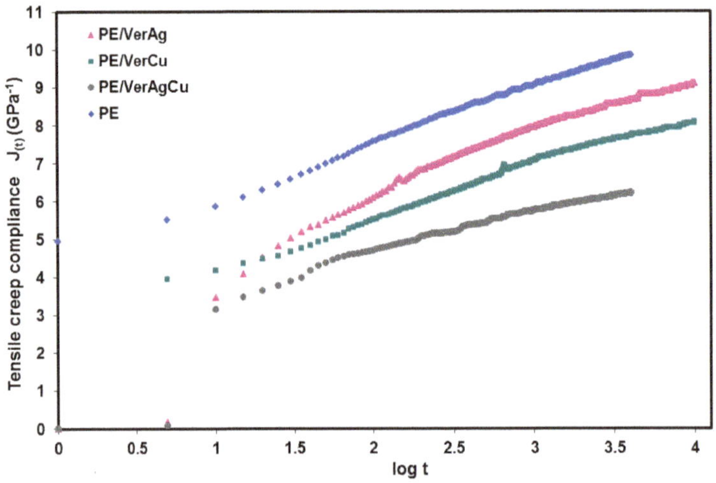

Fig. 6. Tensile creep compliance data for pure PE and PE/ Ver-Ag,Cu samples.

Tab. 1 Values of initial tensile creep modulus $E_{(t)}$ for PE/Ver-Ag,Cu samples.

	PE	PE/VerAg	PE/VerCu	PE/VerAgCu
$E_{(t)}$ (MPa)	202	288	253	317

3.3 Antibacterial test

The results of an antibacterial test in Tab. 2 show MIC values of powder Ver nanofillers. Sample of original Ver had no inhibition effect on bacteria growth. Samples VerAg and VerAgCu showed the MIC value 10% (w/v) after 1 h and 3.33% (w/v) after 24 h. These samples were slightly more effective in comparison to the sample VerCu which showed MIC value 10% (w/v) after 5 h and 3.33% after 48 h. Antibacterial action of Ver-Ag,Cu samples can be attributed to several mechanisms of action of Ag and Cu in samples: damage the bacterial cell membrane, interaction with functional groups in bacteria cell or causing the oxidative stress [21-23].

Tab. 2 Minimum inhibition concentration (MIC) of samples against bacteria *Enterococcus faecalis* after 1h, 5h, 24 h and 48 h.

Sample	MIC % (w/v)			
	1h	5h	24h	48h
Ver	-	-	-	-
VerAg	10	10	3.33	3.33
VerCu	-	10	10	3.33
VerAgCu	10	10	3.33	3.33

The results of antibacterial test in Tab. 3 show that control pure PE exhibited the countless number (CN) of bacterial CFU for 96 h of the tested time. The number of survived bacterial colonies decreased to several hundreds and tens of CFU from initial 1.5×10^8 CFU per ml after 24 h. This represents the decrease about 4-5 orders on surfaces of all samples. Sample PE/VerAgCu inhibited bacterial growth after 24 h and after 96 h no bacteria survived on the surface of this sample. Antibacterial behaviour was result to not only an effect of silver and copper but it was also probably influenced by the surface properties of the plates. Pure PE have smooth surface compared to rougher surface in PE/Ver-Ag,Cu samples. Rougher surface can influence the contact of bacteria with active components in the samples. Also, the charge properties of samples surfaces and bacterial cells could play an important role in antibacterial action, nevertheless, this needs to be better examined.

Tab. 3 Average number of survived colony forming units (CFU) *Enterococcus faecalis* on the three tested composite plates after 24 h, 48 h, 72 h and 96 h. (CN = countless number)

Sample	Average number of CFU *Enterococcus faecalis*			
	24h	48h	72h	96h
PE	CN	CN	CN	CN
PE/VerAg	257	94	49	26
PE/VerCu	71	39	34	13
PE/VerAgCu	40	19	7	0
PE/VerAgCu-b	124	67	44	32

4 CONCLUSIONS

The polyethylene with nanofillers of vermiculite with silver and copper were prepared and characterized. The X-ray diffraction patterns showed that polyethylene was not intercalated into the interlayer space of vermiculite. Further, vermiculite nanofillers were not exfoliated in polyethylene matrix, and only traditional microcomposite structure was formed in the samples. Creep test confirmed increase of rigidity of polyethylene matrix with vermiculite nanofillers compared to the pure polyethylene matrix. The antibacterial test showed good antibacterial effect of powder vermiculite with silver and copper on Gram-positive bacteria *Enterococcus faecalis*. Samples inhibited bacterial growth already after 1 h of bacteria contact with samples. The antibacterial test showed that all prepared samples of polyethylene with nanofillers of vermiculite with silver and copper inhibited the bacterial growth of *Enterococcus faecalis*.

ACKNOWLEDGEMENT

This work was supported by IT4Innovations Centre of Excellence project reg. no.cz.1.05/1.1.00/02.0070; Czech Grant Agency P210/11/2215 and Ministry of Education, Youth and Sport of Czech Republic SP2013/78 and SP2014/82. The authors thank to Ing. D. Kimmer, CSc. and Ing. J. Chmelenský from fy SPUR a.s. for assistance with preparation of PE composites, doc. Ing. P. Slobodian, Ph.D. for creep tests, Mgr. K. Šafářová, Ph.D. for TEM image, M. Heliová and Ing. S. Kozubová for SEM images and elemental maps.

REFERENCES

[1] BRIGATTI, M. F., GALAN, E., THENG, B. K. G. Structures and mineralogy of clay minerals. In: Bergaya F., Theng B.K.G. and Lagaly G. (Eds.), Handbook of Clay Science, Elsevier, Oxford, 2006, pp. 19-86.

[2] LI, B., YU, S., HWANG, J. Y., SHI, S. Antibacterial Vermiculite Nano–Material. *Journal of Minerals and Materials Characterization and Engineering*. 2002, I, pp. 61-68.

[3] DRELICH, J., LI, B., BOWEN, P., HWANG, J.-Y., MILLS, O., HOFFMANN, D. Vermiculite decorated with copper nanoparticles: Novel antibacterial hybrid material. *Applied Surface Science*, 2011, CCLVII, pp. 9435–9443.

[4] VALÁŠKOVÁ, M., HUNDÁKOVÁ, M., MAMULOVÁ KUTLÁKOVÁ, K., SEIDLEROVÁ, J., ČAPKOVÁ, P., PAZDZIORA, E., MATĚJOVÁ, K., HEŘMÁNEK, M., KLEMM, V., RAFAJA, D. Preparation and characterization of antibacterial silver/vermiculites and silver/montmorillonites. *Geochimica et Cosmochimicica Acta.* 2010, vol. LXXIV, pp. 6287–6300.

[5] HUNDÁKOVÁ, M., VALÁŠKOVÁ, M., TOMÁŠEK, V., PAZDZIORA, E., MATĚJOVÁ, K. Silver and/or copper vermiculites and their antibacterial effect. *Acta Geodynamica et Geomaterialia.* 2013, X, Nr. 1 (169), pp. 97-103.

[6] RAO, Y. & POCHAN, J. M. Mechanics of polymer-clay nanocomposites. *Macromolecules.* 2007, XL, pp. 290-296.

[7] GOMES, E. V. D., VISCONTE, L. L. Y., PACHECO, E. B. A. V. Thermal characterization of polypropylene/vermiculite composites. *Journal of Thermal Analysis and Calorimetry.* 2009, XCVII, pp. 571-575.

[8] XUE, B., LI, F., XING, Y., SUN, M., LIU, D., JIANG, Y. Preparation of Cu/OMMT/LLDPE nanocomposites and synergistic effect study of two different nano materials in polymer matrix. *Polymer Bulletin.* 2011, LXVII, pp. 1463-1481.

[9] HONG, S. I. & RHIM, J. W. Preparation and properties of melt-intercalated linear low density polyethylene/clay nanocomposite films prepared by blow extrusion. *LWT-Food Science and Technology*. 2012, XLVIII, pp. 43-51.

[10] ZHANG, J. & WILKIE, CH. A. Preparation and flammability of polyethylene-clay nanocomposites. *Polymer Degradation and Stability*. 2003, LXXX, pp. 163-169.

[11] BRUNA, J. E., PEŃALOZA, A., GUARDA, A., RODRÍGUEZ, F., GALOTTO, M. J. Development of $MtCu^{2+}$/LDPE nanocomposites with antimicrobial activity for potential use in food packaging. *Applied Clay Science*. 2012, LVIII, pp. 79-87.

[12] NIGMATULLIN, R., GAO, F., KONOVALOVA, V. Polymer-layered silicate nanocomposites in the design of antimicrobial materials. *Journal of Materials Science*. 2008, XLIII, pp. 5728-5733.

[13] HAROUN, A. A., AHMED, E. F., EL-HALAWANY, N. R., TAIE, H. A. A. Antimicrobial and antioxidant properties of novel synthesized nanocomposites based on polystyrene packaging material waste. *IRBM*. 2013, XXXIV, pp. 206–213.

[14] VALÁŠKOVÁ, M., TOKARSKÝ, J., ČECH BARABASZOVÁ, K., MATĚJKA, V., HUNDÁKOVÁ, M., PAZDZIORA, E., KIMMER, D. New aspects about vermiculite filler in polyethylene. *Applied Clay Science*. 2013, LXXII, pp. 110-116.

[15] HUNDÁKOVÁ, M., VALÁŠKOVÁ, M., PAZDZIORA, E., SLOBODIAN, P., KIMMER, D. Preparation and some properties of polyethylene/vermiculite nanocomposites. In *4th International Conference NANOCON 2012*, 23-25 Oct. 2012, Brno, Czech Republic. Conference proceedings. Ostrava: Tanger, Ltd., 2013, pp. 86-91, ISBN 978-80-87294-35-2.

[16] KNEIFLOVÁ, J. Hodnocení baktericidní účinnosti dezinfekčních prostředků suspenzní mikrometodou. (Evaluation of the antibacterial effect of disinfectants using a suspension micro-method). *Čs. epidemiologie, mikrobiologie a imunologie*. 1988, XXXVII, pp. 97-104.

[17] WALKER, G. F. Mechanism of dehydration of Mg-vermiculite. *Clays and Clay Minerals*. 1956, IV, pp. 101-115.

[18] MARCOS, C., ARANGO, Y. C., RODRIGUEZ, I. X-ray diffraction studies of the thermal behaviour of commercial vermiculites. *Applied Clay Science*. 2009, XLII, pp. 368-378.

[19] CLEMENTZ, D. M., PINNAVAIA, T. J., PORTLAND, M. M. Stereochemistry of hydrated copper (II) ions on the interlamellar surfaces of layer silicates. An electron spin resonance study. *Journal of Physical Chemistry*. 1973, LXXVII, pp. 196–200.

[20] DAS-GUPTA, D. K. Polyethylene: Structure, morphology molecular motion and dielectric behaviour. *IEEE Electrical Insulation Magazine*. 1994, X, Nr. 3, pp. 5-15.

[21] TONG, G., YULONG, M., PENG, G., ZIRONG, X. Antibacterial effects of the Cu(II)-exchanged montmorillonite on *Escherichia coli* K88 and *Salmonella choleraesuis*. *Veterinary Microbiology*. 2005, CV, pp. 113–122.

[22] LI, W.-R., XIE, X.-B., SHI, Q.-S., ZENG, H.-Y., OU-YANG, Y.-S., CHEN, Y.-B. Antibacterial activity and mechanism of silver nanoparticles on *Escherichia coli*. *Applied Microbiology and Biotechnology*. 2010, vol. LXXXIV 85, pp. 1115–1122.

[23] THURMAN, R. & GERBA, C. The molecular mechanisms of copper and silver ion disinfection of bacteria and viruses. *Critical Reviews in Environmental Control*. 1989, XVIII, pp. 295–315.

RESUMÉ

Vzorky polyethylenu s nanoplnivem vermikulitu se stříbrem a mědí byly připraveny a charakterizovány. Rentgenové difrakční záznamy ukázaly, že nedošlo k interkalaci polyethylenu v mezivrství vermikulitu. Difrakční záznamy také potvrdily, že vermikulitové nanoplnivo nebylo exfoliováno v polyethylenové matrici, tudíž došlo pouze ke vzniku mikrokompozitní struktury. Test krípu potvrdil vyztužení matrice polyethylenu s vermikulitovými nanoplnivy oproti čisté matrici polyethylenu. Antibakteriální test ukázal dobrý antibakteriální účinek práškových vzorků vermikulitů se stříbrem a mědí na gram-pozitivní bakterii *Enterococcus faecalis*. Vzorky inhibovaly růst bakterie již po 1 h kontaktu bakterie se vzorky. Dále, antibakteriální test ukázal, že všechny připravené vzorky polyethylenu s obsahem nanoplniva vermikulitu se stříbrem a mědí inhibovaly růst bakterie *Enterococcus faecalis*.

18

THE ISSUE OF INCOME FROM OPERATING UKRAINE COAL DEEP MINES

Jaroslavav DVOŘÁČEK[1], Elena V. MARTYAKOVÁ[2], Ilona V. KOCHURA[3], Radmila SOUSEDÍKOVÁ[4], Igor SAVIČ[5]

[1] *prof. Ing. CSc., Institute of Economics and Control Systems, Faculty of Mining and Geology, VŠB-Technical University of Ostrava*
17. listopadu 15, Ostrava-Poruba
e-mail: jaroslav.dvoracek@vsb.cz

[2] *prof. DrSc., Production Management Department, Mine Faculty, Donetsk National Technical University*
Artema str. 58, Donetsk 83000, Ukraina
e-mail: emartyakova@gmail.com

[3] *doc. Ph.D., Production Management Department, Mine Faculty, Donetsk National Technical University*
Artema str. 58, Donetsk 83000, Ukraina
e-mail: kochurai@mail.ru

[4] *RNDr. Ph.D., Institute of Economics and Control Systems, Faculty of Mining and Geology, VŠB-Technical University of Ostrava*
17. listopadu 15, Ostrava-Poruba
e-mail: radmila.sousedikova@vsb.cz

[5] *Mgr., Cybex Industrial LTD, C/O Columbus Trading-Partners GMBH*
Riedinger str. 18, 95448 Bayreuth, German
e-mail: igor.savic@cybex-online.com

Abstract

This paper focuses on looking for factors that influence the income of deep coal mining operations in Ukraine. The method of discriminate analysis was employed for assessing economic success of the operations. It could be concluded that mining-geological conditions are the most significant cause of influence; the human factor is of lesser importance.

Abstrakt

Příspěvek se zabývá hledáním faktorů, které ovlivňují dosažení kladného nebo záporného hospodářského výsledku hlubinných uhelných dolů Ukrajiny. Je použitá metoda diskriminační analýzy. Dochází se k závěru, že nejvýznamnější příčinou jsou důlně-geologické podmínky, v menší míře pak vliv lidského činitele.

Key words: coal deep mines, Ukraine, profitable and loss-making mines

1 INTRODUCTION

Ukraine takes the seventh position as regards the world total coal reserves. Until 1989, in analogy to other Comecon countries, the country collieries were owned and operated by the state. The coal was taken as a strategic raw material whose extraction economy was of less importance. The financing of actual mining operations took advantage of redistributing income between open cast and deep mines. The loss-making operation of some collieries was taken for granted and the appropriate planning provisions were made. The mine workers' status and their wages were superior compared to other blue-collar workers.

The progress of restructuring mining industries after 1989 was faster in the Czech Republic than in Ukraine. The restructuring was characterised by privatisation of the industries concurrently with restriction of their activity. At the same time, state subsidy and income redistribution schemes were abolished. The causes of the restructuring consisted in reduced priority of heavy industries, loss of strategic status for many minerals, the need of which could be met by import, and environmental concerns, and other factors inclusive those of politics. The restricted mining industry activities were concentrated and specialised on extracting of choice resources. The mining industry organisational structures simplified.

In such situation, the importance of profitability of individual mining enterprises increased, as loss-making pits were being closed down.

2 INCOME FROM OPERATING COAL DEEP MINES

An economic result of a company, as a difference between revenues and expenses, is a complex indicator of successful or unsuccessful operations. Considering the conditions of deep mine operations, this indicator is influenced by many factors that can be summarized as follows:

- **Natural factors**: Useful seam thickness; seam bedding depth and its dip; coal gas capacity; threats of quakes, coal and gas outbursts; seam tectonics; etc.,
- **Human factors**: Workforce provision as regards quantity and miners' qualification structures; their age, health, work moral, etc.,
- **Economic factors**: Prices of products and production process inputs, ratio of inside and outside capital resources, production quantities, etc.,
- **Organisation and management factors**: Management experience and motivation, mining legislation of work and safety at work, etc.

These factors influence the economic results and cash flows of individual collieries. If it is obvious that factors of negative influence are beyond redress, the colliery is closed down.

Such analyses and processes affected coal (but also other mineral) extraction industries in the Czech Republic in the nineties of the past century. The result was that only half of total production was reached in comparison to a historical maximum. The Ukrainian coal mining industries are at the beginning of the process. The ratio of the state owned and run collieries is still quite high and a lot of them operate in the red.

It is a question, which factors promote chances of success or failure. Obviously, the answer depends on how many factors of the above mentioned items are at disposal for the analysis.

This case investigation took into account monthly statistics of 12 coal deep mines of the Donetsk Region in Ukraine. The input data covered operations between 2000 and 2005. (Kochura, 2012). In this period, all collieries were in the hands of the state. The monthly statistics presented the following variables:

- Coal production [ton/month]
- Direct costs of extraction [Hryvnia/ton]
- Production loss caused by gas capacity[ton/1000 tons of production]
- Production loss caused by geological conditions[ton/1000 tons of extraction]
- Production loss caused by disrupted flows of material supplies [ton/1000 tons of production]
- Workforce provision ratio [ratio to 1, i.e. to full workforce provision rate]
- Workers of basic qualification ratio, i.e. workmen without secondary or higher education [ratio to 1]
- Human factor implied loss of production, i.e. infringement of work safety regulations [tons/1000 tons of production]
- Production loss due to equipment breakdowns [ton/1000 tons of production]
- Increase of production input costs in the last 5 years [%]
- Trading income of the colliery [thousand Hryvnias/month]

The production losses caused by natural and organisational/management factors, as well as a human factor were identified by periods in which the production was disrupted by any of the mentioned factors.

Comparing daily production capacities with the periods of discontinued production, the losses were established in tons per month, which were adapted to 1000 tons of production so as to provide for comparison with other collieries. The input data files were structured into 2 groups, namely: (i) profitable collieries, and (ii) pits operating in the red. The individual indicator (variable) averages of these groups are given in Tab. 1. This table also provides for assessments of statistical significance as regards differences in averages of profitable and loss-making collieries. This statistical significance was evaluated by the Student's t-test at the $\alpha=0.05$ level. (Hindls, 2007). Each group of collieries comprised 32 figures.

The discriminate analysis was used for identifying those variables that are significant for classifying businesses as profitable or loss-making.

Tab. 1 Indicator (variable) averages and statistical significance of their differences

Variable	Average		Difference significance
	Profitable	Unprofitable	
Actual production	63 383.03	25 019.63	Significant difference
Actual direct costs	90.02	136.55	Insignificant difference
Production loss due to coal gas	0.36	1.77	Insignificant difference
Production loss due to geological conditions	6.32	42.35	Insignificant difference
Production loss due to material supply deficiencies	8.33	13.36	Insignificant difference
Workforce qualification structure ratio	0.80	0.81	Insignificant difference
Workforce basic qualification ratio	0.74	0.71	Insignificant difference
Production loss due to human factor	0.71	3.69	Insignificant difference
Production loss due to equipment breakdowns	76.52	90.89	Insignificant difference
Cost increases of material inputs, 5-year period	4.98	8.65	Insignificant difference
Income	+ 685.00	-1259.00	Significant difference

3 DISCRIMINATE ANALYSIS METHOD

3.1 The Method

The discriminate analysis of classification statistically investigates relations between a group of independent attributes, m – discriminators, and a single dependant qualitative variable. The simplest case is represented by a binary variable output of 0 (zero) value if an object belongs to Class I. If an object belongs to Class II, the binary variable output has a value of 1. The objective is to develop a prediction model that would provide for classification of new objects.

The application of the method starts with two classes of objects. Each of them is characterised by its discriminator value, m. The first class comprises objects, n_1, and can be described by an input matrix, $n_1 \, x \, m$. The matrix's lines are formed by discriminator values of specific objects; the columns provide for values of specific discriminators for different objects. The second class comprises objects, n_2, and can be described by an input matrix, $n_2 \, x \, m$. Sample averages, $\bar{x}_1$ and $\bar{x}_2$, are calculated for individual classes. These are vectors, the constituents of which are calculated as discriminator values averages.

Further, a common covariance matrix is formed:

$$S = \frac{(n_1 - 1)S_1 + (n_2 - 1)S_2}{n_1 + n_2 - 2}, \tag{1}$$

where:

S_1, S_2 – covariance matrices of individual classes.

With the aid of the sample averages, $\bar{x}_1$ a $\bar{x}_2$, and the inversion matrix, S^{-1}, as related to the common covariance matrix, S, the vector, a, is calculated (the constituents, of which are the linear discriminate function coefficients), and a coefficient,

$$b = -\frac{1}{2}a^T(\bar{x}_1 - \bar{x}_2) - ln(\frac{\pi_1}{\pi_2}), \tag{2}$$

where:

π_1, π_2 - a priori probabilities that an object pertains to the first or second class.

The simplest assumption is: $\pi_1 = \pi_2 = 0,5$

If new objects, x_0, are classified, the rule is applied that an object is classified as belonging to class I. if $a^T x_0 + b > 0$ (Meloun&Militký, 2004). (3)

3.2 Discriminate analysis application

The linear discriminate analysis was applied to a file, n_1, of profitable collieries, and file, n_2, of loss-making pits. Each colliery is characterised by a set of discriminators, m. The objective is to find a discriminate function that would provide for classifying a specific colliery as profitable or operating in the red. If Class I (profitable pits, n_1) is characterised by mean value vectors,

$x_1 = (x_{11}, x_{12}, \ldots, x_{1m})$ and a covariance matrix, S_1

and Class II. (unprofitable pits, n_2) by mean value vectors,

$x_2 = (x_{21}, x_{22}, \ldots, x_{2m})$ and a covariance matrix, S_2,

than the coefficients, $a_1, a_2, \ldots, a_m$ of the discriminate function,

$$f_i = a_1 x_{i1} + a_2 x_{i2} + \cdots + a_m x_{im}, \tag{4}$$

are calculated by the formula,

$a = S^{-1}(\bar{x}_1 - \bar{x}_2)$(Hebák et al., 2004). (5)

If average value vectors of Classes I and II are substituted in the discriminate function equation, we can calculate average values, $\bar{Z}_1$ and $\bar{Z}_2$.

The optimal threshold value, C, which determines objects as belonging to Class I or Class II, can be calculated:

$$C = \frac{(\bar{Z}_1 + \bar{Z}_2)}{2}. \tag{6}$$

A non-classified colliery, whose linear discriminate function value is in excess of C, can be put in Class I of profitably operating pits. If the linear discriminate function value is less than C, the pit should be qualified unprofitable and as such belongs to Class II. The model efficiency can be assessed by a feedback classification of collieries – or any company for that matter – that provided for the input data files of the model provision, and whose economic performance is known.

The relative significance of individual discriminators can be established by means of standardised discriminate coefficients. The standardized discriminate coefficients can be calculated if the discriminate function coefficients, a_i, are multiplied by standard deviations, s_i, of the respective discriminators. Irrespective of their negative or positive signs, each standardized coefficient represents a relative contribution of its discriminator to the linear discriminate function. The discriminators with standardized coefficients of relatively high values contribute to the linear discriminate function more than those of the lesser values. The negative or positive sign only means that discriminators contribute positively or negatively.

3.3 Discriminate analysis results

The results of the discriminate analysis in a form of standardized discriminate coefficients are provided by Tab. 2.

Tab. 2 Standardized discriminate coefficients

Variable	Value
Actual production	2.460
Actual direct costs	-0.680
Production loss due to coal gas	0.349
Production loss due to geological conditions	-0.499
Production loss due to material supply deficiencies	0.546
Workforce qualification structure ratio	0.228
Workforce basic qualification ratio	1.707
Production loss due to human factor	0.228
Production loss due to equipment breakdowns	-1.215
Cost increases of material inputs, 5 years period	-0.389

As evidenced by Tab. 2, Class I of profitably operating pits is particularly influenced by actual production volumes, and the ratio of the basic qualification workforce. Conversely to this fact, the negative ranking of the Class II of collieries operating in the red is mostly caused by production breakdowns.

By utilising the linear discriminate function – as a result of discriminate analysis application – and the individual pit discriminator values in the respective months of the analysis, a feedback classification was performed, having regarded the input data files of profitable or in the red operating collieries. The ratio of successfully classified pits to their total number in the group (i.e. 32) is a measure of classification success. Concerning 32 profitable and 32 in the red operating pits of each group, the success rate was 93.75 %, and 90.63 % respectively.

4 DISCUSSION

The individual discriminator averages, their quantitative levels, and statistical significance of differences of discriminator averages between profitable and loss-making pits provide for the following characteristic of profitable collieries:

They are clearly major producers of coal, they operate with lower direct costs, and production losses that would be caused by breakdowns, coal gas and geological adversities, as well as increases of input material costs in the last five years are also less significant. The discriminator average differences, as regards production losses due to breakdowns and input material cost increases, are of little statistical consequence. The differences of the other three discriminators are statistically significant.

The variable of actual production can be considered as a factor of major importance. High levels of direct costs – mine pumping-out and ventilation, lighting, and maintenance – are concurrent to any mining operations. If a monthly coal production output is low, the ratio of the actual direct costs per one ton of coal produced can be pretty high. Direct costs further increases due to the fast rise of material input costs. Collieries operating in the red also have higher rates of production losses caused by geological conditions of mining, and production equipment breakdowns. The latter two influences might be considered in their unity, as the equipment breakdowns can be also affected by natural conditions of the equipment operation.

These specific conditions of mining operations imply factors of major significance for classifying the individual pits as profitable or loss-making. Higher production rates not only influence the actual direct costs (fixed cost constituent ratio) but also imply increases of income from coal sales. A rather surprising factor of influence is that of workforce basic qualification ratio. It can be assumed that workers of this category are directly involved in the coal extraction activities so that their higher ratio affect positively the rates of production. The negative effect of equipment breakdowns is easy to grasp – the breakdowns that cause stoppage of the fully mechanised extraction working are clearly of a major impact on the overall coal production because of the production capacities implied in the mechanised working.

5 CONCLUSION

The individual discriminators values of profitable and loss-making collieries of the original input files provide for the following conclusions:

- Lower production losses due to better geological conditions of mining mean that the pits operate in more favourable natural environment.
- Favourable conditions of operation enable utilizing the equipment and machinery both of a larger number and capacity, which reduces a risk of production losses by breakdowns.
- Utilization of the large-capacity equipment and a lesser breakdown risk imply increased production rates, which lowers the ratio of fixed costs per one ton of coal extracted and consequently leads to a reduction of direct costs. These costs may be negatively affected by less favourable geological conditions under which loss-making collieries operate.
- Concerning collieries that operate in the red, the direct costs further increase by the faster growth of material input costs. The material inputs for operating more or less the same levels of machinery are approximately the same. Different material input needs for operating relatively the same machinery signify deficiencies of organization and management of the supply processes. There is a theoretical assumption that low production pits cannot take advantage of bulk discounts, as they buy less and it is more difficult for them to negotiate individual delivery prices.

It can certainly be assumed that natural conditions of mining dominate, concerning profitability of mining. The human factor influence is of lesser importance. Such assumption is in full accordance with the experience of deep mine operations in the Czech Republic, where the process of restructuring was started by closing down pits of adverse natural condition of operation.

Nevertheless, this analysis input data files are not large enough. Their amplification might provide for thoroughgoing investigation of factors that cause or influence the profitability of the Ukraine coal deep mines.

REFERENCES

[1] KOCHURA, I. V. Coal market of Ukraine: Analysis and development background. *Geoscience Engineering*. 2012, LVIII, Nr.1, pp. 17-23, ISSN 1802-5420.

[2] MELOUN, M. a J. MILITKÝ. *Statistická analýza experimentálních dat (Statistical Analysis of Experimental Data)*. Prague: Academia, 2004. 953 pp. 415 pp. ISBN 80-200-1254-0.

[3] HINDLS, Richard et al. *Statistika pro ekonomy. (Statistics for Economists)*. 8. ed. Prague: Professional Publishing, 2007. ISBN 978-80-86946-43-6.

[4] HEBÁK et al. *Vícerozměrné statistické metody (Multivariate Statistical Methods)*. 1. ed. Prague: Informatorium, 2004. 239 pp. ISBN 80-7333-025-3.

RESUMÉ

Restrukturalizace uhelného průmyslu České republiky předcházela těmto procesům v podmínkách Ukrajiny, kde stále zůstává značný podíl hlubinných uhelných dolů v rukou státu. Privatizační procesy a ekonomické změny země se sedmými největšími zásobami uhlí na světě zvyšují význam dosahovaných hospodářských výsledků.

Hospodářský výsledek, tedy zisk nebo ztráta hlubinných uhelných dolů je ovlivňován řadou faktorů – přírodními, lidskými, ekonomickými faktory a faktory organizace a řízení. K analýze vlivu konkrétních faktorů jsme použili výchozí soubor 32 údajů měsíčních hodnot ziskových dolů a 32 údajů měsíčních hodnot ztrátových dolů Doněcké oblasti. Každý měsíční údaj byl charakterizován 10 proměnnými (diskriminátory) zahrnující těžbu uhlí, přírodní, ekonomické a lidské faktory. Dále byl k dispozici údaj o dosaženém hospodářském výsledku. Pro zjištění diskriminátorů, které jsou nejvýznamnější pro zatřídění do skupiny ziskových nebo ztrátových dolů, byla použita diskriminační analýza.

Výsledky ukázaly, že zařazení mezi ziskové doly je nejvíce pozitivně ovlivněno objemem těžby a podílem dělníků se základní kvalifikací. Zařazení mezi ztrátové doly je nejvíce ovlivňováno ztrátami těžby vlivem havárie zařízení.

S ohledem na hodnoty diskriminátorů v obou skupinách dolů, absolutní hodnoty těchto diskriminátorů a statistickou významnost rozdílů průměrných hodnot diskriminátorů mezi ziskovými a ztrátovými doly lze předpokládat, že dominantní vliv na ztrátovost dolů budou mít přírodní podmínky, v menší míře pak lidský činitel.

Rozšíření vstupních souborů a zvýšení počtu diskriminátorů by umožnilo hlubší zkoumání příčin ztrátovosti resp. ziskovosti hlubinných uhelných dolů Ukrajiny.

THE FOGGARA: A TRADITIONAL SYSTEM OF IRRIGATION IN ARID REGIONS

Remini Boualem[1], *Achour Bachir*[2], *Kechad Rabah*[3]

[1]*Department of Water Sciences and Environment, Faculty of Engineering, Blida University, Blida 9000, Algeria, reminib@yahoo.fr*

[2]*Department of Hydraulic, Larhyss laboratory, Biskra University, Biskra, 7000, Algeria*

[3]*Department of Economics, Faculty of Engineering, Blida University, Blida, 9000, Algeria, rkechad@yahoo.fr*

Abstract

This article discusses the traditional irrigation system in the oases of Touat, Gourara and Tidikelt. Since centuries, farmers use the system of foggaras for irrigation of palm groves and gardens.

The results obtained following the inquiries and investigations carried out on the sites of foggaras, showed that since the eighties, drilling works multiplied in the oases of Touat, Gourara and Tidikelt. However, despite the application of these new techniques of water catchment, farmers still irrigate their gardens by traditional methods. For social, economic and environmental problems, foggaras degrade more and more; 50% of the foggaras decayed. Currently, the discharge of foggaras in service greatly diminished. The adoption of modern catchment techniques entails to the lowering of the water table.

Keywords: water, foggara, oasis, irrigation system, Sahara

1 INTRODUCTION

The Algerian Sahara is a hyper-arid region where rainfall does not exceed 100 mm/year. The temperature can reach 50 °C during summer. Evaporation exceeds the threshold of 2 m/year. This dry environment with bare soil characterized by the absence of surface water pushed farmers to exploit the waters hidden in the soil. Thanks to the knowledge of farmers on hydraulics and engineering, hundreds of foggaras were developed in the regions of Touat, Gourara and Tidikelt. This ancestral water catchment system has been developed in the ancient Iran for over 3000 years (Goblot, 1963; Goblot, 1979; Kazemi, 2004). Called Qanat in Iran and Khettara in Morocco, foggaras have spread to over 30 countries around the world (Hofman, 2007). Just through this ingenious hydraulic system, thousands of hectares of land have been earmarked and promoted. Hundreds of hectares of palm groves are irrigated by foggaras. Several oases were developed in southwest Algeria.

The objective of this paper is to study the effectiveness of irrigation by foggaras during the last ten centuries. The impact of new technologies of water catchment on the degradation of foggaras is another subject of this study.

2 REGION OF STUDY AND INVESTIGATIONS

2.1 Location and characteristics of the region

The regions of Touat, Gourara and Tidikelt known as the country of foggaras are located more than 1200 km to the southwest of Algiers (Fig. 1). About 1400 foggaras have been dug for more than 10 centuries. Today, there are not more than 700 operating foggaras.

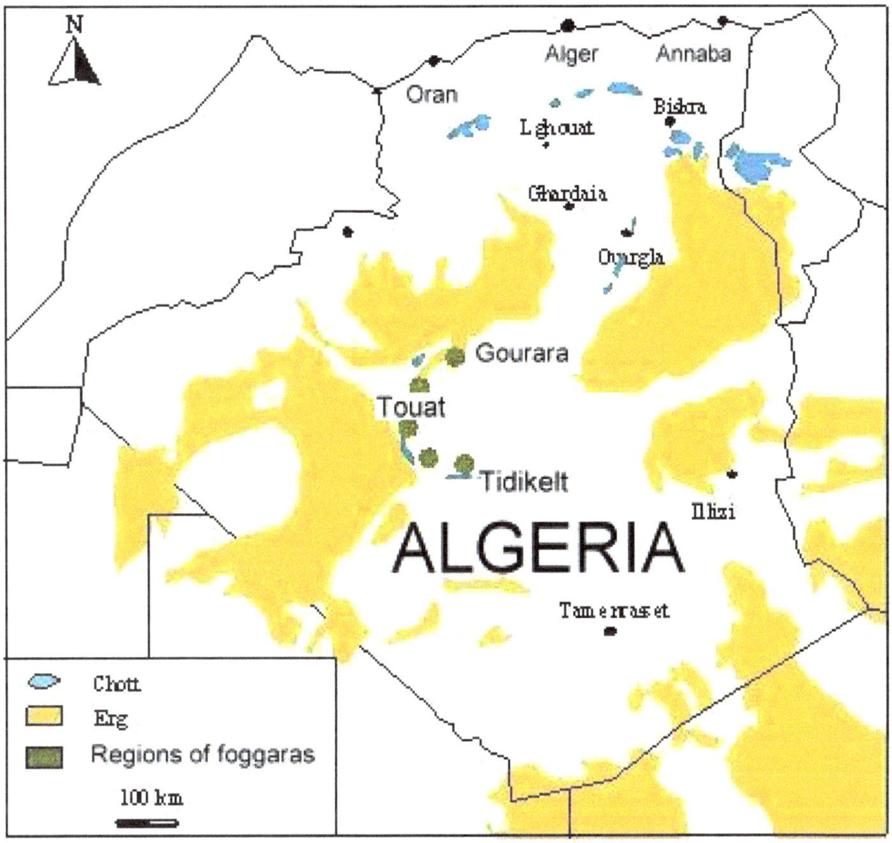

Fig. 1 Location of study areas

2.2 Missions and investigations

We conducted several missions in the oasis of Touat, Gourara and Tidikelt. Surveys were performed with local people and owners of foggaras. Visits were carried out to twenty foggaras to detect the real problems of degradation.

3 RESULTS AND DISCUSSION

3.1 Irrigation by foggaras

The foggara is an underground tunnel, one hundred meters in length, equipped with several air shafts for maintenance and ventilation of the gallery. To ensure gravity flow, the gallery is slightly inclined. The length of the gallery ranges from 100 to 14,000 m (Fig.2 (a, b, c)). The number of aeration wells varies from 10 up to 500 with depths ranging from 8 m to 40 m (Fig.3). The foggaras in the regions of Touat, Gourara and Tidikelt exploit the water of Intercalary Continental series. Considered one of the largest tablecloths of the planet, the Intercalary Continental series is a fossil aquifer. The gallery collects and drains the water from the underground to the ground surface. As the foggara is a collective structure, the water outlet of the gallery is shared between the owners of the foggara. A small pond triangular ending by a diverter at its base that takes the form of a comb is called kasria (Fig.4). The largest kasria (main kasria) is placed at the outlet of the gallery which serves for sharing water among the owners of the foggara. The portion of water depends on the contribution of each owner. From the main kasria, each part of water is fed by a channel (seguia) to a storage basin called madjen which is filled in 24 h (Fig. 5). From this madjen, the water flows in the small seguias embedded in the garden (Guemoun) to irrigate palm trees and other plants (Fig. 6). Generally, in the irrigation channel from the main kasria a portion of water for one family (a group of owners) flows that ends in a secondary irrigation channel. Once the water sharing is established, the water flows in seguias low sections to reach the secondary kasria. Shared by the secondary kasria, the water flows in small seguias to run into tertiary kasrias. Each owner then receives a share of water in terms of his contribution to the development and maintenance of the foggara.

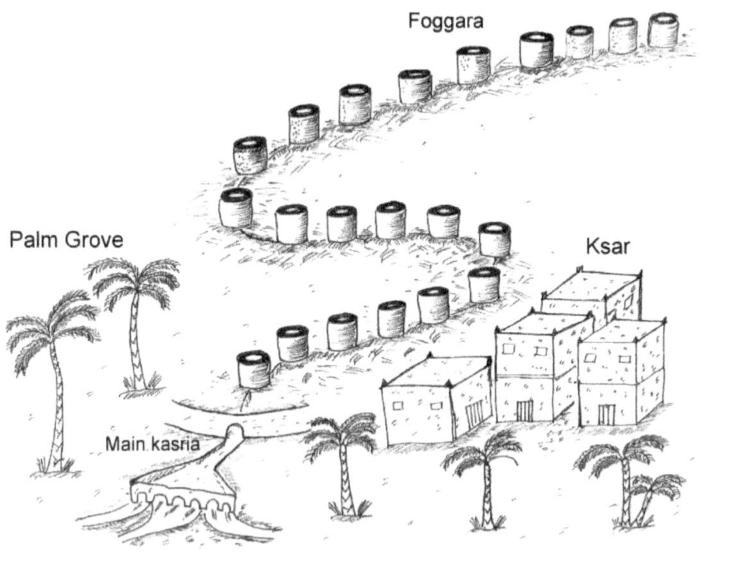

a) An overview of the foggara

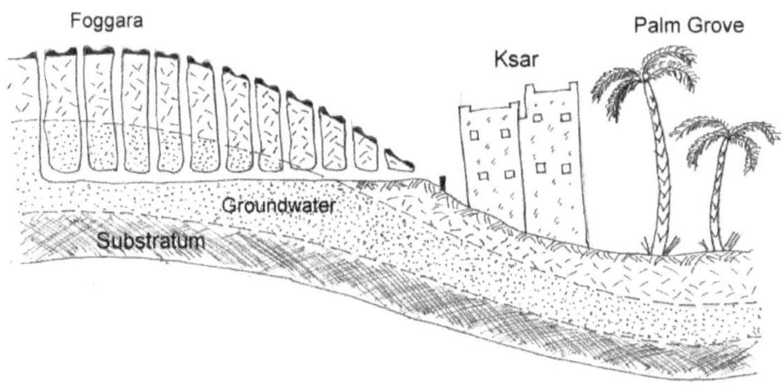

b) A logitudinal cup of the foggara

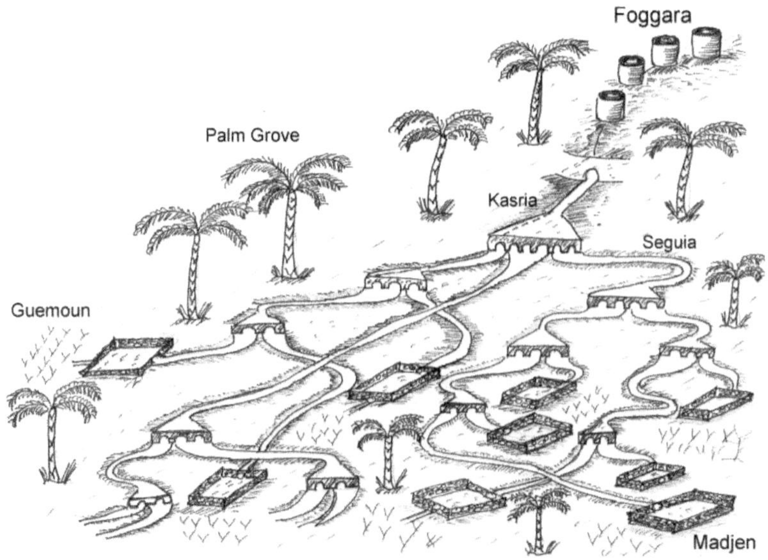

c) A distribution network of the foggara

Fig. 2 Schematic diagram of the foggara

Fig. 3 An airshaft at the foggara of Gourara

Fig. 4 The secondary Kasria at the foggara of Gourara

Fig. 5 The seguia in an oasis of Gourara

Fig. 6 Guemoun (garden) in an oasis of Touat

3.2 Relationship between the palm grove and the foggara

In the oases of Touat, Gourara and Tidikelt, there is a relationship between the foggara and the palm grove. Indeed, it is the flow of the foggara which determines the size of the palm grove. As soon as the palm grove area increases by implanting a new palm trees, the farmers increase the discharge of the foggara or they get digging a second foggara. To support this idea, we show in Fig. 7(a, b) the discharge of the foggara depending on the size of the palm grove of oases of Touat and Gourara using the inventory data on foggaras from 1962 (Arrus, 1985).

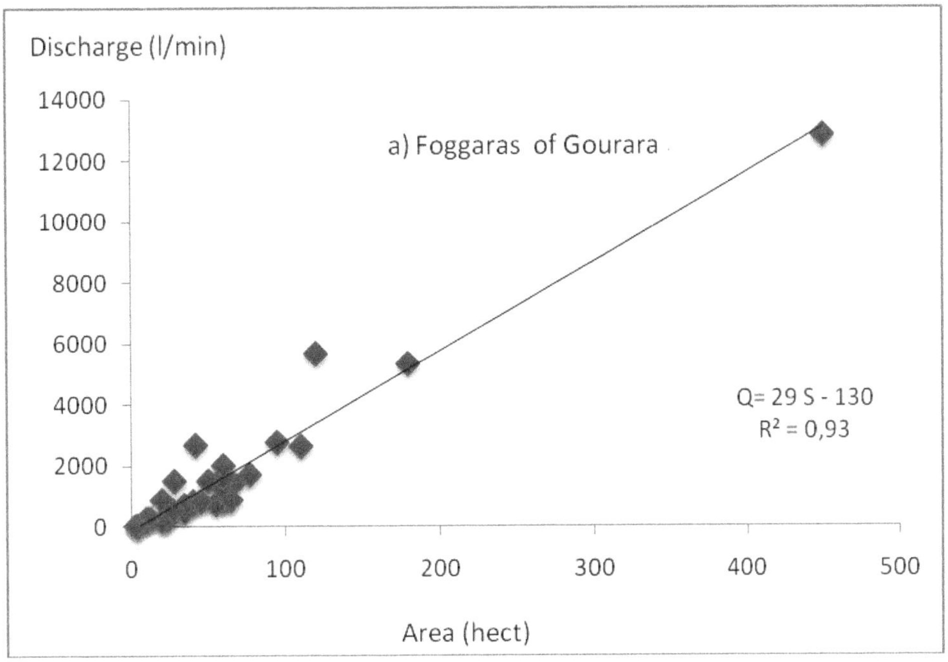

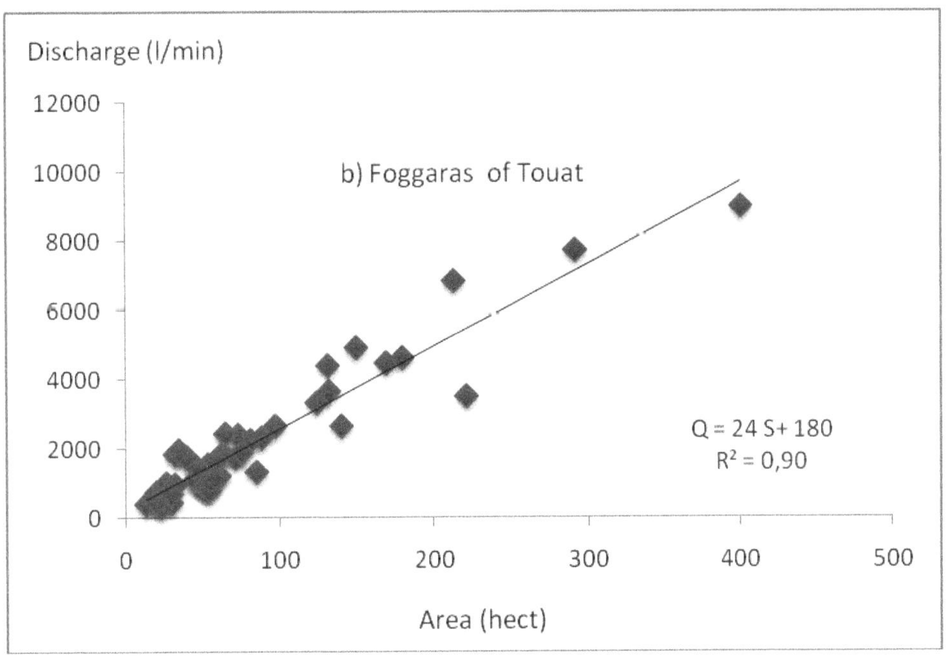

Fig. 7 Evolution of the foggara flow rate depending on the surface area of the palm grove (Data: Arrus, 1985)

It is interesting to note that there is a clear correlation between the throughput of the foggara and the area to irrigate. It consists in that there is no surplus or deficit of water. It is the know-how and engineering of farmers demonstrating that the irrigation of the palm grove was well studied by farmers of Touat, Gourara and Tidikelt. By highlighting new agricultural land, the farmers increase the discharge of the foggara by extending its gallery. Increasing the length of the gallery, automatically the number of aeration wells increases. To verify this hypothesis, the figure (Fig. 8 (a, b)) shows that there is indeed a relationship between the discharge and the length of the gallery of the foggara.

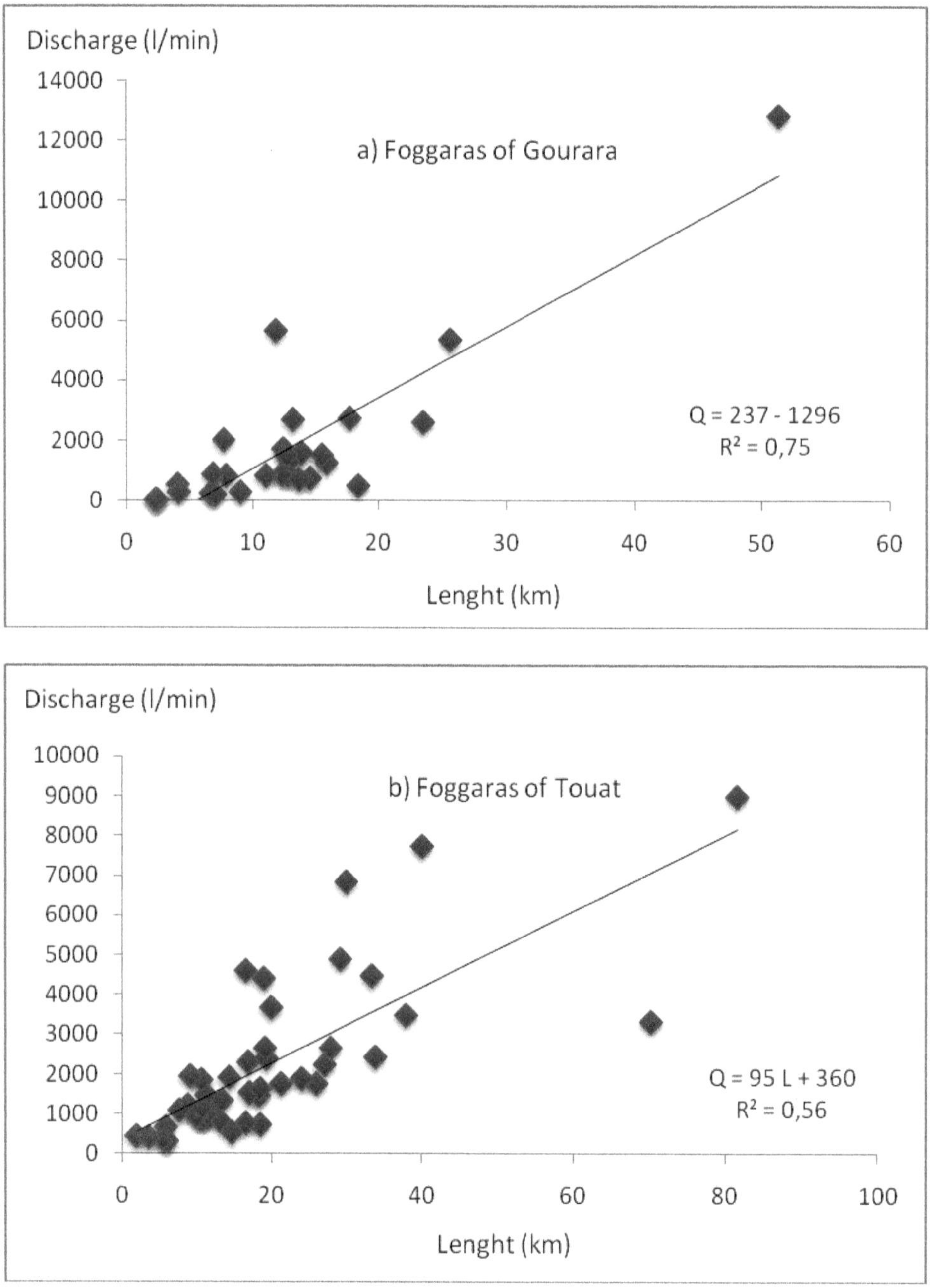

Fig. 8 Evolution of the foggara discharge depending on the length of the gallery (Data : Arrus 1985)

3.3 The contribution of drilling and degradation of the foggara

Dry environments are very fragile ecosystems. So the slightest disturbance can cause harmful consequences to the environment. For over 10 centuries, the farmers of Touat, Gourara and Tidikelt irrigate their gardens by the foggaras without causing any irregularity in the environment. No excess water was recorded during the operation of foggaras which avoided the lowering of the water discharge. The application of modern technology in the area during the fifties and the proliferation of boreholes from the eighties have greatly reduced the discharge of foggaras. On the other hand, given the good performance of a drill, farmers prefer irrigation drilling at the expense of foggaras.

For example, one foggara gives on average one liter of water per second to a 2 km long draining gallery, while the drilling method in the same fields with a well of a few meters provides a ten times higher flow rate (Bisson, 1991). This difference that allows the Sahara to get a water flow without much greater physical energy encourages the oasis to abandon the foggaras whose number continuously declines. We no longer maintain the old galleries and build new drains. Maintaining a foggara is much more complicated than its realization. Today, it is difficult to proceed with the construction of new foggaras. There were more than 900 foggaras in the early 20th century, against 650 in 1996 (Kassah, 1998) and about 600 foggaras productive today (Kaoula, 2002). The total available discharge does not exceed 3 m^3/s.

4 CONCLUSION

As we mentioned above, the foggara is the oldest irrigation technique which remains in service today in the areas of Touat, Gourara and Tidikelt. Despite its age of 10 centuries, the foggara irrigates thousands of acres of a palm grove with the best dates in the world. Unlike drilling, the discharge of the foggara is directly related to the irrigated area. There is no deficit or surplus of water in the gardens. Despite the competition with drilling, the foggara continues to irrigate thousands of palm trees. It is virtually impossible to dig new foggaras, however we can maintain and protect the foggaras in service.

REFERENCES

[1] Arrus, R. 1985: Water in Algeria, University Publications Office, 388 p.

[2] Bisson J. 1991: The Sahara in the development of North African states. Arab Maghreb Mashreq No 134, October - December, Studies, pp. 3-17.

[3] Goblot H., 1979. Qanats: an acquisition water technique Mountain Edition, Paris, 231 p.

[4] Goblot H., 1963. In ancient Iran, the techniques of water and great story. Annales, Vol. 18, No 3, pp. 499-520.

[5] Hofman A., 2007. Traditional water management by qanat in Iran is compatible with the concept of "GIRE"? Technical Summary, February, Engref (Montpellier France), 17 p.

[6] Kaoula A. 2002: Foggaras. Environmental Atlas of Algeria. Symbiosis Edition, 105 p.

[7] Kazemi G.A., 2004. Temporal changes in the physical properties and chemical composition of the municipal water supply of Shahrood, north-eastern, Iran. Hydrogeology Journal, No12, pp. 723-734.

[8] Kessah A. 1998: Water and agricultural development in Maghreb Sahara: challenges, conflicts and tradeoffs. "Revue Secheresse", Vol. 9, No 2, June, pp.95-102.

20

THE USE OF FILTRATION MATERIALS TO REMOVE AMMONIA FROM WATER

Anna HANUSOVÁ [1]

[1] *Ing, Department of Sanitary and Environmental Engineering, Faculty of Civil Engineering, Slovak University of Technology*
Radlinského 11, 813 68 Bratislava
anna.hanusova@stuba.sk

Abstract

Ammonia nitrogen is the primary product of the decomposition of plant and animal organic nitrogenous materials. Ammonium salts are part of some nitrogen fertilizers and are getting into ground and surface waters from agricultural areas. Secondary ammonium compounds can be formed directly in waters by reducing nitrates. This article is focused just on removing ammonia from drinking water, while the current research emphasizes the use of natural materials, especially for the removal of inorganic contaminants from water. The aim of this work consists in verifying the possibility of using natural filter materials - zeolites in ammonia removal and other filter materials, verify sorption (ion exchange) properties and efficiency of available types of materials under laboratory conditions.

Abstrakt

Amoniak je primárnym produktom rozkladu rastlinných a živočíšnych organických dusíkatých látok. Amónne soli sú súčasťou niektorých dusíkatých hnojív a z poľnohospodárskej plochy sa dostávajú do podzemných a povrchových vôd. Sekundárne amóniové zlúčeniny môžu byť vytvorené priamo vo vodách redukciou dusičnanov. Tento článok je zameraný práve na odstránenie amoniaku z pitnej vody, pričom súčasný výskum kladie dôraz na použitie prírodných materiálov, najmä pre odstránenie anorganických kontaminantov z vody. Cieľom tejto práce bolo overiť možnosti použitia prírodných filtračných materiálov - zeolitov a iných filtračných materiálov pri odstraňovaní amoniaku z vody, overiť sorpčné (ionovýmenné) vlastnosti a účinnosť dostupných typov materiálov v laboratórnych podmienkach.

Key words: ammonia removal, zeolite, lewatit, filtralite, klinomangan, Read-As

1 INTRODUCTION

Considering currently legal requirements to ensure the supply of "safe drinking water" in terms of water treatment technology, it is necessary to optimize the water treatment process, thus to use modern technologies which can cope with changing water quality. In many cases, water treatment is the only possible way to supply the population with quality drinking water. The issue of water treatment, however, concerns of different industries, not only the problems of drinking water, but also water for industrial, pharmaceutical and other purposes. The article focuses on verifying the possibility of using selected filter materials in removing ammonia from water.

The study by Woo-Hang deals with the evaluation of the feasibility of ammonia removal by zeolite adsorption in drinking water treatment process. They tested the feasibility using two methods: powdered zeolite dosing to coagulation tank and substituting sand for granular zeolite in a sand filter. When the zeolite dosage was above 100mg l, the ammonia concentration decreased below 0.5mg l of NH_4^- -N in a powdered zeolite test, but the turbidity increased to 30NTU. In a granular zeolite test, ammonia was not detected in treated water until 8 days. This result suggest that the use of granular zeolite in sand filter could remove ammonia in winter [1].

Balci and DinÇel examined the ammonium removal from the solution using Turkish sepiolite having an average particle size of 2.05 mm. Ammonium uptake experiments for two different adsorbent quantities were performed for solutions with initial solution concentrations in a range of 8.333–222.23 mmol NH_4^- /l, and equilibrium isotherms were performed at 25°C. The removal of ammonium was observed around 60l , 90l of which were achieved within 500 s [2].

The objective of a Chmielewská study consisted in determining the feasibility at the laboratory and to evaluating a pilot potential in field operation for ammonium removal from drinking water with enhanced ammonium concentration by applying Slovakian clinoptilolite-rich tuff deposited nearby Nižný Hrabovec, a village in the Eastern Slovakia. Detailed technological parameters (specific and shipping weights, porosity, sphericity, abrasion) were determined for the Slovakian clinoptilolite-rich tuff and compared with some other foreign samples including common water treatment adsorption materials like active coke and silica sand [3].

It is necessary to note the researches dealing with wastewater treatment who work with higher concentrations of ammonia nitrogen, but with similar materials and the same principle. Several works related to the wastewater treatment using bentonite, sepiolite and clinoptilolite from many different deposits are cited in the literature. Suzuki and Ha [4], investigated the ion exchange and the equilibrium for NH_4^-–Na, NH_4^-–K, and NH_4^-–Ca_2 systems in the channels of clinoptilolite at room temperature. It was observed that clinoptilolite has a higher selectivity coefficient for ammonium relative to sodium and calcium ions. The effective diffusion coefficient for ammonium and sodium ions in clinoptilolite was observed at around 4–5×10−12 m^2 s. Bernal and Lopez [5] studied ammonia and ammonium removal using clinoptilolite and sepiolite. They showed that sepiolite also had a comparable ammonium removal capacity. Booker et al. [6] observed that the NH_4–N concentration decreased below 1 mg l during the treatment of sewage with the concentration ranging from 25 to 50 mg NH_4–N l using natural zeolite. Veigh and Weatherley [7] observed around 90l removal of influent ammonia in the continuous packed bed operation using clinoptilolite. Rozic et al. [8] obtained around 50l ammoniacal nitrogen removal at a low solution concentration using zeolite and natural clays. They also observed a decrease of removal efficiency upon increasing initial concentration.

The main aim of this work consists in verifying the possibilities of using natural filter materials - zeolites in ammonia removal and other filter materials, verifying sorption (ion exchange) properties and efficiency of available types of materials which were originally developed to remove other undesirable substances in the process of water treatment such as arsenic, heavy metals, reduction of water hardness under the laboratory conditions.

1.1 Occurrence and properties of ammonia

Simple ammonium salts do not form minerals in nature. They mostly come from secondary sources as wastes and from farming. Ammonia nitrogen is a primary product of the decomposition of animal and plant organic nitrogenous materials. Ammonium compounds can be formed secondarily in natural waters by a chemical reaction of nitrates and biochemical decomposition.

Its concentrations in natural water are below 0.2 mg.l^{-1}. Groundwaters associated with petroleum waters sometimes contain more than 100 mg.l^{-1} of ammonia nitrogen. This is explained by the decomposition of organic matter under anoxic conditions where nitrification processes cannot take place.

The term ammonium includes both unionized (NH_3) and ionized (NH_4) forms. Ammonia dissolution in water gives rise to hydrate $NH_3.H_2O$ which directly dissociated to NH_4^- and OH^- ions. In groundwater evaluation, ammonium is an indicator of possible groundwater pollution. Its values are particularly significant in association with a demonstrated bacteriological pollution, but it is very important to rule out in advance its inorganic origin or decomposition of nitrogenous organic matter of vegetable origin. Toxic effect of ammonia nitrogen on fish are considerably controlled by water pH, because it is not the NH_4^- ion which has toxic effect, but the undissociated ammonia which more easily penetrates through cell membranes. Ammonium is the principal component of mammal metabolism. Toxic effects have been noted only if its content attains about 200 mg.kg^{-1} of body weight. [9]

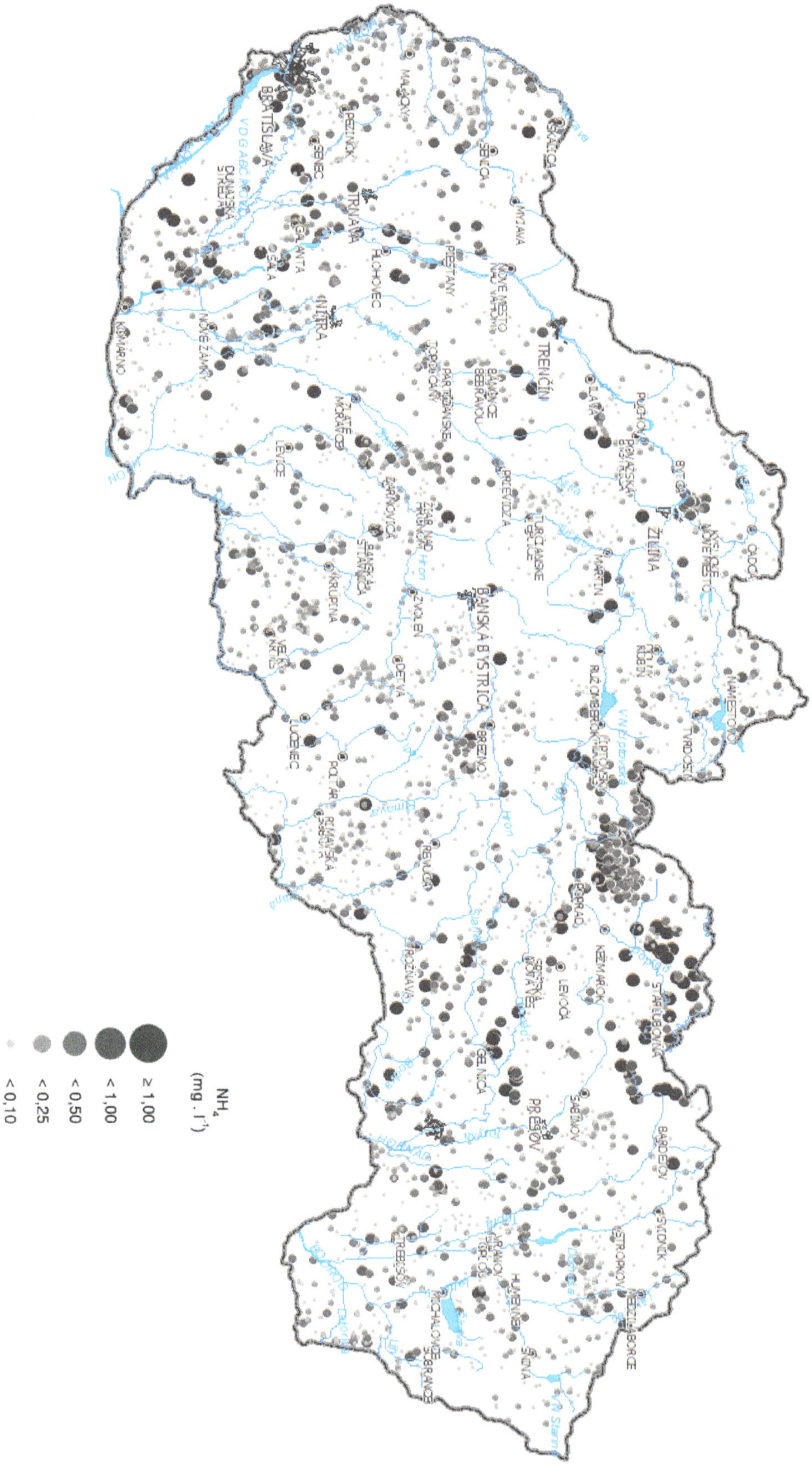

Fig. 1 Occurrence of ammonia SR [9]

WHO does not give its limit value. SR Government Regulation no. 496 2010 Coll. about *Drinking Water* permits the maximum NH_4^- concentration of 0.5 $mg.l^{-1}$. An average NH_4^- content in groundwater samples is 0.075±0.53 $mg.l^{-1}$ as shown in Fig.1. Ammonium in groundwater is fairly evenly distributed, 98l of samples contain it in an amount below the limit value permitted by the SR Government Regulation no. 496 2010 Coll. In high-altitude areas, called headwaters, notably in their summit section with little soil and biomass, winter precipitation is the chief source of NH_4^-. Elsewhere, however, the vast majority of occurrences are man-made-agriculture and missing sewerage are their main sources [9].

2 MATERIALS AND METHODS

For this experiment was produced the stock solution with the concentration of NH_4^- 6.8 $mg.l^{-1}$ at pH 7.84. To each beaker was dosed 60 ml the stock solution (model raw water) and 0.6 g of each material. The experiment was conducted in a stationary mode with occasional stirring at laboratory temperature 23.4 °C. The samples was collected after 1, 4, 12 and 24 hours. Based on previous research were selected following available materials with proven efficiency for removing ammonia nitrogen and other undesirable substances:

1. Zeolite - mined in Nizhny Hrabovec, supplier Zeocem, a.s. Bystré, Slovakia; particle size: 0.5 to 1.0 mm (Fig. 2)
2. Filtralite - developed in Norway, supplier Envi-Pur, s.r.o. Prague (Fig. 3)
3. Cation exchange resin - Ostion KS (Fig. 4)
4. Read-As - supplier Global Water Kft., Hungary(Fig. 5)
5. Zeolite - mined in Nizhny Hrabovec, supplier Zeocem, a.s. Bystré, Slovakia; particle size: 1.0 to 2.5 mm (Fig. 6)
6. Lewatit TP 207 - supplier SAMPLE Bratislava, s.r.o., Slovakia, producer Lanxess AG (Fig. 7)
7. Klinomangan -BBC-NA1000-0.5-1.2 - producer Geoproduct Kft., Hungary (Fig. 8)
8. Lewatit S 108 - supplier SAMPLE Bratislava, s.r.o., Slovakia, producer Lanxess AG (Fig. 9)

Fig.2 Zeolit 0,5-1,0 mm

Fig.3 Filtralite

Fig.4 Ostion KS

Fig.5 Read-As

Fig.6 Zeolit 1,0-2,5 mm

Fig.7 Lewatit TP 207

Fig.8 Klinomangan

Fig.9 Sk Lewatit S 108

Kinetics of adsorption was carried out in a stationary mode with occasional mixing at a constant room temperature of 23 °C, and was used 0.6 g of adsorbent and 60 ml of a model solution with the initial ammonia concentration of 6.8 mg .l^{-1}.

Zeolite (ZeoCem Eco) is used for drinking water treatment as a filter inset. It reduces material cost by saving the costs for chemicals. It reduces iron and manganese, antimony. Zeolites also effectively remove ammonium ions from drinking water and wastewater so they can be used both for chemical filtration and as a substrate for biological treatment. As a biofilter for water purification, they provide an appropriate environment for the growth of bacterial cultures which oxidize NH_4 to NO_3^-. The structure of clinoptilolite includes a three-dimensional grid which consists of Silicate tetrahedrite $(SiO_4)^{4-}$ each interconnected via oxygen atoms; the atoms of silicon is replaced by aluminum $(AlO_4)^{5-}$ way, creating a characteristic spatial structure with a significant incidence of cavities interconnected by channels, in which metal cations, or water molecules are stored. The total volume of these cavities is from 24 to 32 l [10].

Filtralite filter media is made by burning of clay at about 1200° C, followed by crushing and sieving. Filtralite are inert ceramic particles with a dense shell surrounding a porous core. The chemical and physical composition gives the beads several properties used for many different purposes included filter materials. Filtralite's large pore volumes and surface areas make it an ideal media for biological treatment of drinking water. Biofilm processes are excellent for treating raw water containing ammonia, manganese, iron etc. Tests have shown that Filtralite works perfectly in these kinds of processes. In conventional filtration processes Filtralite can be used in different ways. In dual media filters Filtralite can be used in combination with sand. Tests have shown that Filtralite operates as well as or better than traditional filter media in these types of filters. Filtralite can not only be used in traditional drinking water treatment, but also in treatment of saline water. In desalination plants one of the most important parts is the pretreatment process. Filtralite is an excellent filter media for use in pre-treatment filters in desalination plants [11].

READ-As was provided by the company Global Water KFT. The material, granulated ceric oxide, has been developed by the Nihin Kaisui Co., Ltd. Company in Japan. READ-As is a material produced for removal of arsenic from water under a large scale of conditions and it is sorbs efficiently both arsenate and arsenate (III). Neither oxidation of arsenate (III) to arsenate, nor adjusting the pH of water before and after sorption is needed. It is possible to regenerate this material by adding natrium hydroxide followed by sodium hypochlorite and washing it with water at the end. The regenerated material needs to be neutralized by hydrochloric acid and must be washed with water before the next use. Depending on the amount of sorbent and a chemical composition of water, the regeneration should be performed after 4-12 months. After 7-10 years, it is necessary to change the whole filling of the filter as the material could be dangerous [12].

Klinomangan-BBC-NA1000-0.5-1.2 is a natural filtration medium with high content of dispersed manganese dioxide. A special technique was used for binding of manganese oxide to the surface and in micropores of zeolite, thanks to which only a low concentration of colloidal particles is washed out while the medium is in water. Manganese is present in Klinomangan in two forms: in from of manganese oxide and in ionic form in the crystal structure of zeolite. The manganese ions bound in zeolite are replaceable, which means that they can be replaced by alkaline cations. They may occure in filtered water, if the salt concentration of the water is very high. Klinomangan is an excellent adsorbent of ammonium ions. The elimination of the ammonium ions, iron and manganese runs simultaneously in one technological step. The regeneration solution, NaCl (5l) is used. The regeneration by means of NaCl restores only the absorption capacity of ammonium ions. Then it should be followed by regeneration using the Perma preparation, which restores the original oxidative capacity of the medium [13].

Ion exchange resins (Lewatit) are used to remove dissolved particles from liquids. Therefore, their functional groups are bonded to polymer beads. These functionalized polymers adsorb particular anions or cations from the liquid and discharge others. Water treatment is the best-known and biggest field of application of ion exchange resins. In household, such applications are used, among others, to soften water. They ensure that the calcium ions responsible for water hardness are replaced by sodium ions. Ion exchange resins are also used in industry, for example in power generation. There, Lewatit is used in the production of ultra-pure boiler feed water and steam to avoid incrustations and corrosion. This improves the efficiency, operating reliability and lifetime of power plants. In the cleaning of industrial effluent and treatment of groundwater, the removal of toxic substances plays a key role. A range of special Lewatit resins are able to selectively remove, for example, heavy metal ions and organic pollutants from (waste)water. Other fields of application for ion exchange resins include semi-conductor production, galvanic industry (recovery and treatment of metals such as gold, copper, nickel or cobalt in hydrometallurgy), pharmaceuticals, and the food and beverage industry. Lewatit ion exchange resins are also reliable catalysts which are crucial for many chemical reactions [14].

3 RESULTS AND DISCUSSION

The experimental work weas focused mainly on the removal of ammonium ions from drinking water; the limit value for drinking water is 0.5 $mg.l^{-1}$. The results achieved by volumetric analysis method are presented in tabular form (Tab. 3.1-3.9).

Tab. 3.1 Concentration of ammonia after the adsorption in mg/l

Material	Ammonia [mg l]			
	1 h	4 h	12 h	24 h
Zeolite 0.5-1.0 mm	6.40	4.90	3.05	2.15
Filtralite	6.80	6.70	6.75	6.75
Ostion	3.65	1.45	1.25	0.50
Read-As	6.80	6.75	6.50	6.20
Zeolite 1.0-2.5 mm	6.55	5.25	4.80	3.70
Lewatit TP 107	6.80	6.60	6,.35	6.30
Klinomangán	6.60	5.10	2.85	2.10
Lewatit S 108	6.05	3.90	1.50	1.35

Tab. 3.2 Removal of ammonia from water by Zeolite 0.5-1.0 mm

Variable	Unite	*Zeolit 0,5-1,0 mm*			
t	[h]	1	4	12	24
c_m	[mg l]	6.40	4.90	3.05	2.15
η	[l]	5.88	27.94	55.15	68.38
a_t	[μg g]	40	190	380	470

Tab. 3.3 Removal of ammonia from water by Filtralite

Variable	Unite	*Filtralite*			
t	[h]	1	4	12	24
c_m	[mg l]	6.80	6.70	6.75	6.75
η	[l]	0.00	1.47	0.74	0.74
a_t	[μg g]	0.00	10	0.00	0.00

Tab. 3.4 Removal of ammonia from water by Ostion KS

Variable	Unite	*Ostion KS*			
t	[h]	1	4	12	24
c_m	[mg l]	3.65	1.45	1.25	0.50
η	[l]	46.32	78.68	81.62	92.65
a_t	[μg g]	320	540	560	630

Tab. 3.5 Removal of ammonia from water by Read-As

Variable	Unite	*Read-As*			
t	[h]	1	4	12	24
c_m	[mg l]	6.80	6.75	6.50	6.20
η	[l]	0.00	0.74	4.41	8.82
a_t	[μg g]	0.00	0.00	30	60

Tab. 3.6 Removal of ammonia from water by Zeolite 1.0-2.5 mm

Variable	Unite	*Zeolit 1.0-2.5 mm*			
t	[h]	1	4	12	24
c_m	[mg l]	6.55	5.25	4.80	3.70
η	[l]	3.68	22.79	29.41	45.59
a_t	[μg g]	30	160	200	310

Tab. 3.7 Removal of ammonia from water by Lewatit TP 107

Variable	Unite	*Lewatit TP 107*			
t	[h]	1	4	12	24
c_m	[mg l]	6.80	6.60	6.35	6.30
η	[l]	0.00	2.94	6.62	7.35
a_t	[μg g]	0,00	20	50	50

Tab. 3.8 Removal of ammonia from water by Klinomangan

Variable	Unite	*Klinomangán*			
t	[h]	1	4	12	24
c_m	[mg l]	6.60	5.10	2.85	2.10
η	[l]	2.94	25.00	58.09	69.12
a_t	[μg g]	20	170	400	470

Tab. 3.9 Removal of ammonia from water by Lewatit S 108

Variable	Unite	*Lewatit S 108*			
t	[h]	1	4	12	24
c_m	[mg l]	6.05	3.90	1.50	1.35
η	[l]	11.03	42.65	77.94	80.15
a_t	[μg g]	80	290	530	550

The instantaneous adsorption capacity at (1) and adsorption efficiency η (2) is calculated according to the formulas:

$$a_t = \frac{(c_o - c_m)V}{m} \qquad (1)$$

$$\eta = \frac{(c_o - c_m)100}{c_o} \qquad (2)$$

where:

c_o – concentration of ammonia before adsorption (6.8 mg l) $\left[\frac{mg}{l}\right]$,

c_m – concentration of the ammonia after adsorption at the time t $\left[\frac{mg}{l}\right]$,

V – volume of the aqueous solution (0.060 l) $[l]$,

m – weight of sorbent material (0.6 g) $[g]$,

η – adsorption efficiency [l],

a_t – instantaneous adsorption capacity; adsorbed amount per unit weight of sorbent material at time t $\left[\frac{mg}{g}; \frac{\mu g}{g}\right]$,

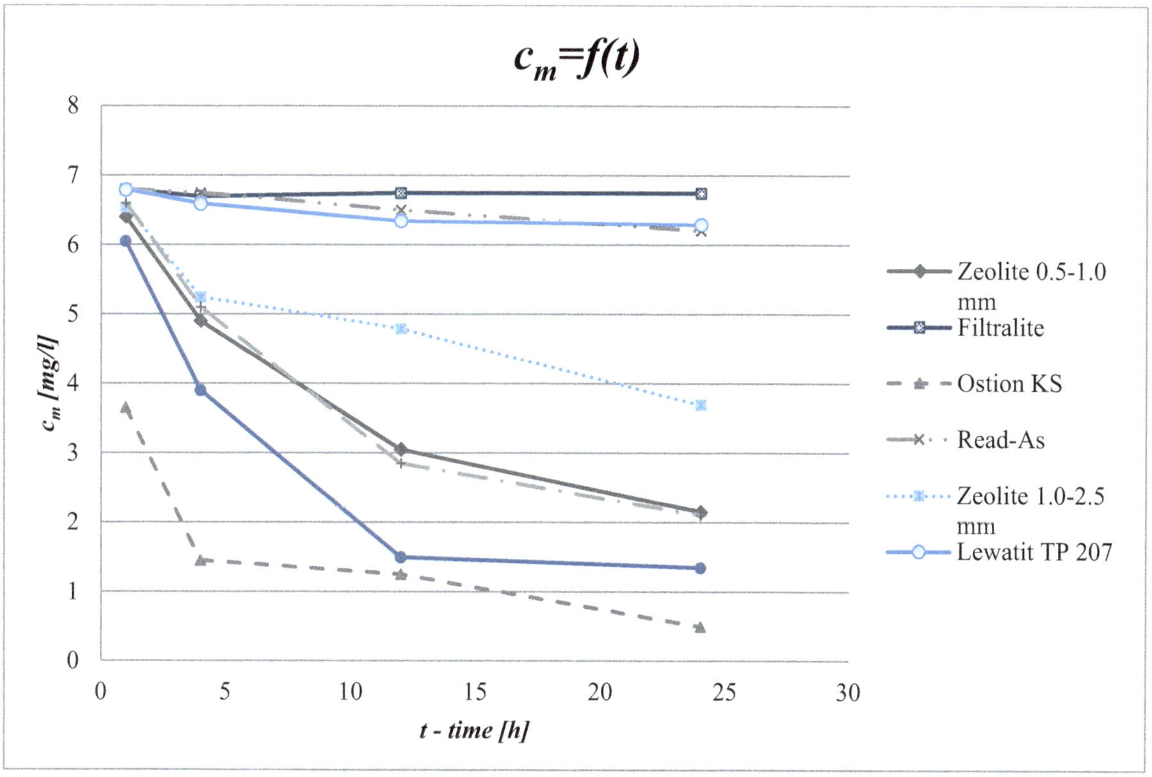

Fig.10 Dependence of ammonia concentration from adsorption time

As shown in Tab.3.4 and Tab.3.9, the highest adsorption capacity for ammonia removal was achieved by the cation exchangers Ostion KS and Lewatit S 108 which presented their ion exchange properties as they are high molecular compound (polymer resin) able to exchange sodium for different cation with high efficiency.

Klinomangan and natural adsorbent zeolite clinoptilolite - mined in Nižný Hrabovec, are the second most effective adsorbent depending on their particle size; a finer particle size has higher efficiency (Tab. 3.2, Tab. 3.8). Broad application possibilities of zeolites (e.g. in construction, agriculture, water purification and treatment, pharmacology) are the result of their unique physicochemical properties such as ion exchange and sorption capacity, relatively large specific surface or economic access. The three-dimensional network of the tetrahedron structure of SiO_4^{4-} and AlO_4^{5-} and the permanent negative charge in the structure makes it possible to bind exchangeable cations [15, 16, 17].

The graph (Fig.10) shows the change in ammonia concentration with adsorption time, which means that the concentration of both substances decrease with time during the ongoing adsorption. It is expected that under a longer influence of adsorbent in a model solution, the ammonia concentration will not significantly decrease in the case of non-active materials (it follows from the plot c_m^{-} f (t)) and sorption properties of these selected materials will be exhausted under the given laboratory conditions. In case of an active material, the dependence c_m^{-} f (t) shows a constant decrease in the ammonia concentration unless the sorption properties of the material are exhausted.

It is necessary to emphasize the inclusion of the cation exchangers in the experiment as a reference material due to its ion exchange properties and proven efficiency. From previous research and from the actual experiment, it is clear that the use of zeolites and their modified forms achieves high efficiency in the removal of ammonia (and other undesirable substances). In the case of materials, primarily developed for the removal of different undesirable substances from water than ammonia, their inclusion in the experiment didn't prove validity.

4 CONCLUSIONS

The study of adsorption properties of selected natural and synthetic adsorbents was performed in order to determine and compare the adsorption efficiency of the NH_4^{-} removal from a model solution in the stationary laboratory experiment. It is clear, from the obtained results, that several of the materials can be used for ammonia removal; natural materials - clinoptilolite zeolites are able to remove NH_4^{-} with an efficiency of about 45-68l , klinomangan with efficiency of about 69l and cation exchangers with their ion exchange properties with the efficiency of about 80-92l , however not to the legislative required value 0.5 mg l^{-1}. Other tested materials did not demonstrate the efficiency.

The next task will be to study the influence of selected parameters on the adsorption process in water treatment and transformation of the stationary experiments into dynamic regime consisting of a set of experimental columns, 5 cm in diameter.

ACKNOWLEDGEMENT

Experimental measurements were carried out with the financial support of the VEGA 1 1243 12 research project from the Scientific Grant Agency of the Slovak Ministry of Education, Science, Research and Sport.

REFERENCES

[1] WOO-HANG, K., CHUNG-HWAN, K. *Zeolite filtration for ammonium nitrogen removal in drinking water treatment.* Journal of environmental science international, Vol.12, Issue 3, 2003, pp.281-286. ISSN: 1225-4517

[2] BALCI, S., DINÇEL, Y. *Ammonium ion adsorption with sepiolite: use of transient uptake method* Chemical Engineering and Processing 41, 2002, pp. 79-85

[3] CHMIELEWSKÁ, E. *Designing clinoptiloliterich tuff columns for adsorptive filtration of water with enhanced ammonium concentration* Fresenius Environmental Bulletin, 23 (5), 2014, pp. 1277-1283.

[4] SUZUKI, M., HA, K.S. *Equilibrium and rates of ammonium ion exchange by clinoptilolite.* Jpn. J. Chem. Eng. 17, 1984, pp. 139-144.

[5] BERNAL, M.P., LOPEZ-REAL, J.M. *Natural zeolites and sepiolites as ammonium and ammonia adsorbent materials.* Bioresour. Technol.43, 1993, pp. 27-33.

[6] BOOKER, N.A., COONEY, E.L., PRIESTLEY, A.J. *Ammonia removal from sewage using natural Australian zeolite.* Water Science Technology 34, 1996, pp. 17-24.

[7] VEIGH, R.M., WEATHERLEY, L. *Ammonium ion (NH_4^+) removal from secondary effuluent through ion-exchange: the effect of biological activity and the presence of other cations.* Water Science Technology 40, 1999, pp. 143-149.

[8] ROZIC, M., CERJAN-STEFANOVIC, S., KURAJICA, S., VANCINA, V., HODZIC, E. *Ammoniacal nitrogen removal from water by treatment with clays and zeolites.* Water Resource 34, 2000, pp. 3675-3681.

[9] RAPANT, S., VRANA, K., BODIŠ, D. *Chemical atlas of Slovakia - Groundwater.* Bratislava: SLOVSPOL Košice, a.s., Geologická služba Slovenskej republiky, 1996, pp.22-23. ISBN 80-85314-67-3.

[10] http: www.zeocem.com about-zeolite

[11] http: www.filtralite.com 26225.html

[12] http: globalwater.hu index.php globalwater oldal termekek.html

[13] http: www.zeoclay.eu domain zeoclay files klinomangan-1000-info.pdf

[14] http: lewatit.com

[15] HODOSSZOVÁ, R., CHMIELEVSKÁ, E., GÁPLOVSKÁ, K. *Adsorpcia fosforečnanov z modelových roztokov na vybraných prírodných a syntetických adsorbentoch*, Waste Forum, pp. 126 – 131.

[16] KOZÁČ, J. *Výskum základných vlastností a úpravy zeolitového tufu Nižný Hrabovec*. Košice: Výskumná správa GS AT NS, 1989.

[17] SABOVÁ, L., CHMIELEVSKÁ, E., GÁPLOVSKÁ, K.. *Príprava a využitie kombinovaných adsorbentov na zeolitovej báze pri odstraňovaní oxy-aniónových polutantov z vôd.* Chemické listy 104, 2010. pp. 243 – 250.

[18] http: bigcichlids.6f.sk voda-chemicke-zlozenie-vody.p93.html

RESUMÉ

Vzhľadom na dnes oprávnenú požiadavku zaistenia dodávky „bezpečnej pitnej vody" z hľadiska technológie úpravy vody, je potrebné optimalizovať proces úpravy vody, resp. začať prevádzkovať nové moderné technológie, ktoré sa vedia s meniacou sa kvalitou vody vysporiadať. V mnohých prípadoch je úprava vody jediným možným spôsobom, ako zásobovať obyvateľstvo kvalitnou pitnou vodou. Problematika úpravy vody sa však týka rôznych odvetví, nielen problematiky pitnej vody, ale aj vody pre priemyselné, farmaceutické a iné účely. Tento článok je zameraný na overenie možnosti využitia vybraných filtračných materiálov pri odstraňovaní amoniaku z vody.

Amoniakálny dusík je primárnym produktom rozkladu rastlinných a živočíšnych organických dusíkatých látok. Amónne soli sú súčasťou niektorých dusíkatých hnojív a z poľnohospodárskych plôch sa dostávajú do podzemných a povrchových vôd. Sekundárne môžu amónne zlúčeniny vznikať priamo vo vodách redukciou dusičnanov. Tento jav nastáva v podzemných vodách vo väčších hĺbkach, ktoré obsahujú zvýšené koncentrácie železa a mangánu, prípadne aj sulfán a jeho iónové formy. Amoniakálny dusík obsahujú odpadové vody z plynární, koksární a galvanizovní. Môžu ho obsahovať aj pitné vody dezinfikované chlóramináciou. Do atmosférických vôd sa dostáva z priemyselných exhalátov. Pri rozpúšťaní amoniaku vo vode vzniká hydrát $NH_3.H_2O$, ktorý priamo disociuje na ióny NH_4^+ a OH^-. Zvyšujúca teplota podporuje disociáciu NH_4^+ na NH_3. Podzemné vody a čisté povrchové vody majú koncentráciu dusíka ($NH_3 + NH_4^+$) asi do 0,1 $mg.l^{-1}$. Väčšie koncentrácie dusíka v podzemných vodách sa vysvetľujú chemickou redukciou dusičnanov, pokiaľ nie sú organického pôvodu. [18]

Amoniak môže byť detegovaný v pitnej vode prostredníctvom série dobre zavedených metód. Voľba použitej metódy bude závisieť od prítomnosti interferencií. V komunálnych odpadových vodách, môže byť amoniak odstránený pomocou biologického čistenia (riadená nitrifikácia) a fyzikálno-chemických procesov, ako je breakpoint chlorácia, iónová výmena a membránová filtrácia.

V rámci experimentu boli testované viaceré materiály: zeolity (Zeocem, a.s. Bystré) Filtralite (Envi-Pur, s.r.o.), ionexy Lewatit (Lanxess AG), Read-As (Global Water Kft.) a Klinomangán (Geoproduct Kft.). Štúdium adsorpčných vlastností na vybraných prírodných a syntetických adsorbentoch bolo vykonané s cieľom zistiť a porovnať ich adsorpčné účinnosti pri odstraňovaní NH_4^+ z modelového roztoku v laboratórnych podmienkach, zatiaľ len stacionárnym pokusom. Zo získaných výsledkov je zrejmé, že niektoré materiály sú použiteľné pri odstraňovaní amoniaku; prírodný materiál - zeolity typu klinoptilolit sú schopné odstraňovať NH_4^+ s účinnosťou cca 45-68%, klinomangán s účinnosťou cca 69% a katexy so svojimi ionovýmennými vlastnosťami s účinnosťou cca 80-92%, avšak nie na legislatívne požadovanú hodnotu 0.5 $mmol.l^{-1}$. Ostatné testované materiály nepreukázali účinnosť.

EVALUATION OF SURFACTANT PERFORMANCE FOR ENHANCED OIL RECOVERY

Michal PORZER [1], Petr BUJOK [2], Martin KLEMPA [3], Petr PÁNEK [4]

[1] *Ing, Institute of Geological Engineering, Faculty of Mining and Geology, VŠB – Technical University of Ostrava*
17. listopadu 15, Ostrava-Poruba
e-mail michal.porzer@vsb.cz

[2] *prof. Ing. CSc. Institute of Geological Engineering, Faculty of Mining and Geology, VŠB – Technical University of Ostrava*
17. listopadu 15, Ostrava-Poruba
e-mail petr.bujok@vsb.cz

[3] *Ing, Institute of Geological Engineering, Faculty of Mining and Geology, VŠB – Technical University of Ostrava*
17. listopadu 15, Ostrava-Poruba
e-mail martin.klempa@vsb.cz

[4] *doc. Ing. CSc. Department of Chemistry, Faculty of Metallurgy and Materials Engineering, VŠB – Technical University of Ostrava*
17. listopadu 15, Ostrava-Poruba
e-mail petr.panek@vsb.cz

Abstract

This paper focuses on the field of enhanced oil recovery by means of a chemical flooding of oil deposit especially a surfactant flooding method. The main objective is the application of the aforementioned method at the Czech oil deposit Ždánice – Miocene which bears the crude oil of significant viscosity and gravity that does not allow conventional production methods to be used. We evaluated the performance of various surfactants in the laboratory environment by simulating oil recovery processes.

Abstrakt

Tento výzkum se zabývá problematikou zvyšování výtěžnosti ropy pomocí rozšířeného zavodňování chemikáliemi, z nichž je věnována největší pozornost povrchově aktivním látkám. Dlouhodobým hlavním cílem tohoto výzkumu je aplikace těchto metod v podmínkách české republiky, konkrétně na ložisku Ždánice – miocén, kde vlastnosti obsažené ropy, jako je viskozita a objemová hmotnost, neumožňují použití klasických těžebních metod. Pro vhodnost použití jsme v laboratorních podmínkách simulovali procesy probíhající při těžbě ropy za ložiskových podmínek a hodnotili vliv jednotlivých povrchově aktivních látek na výtěžnost těžké ropy.

Key words: enhanced oil recovery, surfactant flooding, heavy crude oil, Ždánice – Miocene

1 INTRODUCTION

The industry of oil and gas production experiences a significant evolution due to the last decade's improvements in the exploration for new resources, the characterisation of hydrocarbons reservoirs, the completion of production wells, and the production engineering itself. Mathematical modelling and laboratory testing methods are more frequently utilized for success assurance and risk reduction during the production projects. [1]

Significant advances were achieved also in the production of previously hard-to-get oil, i.e. very viscous oil and unrecoverable remaining oil in place. Currently, the oil recovery varies from 30 to 40 % in case of a convenient reservoir and geological conditions. With more challenging geological features in the reservoir or unfavourable formation fluid properties, the oil recovery decreases dramatically to the 10 – 5 % or even less. The improved and enhanced oil recovery methods (IOR and EOR) are in the main focus of production engineers. [2]

Vast majority of conventional and convenient oils, in terms of production difficulty, is already explored, produced, or also depleted, thus the main focus of production companies inclines towards offshore projects, unconventional resources and previously unrecovered oils, the resources of which are considered to be of great importance. Current economical market conditions with the last 3-year growth of price per barrel enables to increase investments in production projects, and therefore the previously expensive technologies like the chemical flooding become to be implemented on a regular basis.

At our facilities of the Department of Geological Engineering, the heavy oil production and enhancement of oil recovery research is based on a long-term cooperation agreement with the Czech oil company Moravské Naftové Doly, a.s. (MND). The hydrocarbons located at the deposit Ždánice – Miocene which is operated by MND are considered to be very promising, especially the 1st lower miocene horizon where the greatest volume of oil and gas in place (ca 60 % of Ždánice – Miocene deposit) is contained. The production attempts at this particular horizon, however, showed significant difficulties caused by the formation crude oil properties. This fact brought the issue of the proper production method possibilities under the discussion. [3, 4]

The three hydrocarbon deposit horizons were located in the lower Miocene sediments and were named the 1st, the 2nd and the 3rd horizon. The fundament of the Neogene formations belongs to the Brunovistulian terrane and is formed by fractured granitoids which are locally mylonitized. The deposits of the 1st and the 2nd horizons are isometric with a maximum length of ca 2 km and 1.2 km in width. The depth of these horizons varies from 700 to 1000 m under the ground surface (Fig.1). [5]

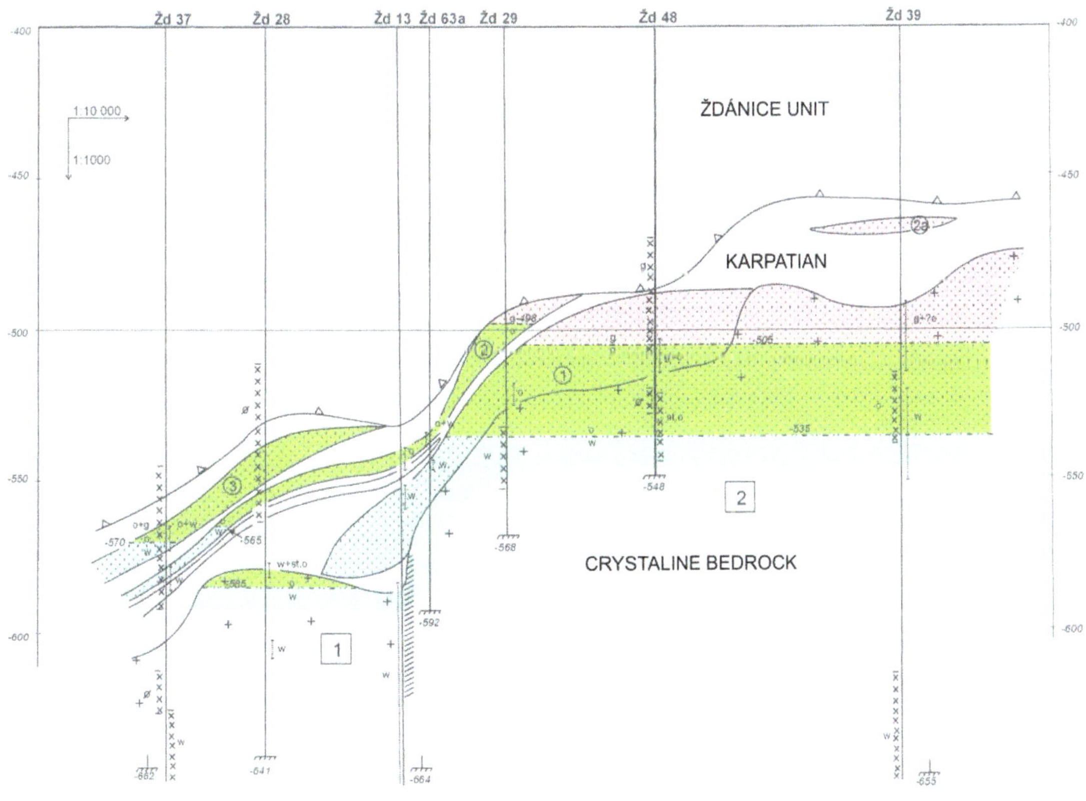

Fig. 1 The Ždánice Miocene deposit (circled numbers 1-3 mark the 1st to the 3rd lower Miocene horizons) [3]

The oil, which the 1st lower Miocene horizon bears, is classified as paraffinic-napthetic, very viscous (ca 250 mPa·s at 20 °C) with a heavy fraction content and with the gravity of about 21° API. The oil bearing zone in the fractured crystalline fundament is about 30 m thick. The reservoir formations of basal Miocene clastic sediments which are of sandstone to conglomerate character have the porosity of 10 – 20 % and the permeability ranging from 81.06 to 202.65 mD. The formation brine is classified as strongly mineralized with an average mineralization of 11996.87 mg/l. [3, 6]

The original formation pressure on the gas – oil contact (-505 m) was 7,088 MPa and the formation temperature was 29.2 °C. The original oil in place was assessed to be 8.944 million bbl; associated gas: 30.29 million m^3; natural gas in the gas cap: 63.604 million m^3. [3, 6]

Conventional methods for producing the crude oil (i.e. downhole pumps, etc.) are considered ineffective for the cases of very viscous oil recovery. Therefore, the utilization of EOR methods seems appropriate. Extensive mathematical modelling was applied to the evaluation of various EOR methods effectiveness at the 1st lower Miocene horizon. Predictive modelling was conducted for following production scenarios: water flooding, application of polymers, application of chemicals (i.e. surfactants, alkali, acids, and solvents), steam injection, *in situ* combustion, and carbon dioxide injection. [7]

The aforementioned predictive modelling evaluated the application of surfactants during the water flooding as the most effective production scenario [7]. This EOR method was previously the research objective at the Section of Exploration Technology (SET) under the Department of Geological Engineering, Faculty of Mining and Geology, VŠB – Technical University of Ostrava. Laboratory experiments in this field of interest were conducted on the MAF (Filtration Measurement Apparatus).

The analysis of the surfactant utilization in the EOR for the Ždánice crude oil case was proposed to be conducted at facilities of the SET and the Institute of Clean Technologies and Raw Materials Utilization (ICT) after the consultation with MND employees.

2 BASIS OF CHEMICAL FLOODING EOR METHOD

The history of oil recovery by means of chemical flooding reaches as far as 70 years of today with the first study patented by Detling (Shell) in 1944. The basic principle is the dynamic interaction of flooding agents with crude oil which is generally seized by the porous media of reservoir formation. The success of flooding operation depends especially on reservoir formation properties (i.e. anisotropy, wettability, porosity, permeability, fracture/pore geometry, etc.), formation fluids properties, injected slug characteristics. The properties of injected slugs are modified according to the reservoir conditions by a wide variety of chemicals. [1]

The chemicals used for flooding slug design are of a various character: surfactants, polymers, alkali, foaming agents, and acids. According to the aforementioned reservoir conditions, the injected fluid could be designed as a single component or multiple component solution. The surfactant flooding slug generally consists of water, surfactant, electrolyte, and solvents. The additional oil recovery achieved by this method could be up to approximately 30 % of residual oil in place. [8]

The main processes which enable additional oil recovery during chemical flooding are: 1) reduction of fluids interfacial tension; 2) oil dissolving; 3) crude oil emulsification into a mobile fluid phase; 4) formation rock wettability alteration; 5) improvement of water – oil mobility ratio. [8]

2.1 Monitoring effectiveness of the flooding agents for EOR under laboratory conditions

Laboratory experiments as well as mathematical modelling implemented during the search for a solution of the real field problems tend to yield not only important information, but also a certain degree of uncertainty caused by the necessary simplification of the real process as delivering the actual physical or mathematical model representing the real conditions is nearly impossible. The effort should be, therefore, made towards monitoring the single factor at the time instead of monitoring the whole heterogeneous and complex system at once. With this thought in mind, the laboratory simulations of surfactant flooding process were conducted.

In order to properly conduct the experiments, the following objectives and order of tasks were set:

- Selecting proper porous media representing the reservoir formation rock and selecting suitable fluids for injection slug design
- Conducting gravitation filtration experiments for further selecting applicable fluids
- Extended filtration experiments with an induced pressure thanks to the MAF VII apparatus and variety of fluids (i.e. brine, surfactant)
- Filtration EOR experiments simulating reservoir pressure and temperature conditions thanks to the BRP 350 apparatus
- Evaluating the best performing fluids

2.2 Selection of oil bearing formation rock model and selection of fluids

The real formation core sample of the 1st Miocene horizon was not obtainable; therefore, the rock model had to be utilized. The main requirements were the effortless obtainability, intergranular hydraulically connected porosity and chemical inertness to the process fluids. For the first unconsolidated rock model design, we selected three fractions of silica sand (a grain diameter of 0.8-1.2 mm, a porosity of 36.2 %).

The second consolidated sample type was decided to be of Hořice sandstone (permeability of 213.04 mD at a 15.9MPa overburden pressure; a 7MPa pore pressure and a porosity of 27.83 % at a 15.9MPa overburden pressure). This sandstone is formed mainly by quartz clasts, with slight amount of feldspar clasts and clay grain cement. The grain diameter varies from 0.07 up to the 0.38 mm. [9, 10]

The base fluids for EOR experiments were the crude oil and brine provided by MND, water, and as the process fluids, four types of surfactants with process concentrations in water of 5 and 15 g/l (Tab. 1) were selected.

Tab. 1 The selected process surfactants and their abbreviations (* commercially distributed surfactants were selected intentionally to assess their effect compared to Slovanik 3040 which is utilized in the oil industry)

Surfactant	Abbreviation
Slovanik 3040	PAL 1
Henkel – Pur*	PAL 2
Jar*	PAL 3
Tesco*	PAL 4

3 FILTRATION EXPERIMENTS EVALUATION

The vast number of surfactants and their concentrations in process slug had to be reduced by inexpensive, less time consuming means as all slugs could not be tested in induced filtration experiments under reservoir conditions. As a first screening option, the gravitational filtration experiment was selected.

3.1 Gravitational filtration experiments for surfactant evaluation

The measurement consisted of placing the unconsolidated rock model into the gravitational filtration testing apparatus (TARGF) (Fig. 2). The rock model was saturated with crude oil provided by MND. Then the secondary method of water flooding with the brine was simulated. Subsequently, the surfactants (PAL 1 – 4) were used for oil displacement from the rock model.

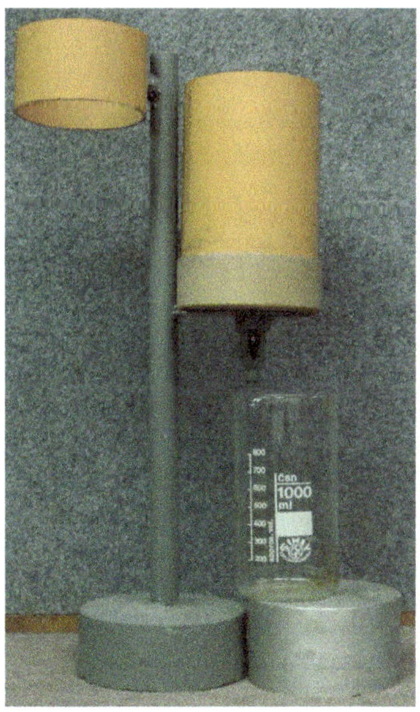

Fig. 2 Gravitational filtration testing apparatus (TARGF)

The resulting filtration curves of recovered oil after the certain volume of slug injected were produced. The residual oil saturation and the primary recovery effectiveness was used to set succeeding steps. This experiment reduced the amount of surfactants to the two with the best performance which were selected for additional testing. The best oil recovery after the injection of 3 pore volumes of process fluid was held by the surfactants PAL 1 and PAL 3 with 89.2 % and 76.6 % respectively. The worst performing by means of oil recovery was showed by the brine (9 %) (Fig. 3).

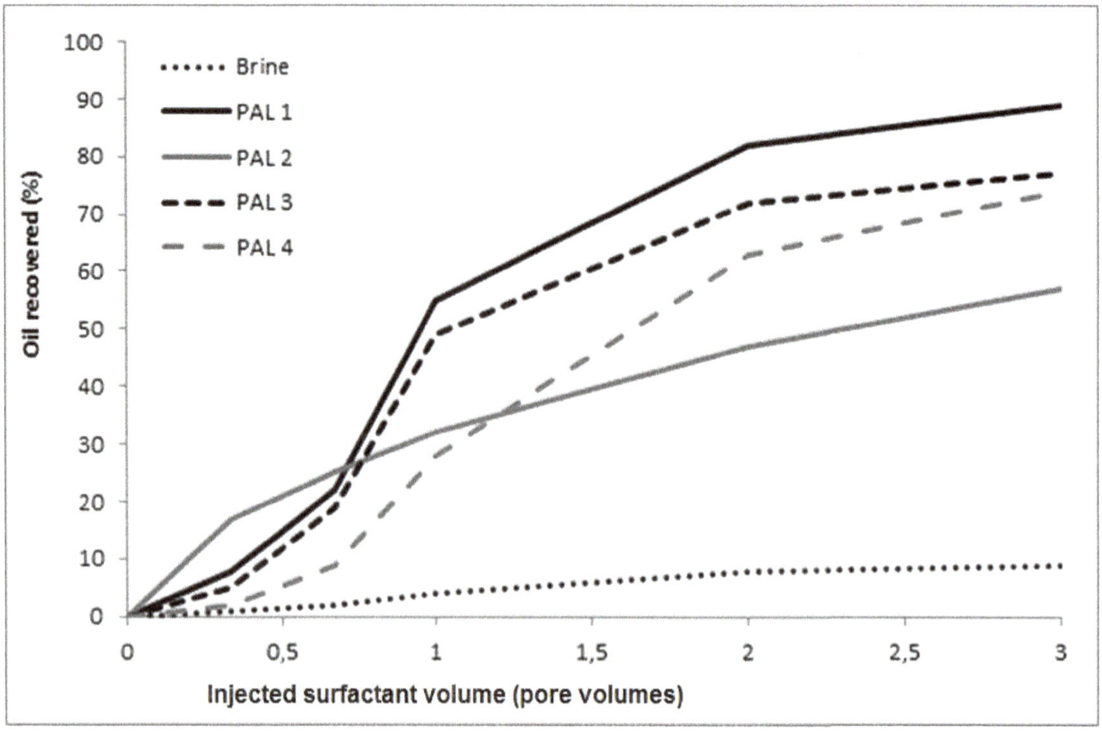

Fig. 3 Graphical record of oil production to the injected fluid volume for PAL 1 – 4 and brine

3.2 Surfactant evaluation by means of experiments under reservoir conditions

One of the previously best performing surfactants PAL 3 was selected for oil displacement filtration under the reservoir conditions thanks to the BRP 350 apparatus (Fig. 4) at the facilities of the Laboratory of Wells and Hydrocarbon Deposits Stimulation, under the ICT. The second type of rock model was utilized in the form of the core plug which has 37.4 mm in diameter and 54.1 mm in length. The sample was originally saturated with brine and subsequently put under a confining pressure of 15.9 MPa and a pore pressure of 7 MPa at a laboratory temperature of 23 °C. The drainage of brine with oil was conducted afterwards to put the sample into the most representative saturation state. The PT and saturation conditions very well represented the in situ conditions of the 1st lower Miocene horizon of the Ždánice Miocene deposit.

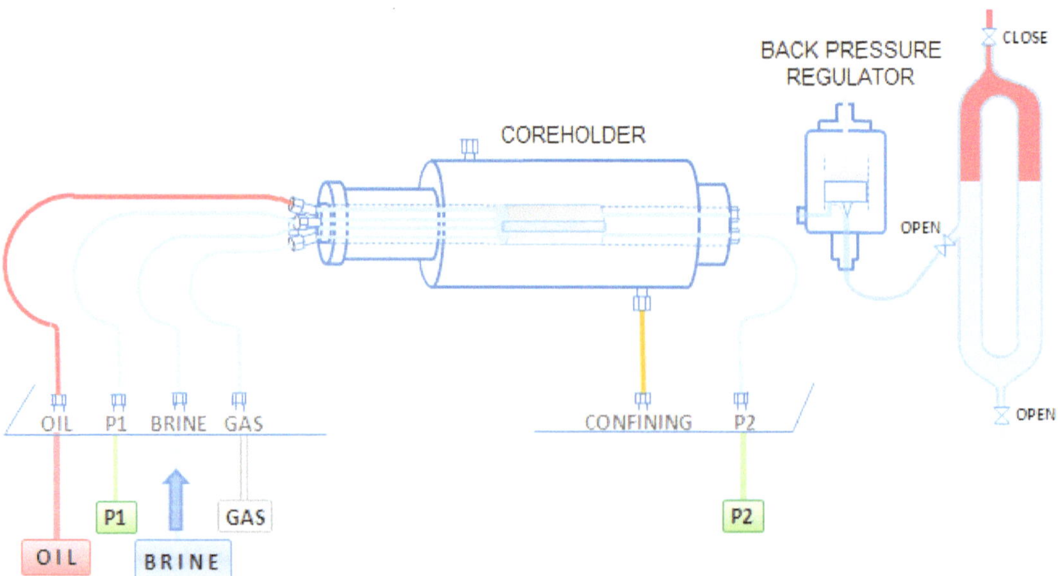

Fig. 4 Scheme of the BRP 350 apparatus (OIL, BRINE, GAS – lines for injection; P1, P2 – lines for inlet and outlet pressure measurements respectively; CONFINING – line for confinement of core sample; BACK PRESSURE REGULATOR – for regulating pore pressure)

The imbibition by brine injection was then performed to simulate a secondary water flooding method of production. The oil recovery at exact pore volume of brine injected was measured, and after the oil volume level stabilization in measurement burette, the recovery by the brine injection was considered done. The oil recovery by brine imbibition was 23.87 %, if the initial oil saturation after the 3.8 pore volumes of brine were injected.

Subsequently, the imbibition by surfactant slug was conducted and the produced oil, in other words the additional oil recovery, was measured. The resulting filtration curves of recovered oil after the certain volume of slug injected were produced (Fig. 5). The slug consisted of brine and 15 g/l PAL 3 concentration held the additional oil recovery at 30.23% residual oil saturation, which means 54.10% recovery of initial oil in place.

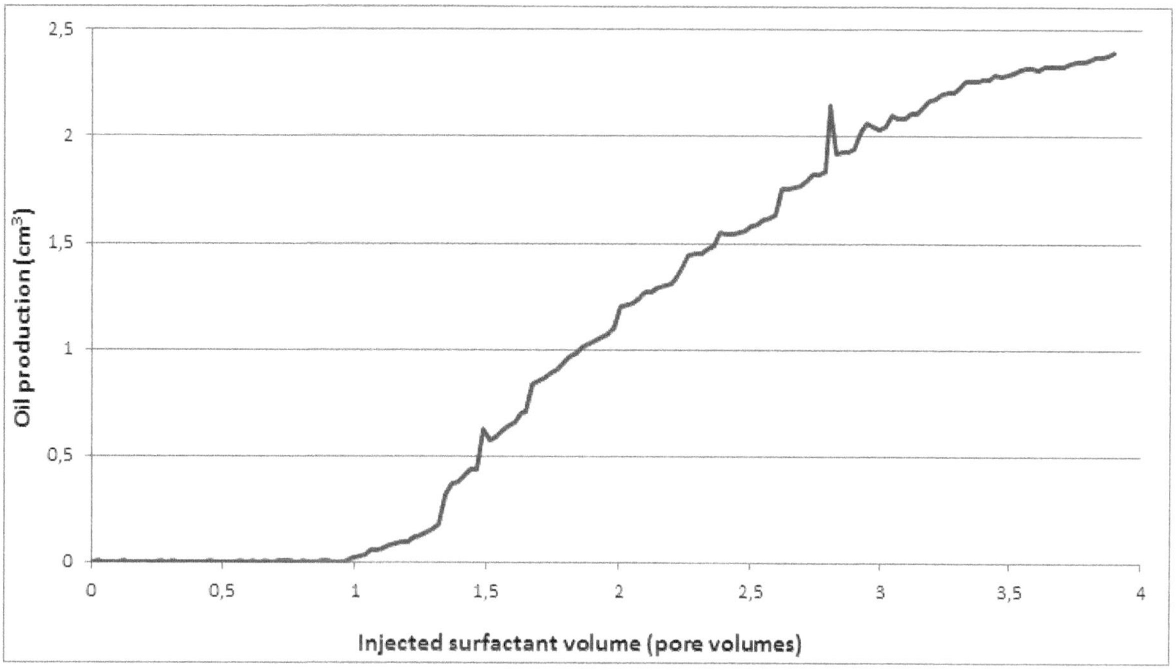

Fig. 5 The filtration curve of recovered oil after the certain volume of slug injected

4 CONCLUSION

The percentage of recovered oil by surfactant injection is quite high and tempting, therefore, this EOR method seems appropriate, however much more testing has to be done in this particular case. The obtained results so far will be used as the screening method for reducing possible injection fluids variety to the most effective one. The surfactants PAL 1 and PAL 3 will be evaluated by the filtration experiments under the reservoir conditions by the utilization of the MAF and BRP 350 apparatuses.

ACKNOWLEDGEMENT

The article has been done in connection with the project of the Institute of Clean Technologies for mining and utilizing raw materials for energy use, reg. no. CZ.1.05/2.1.00/03.0082 supported by the Research and Development for Innovations Operational Programme financed by the Structural Founds of the European Union and from the means of the state budget of the Czech Republic.

REFERENCES

[1] GREEN, D. W. WILLHITE, G. *Enhanced Oil Recovery*. 2st ed. Texas, 2003. 545 pp. ISBN 1-55563-096-0.

[2] HIRASAKI, G. & MILLER, C. & PUERTO, M.. *Recent Advances in Surfactant EOR*. SPE Journal. 2011, vol. 16, issue 4, s. 889-907. DOI: 10.2118/115386-PA.

[3] BOŠÁCKY, P. *Ložisková simulácia – ťažby na ložisku Ždánice*. Bratislava, SPP 02 Výzkum a vyhledávanie nafty a plynu, Bratislava, 1997.

[4] DOHNAL, K. & HAMRŠMÍD, B. & ZÁDRABA, M. Intraformační slepenec v sedimentech spodního miocénu v oblasti Ždánického lesa. In *Nové výsledky v terciéru Západních Karpat II*. Knihovnička zemního plynu a nafty 1995.

[5] DUDEK, A. *The crystalline basement block of the Outer Carpathians in Moravia: Bruno-Vistulicum*. Praha : Čs.Akad. Věd, 1980.

[6] BOŠÁCKY, P. *Ložisková simulácia primárnej ťažby a aplikácia druhotných ťažobných metód na ropných ložiskách Ždánice – miocén a Kloboučky*. Bratislava, SPP 02 Výzkum a vyhledávanie nafty a plynu, Bratislava, 1997.

[7] BOŠÁCKY, P. *Optimalizácia ťažby ropy na ložisku Ždánice 1. spodní miocén*. Final Report, EURO-GEOLOGIC, a.s., Bratislava, 1997.

[8] SHENG, J. *Modern Chemical Enhanced Oil Recovery: Theory and Practice*. 1st ed. Burlington : Gulf Professional Publishing, USA 2011. ISBN 978-1-85617-745-0.

[9] HALATOVÁ, M. *Laboratorní stanovení petrofyzikální charakteristiky vybraných vzorku vrtného jádra*. Bachelors thesis.. Ostrava : VŠB-TU Ostrava, 2013.

[10] VONDROUŠOVÁ, S. *Petrografie a využití hořických pískovců*. Bachelors thesis.. Ostrava : VŠB-TU Ostrava, 2009.

RESUMÉ

Tento výzkum se zabývá problematikou zvyšování výtěžnosti ropy pomocí rozšířeného zavodňování chemikáliemi, z nichž je věnována největší pozornost povrchově aktivním látkám. Dlouhodobým hlavním cílem tohoto výzkumu je aplikace těchto metod v podmínkách české republiky, konkrétně na ložisku Ždánice – miocén, kde vlastnosti obsažené ropy, jako je viskozita a objemová hmotnost, neumožňují použití klasických těžebních metod. Pro vhodnost použití jsme v laboratorních podmínkách simulovaly procesy probíhající při těžbě ropy za ložiskových podmínek a hodnotili vliv jednotlivých povrchově aktivních látek na výtěžnost těžké ropy.

Laboratorní výstupy navazují na prediktivní modely společnosti MND, a.s., které vyhodnotily nejperspektivnější metody těžby na ložisku Ždánice – 1. spodnomiocenní obzor. Matematické modelování považuje použití rozšířeného zavodňování s chemikáliemi, jako nejlepší variantu z běžně užívaných metod zvyšování výtěžnosti ropy (EOR). V laboratořích Institutu čistých technologií těžby a užití nerostných surovin navazujeme na náš předešlý výzkum, který hodnotí řadu povrchově aktivních látek (PAL), z hlediska jejich vhodnosti jako příměs pro EOR v podmínkách Ždánického ložiska. Byly vybrány PAL, které se jevily nejefektivnější a byly provedeny testy gravitační filtrace, nucené filtrace a nucené filtrace za navozených ložiskových podmínek v laboratorním prostředí. Použitá metoda vykazuje velmi slibné výsledky a bude nadále zkoumána formou testování nejvhodnějších PAL a jejich kombinaci s dalšími vhodnými tekutinami, jako jsou polymery, etc.

Článek byl vypracován v rámci projektu Institut čistých technologií těžby a užití energetických surovin, reg. č. CZ.1.05/2.1.00/03.0082 podporovaného Operačním programem Výzkum a vývoj pro Inovace, financovaného ze strukturálních fondů EU a ze státního rozpočtu ČR.

22

ANALYSIS OF SOCIAL AND HEALTH REIMBURSEMENTS FOR MINERS IN TEMS OF THE DECLINE OF MINING ACTIVITY IN THE CZECH REPUBLIC FUNDED FROM THE STATE BUDGET

Jana MAGNUSKOVÁ[1], Jana BARTOŇOVÁ[2], Lucie KRČMARSKÁ[3], Dagmar LÉTAVKOVÁ[4]

[1]*Ing. Ph.D., Institute of Economics and Control Systems, Faculty of Mining and Geology, VŠB – Technical University of Ostrava 17. listopadu 15/2172, Ostrava*
e-mail: jana.magnuskova@vsb.cz

[2]*Ing., Institute of Economics and Control Systems, Faculty of Mining and Geology, VŠB – Technical University of Ostrava 17. listopadu 15/2172, Ostrava*
e-mail: jana.bartonova@vsb.cz

[3]*Ing. Ph.D., Institute of Economics and Control Systems, Faculty of Mining and Geology, VŠB – Technical University of Ostrava 17. listopadu 15/2172, Ostrava*
e-mail: lucie.krcmarska@vsb.cz

[4]*Ing. Ph.D., Institute of Economics and Control Systems, Faculty of Mining and Geology, VŠB – Technical University of Ostrava 17. listopadu 15/2172, Ostrava*
E-mail: dagmar.letavkova@vsb.cz

Abstrakt

The present article deals with the area of social and health welfare of former and current employees of the abolished, privatized or otherwise affected by reorganization mining enterprises in connection with the decline of uranium, ore and coal mining industry in the country. It describes a thorough analysis of the legislation governing the given area from the start of the decline in mining activity in 1990 to the present. Subsequently, relying on real data from practice, the article shows elaborated overviews of trends in the number of beneficiaries of all kinds of provided mandatory and decline-based social and health reimbursements in the years 2001-2012, including the total cost financed in a form of subsidies from the state budget. The forecast of trends of these monitored indicators, processed by the age of beneficiaries of compensations, type of work and other practice-verifiable data, suggests a sharp reduction in the number of recipients of mandatory compensations in the year 2015, the complete termination of payment of a special health benefit for miners by 2025, compensations of accidents at work, occupational diseases and allotments of coal and wood by the year 2035. The payment of the special benefit to miners, a claim for which occurred after 31 December 1992, will have finished by the year 2040.

Abstrakt

Předložený článek se zabývá oblastí sociálně zdravotního zabezpečení bývalých a současných zaměstnanců rušených, privatizovaných či jinak reorganizací dotčených hornických podniků v souvislosti s útlumem uranového, rudného a uhelného hornictví v ČR. Je provedena důkladná analýza legislativy upravující danou oblast od zahájení útlumu hornické činnosti v roce 1990 až po současnost. Následně jsou na základě reálných údajů z praxe zpracovány přehledy vývoje počtu poživatelů všech druhů poskytovaných mandatorních i útlumových sociálně zdravotních náhrad v letech 2001-2012, včetně celkových nákladů, financovaných formou čerpání dotací ze státního rozpočtu. Prognóza vývoje těchto sledovaných ukazatelů, zpracovaná na základě věku příjemců dávek, druhu práce a dalších praxí ověřitelných údajů naznačuje, že k prudkému poklesu počtu příjemců mandatorních dávek dojde kolem roku 2015, k úplnému ukončení výplaty zvláštního příspěvku horníkům zdravotního do roku 2025, k odškodnění pracovních úrazů, nemoci z povolání a deputátů do roku 2035. Výplata zvláštního příspěvku horníkům, na který vznikl nárok po 31. 12. 1992, skončí do roku 2040.

Key words: decline in mining activity, mandatory compensations, decline-based compensations

1 INTRODUCTION

Decline in mining activity in the Czech Republic caused by political and economic changes began in 1990 in the ore mining, and in 1992 was followed by the decline in coal mining. The decline in manufacturing, strong pressure to improve the environment, and economic changes meant a gradual process of reducing the level of mining, closure of unpromising operations and reducing the number of workers in the sector. The very process of mining decline would not be possible without the state participation which was absolutely essential for its implementation. It was not just about a legislative endorsement of the entire process, but especially about its financial resources. The state is responsible for implementing and financing the decline, remediation of mining activity and removal of old environmental burdens, including state subsidies for the social programme (both mandatory and obligatory expenses) for workers affected by mining downturn.

As is clear from the information of the Ministry of Industry and Trade (MIT), more than CZK 78 billion was released from the state budget for the decline of mining since its inception in 1992 until the end of 2010. Hereof more than CZK 30 billion was drawn for the payment of social and health welfare of miners which is an integral part of the decline in mining. The subsidy provided for ensuring the payment of mandatory compensations as of 31 December 2011 amounts to CZK 1 149.6 million. Pending the completion of mining downturn in 2040 it should be another CZK 130 billion; thereof CZK 37.2 billion should be the cost of social and health welfare of miners. [1]

The decline of mining induced by the change in economic conditions is not a problem just in the Czech Republic. Coal mining in Western Europe has begun to decrease since around 1960, and the decline continues to this day. To prevent social and political impacts, governments of these countries adopted the measures with significant financial state subsidies. In all these economically developed countries with well-developed coal industry, the governments organized and managed the process of decline and addressing the consequences, either directly or through state organizations, regardless of whether the coal industry was in public or private ownership.

2 LEGISLATION

Despite different conditions of Czech mining compared to the Western European mining, the restructuring processes have common ground. The restructuring systems were built on a clear definition of the role of the state and mining organizations; herewith managing the entire process is supported by binding legislative acts.

First, a decline of Czech uranium mining commenced; it was on 1 January 1990 based on **Government Resolution of the then ČSSR No. 94 of 19 October 1989, on the concept of reduction of uranium mining loss-making in the Czechoslovakia.** [2] At the same time, the financing of some related activities on the part of the state, including social measures, could also start. At that time, it was about three benefits – a wage compensation, special health compensation for miners and special decline-based compensation for miners.

The scope of rights to social measures, however, expanded with the increasing decline. **Government Resolution No. 894 of 17 December 1990, on the change in the concept of the decline of uranium mining in the ČSSR in relation to the needs of nuclear energy**, was broadened to the scope of subsidized social actions outside of the above mentioned compensations even by a grant for retraining and severance when meeting the highest permissible exposure (HPE) which was established by the Chief Health Officer with effect from 1 December 1990 for underground work with a ionizing radiation risk for 2,100 shifts. [3]

Since 1993, **Government Decree of the CR No. 429/1993 amending the concept of decline of uranium mining and preservation of the Hamr I** has been a basic document. [4] The state budget has covered these social and health benefits:

- **special benefit for miners;**
- **wage compensation for transfer to another job after meeting the HPE;**
- **compulsory severance pay to meet the HPE;**
- **severance pay on termination of employment due to organizational reasons;**
- **retraining;**
- **compensations for accidents at work and occupational diseases, including allowances;**

- **loyalty bonus for miners;**
- **single compensation of loyalty bonus for miners;**
- **reconditioning and rehabilitation programmes;**
- **social insurance and health insurance (if transferred from certain benefits).**

On 31 May 1994, the MIT issued an opinion stating that *"... the decline of uranium mining affected all sites that are covered by Government Resolutions, and that the decline of mining affected both the current workers of the DIAMO state enterprise and its former employees whose employment in the state company was terminated for some reason giving the entitlements and fulfilment given in Tab. 2 of Annex 1 of the cited Government Resolution No. 429/1993 ... "*. [5] All the paid out social and health benefits were therefore considered as decline-based benefits and were financed from the state budget.

In 2002, however, a significant change in their funding occurred. **Act No. 154/2002 Coll., on the interim financing of certain social and health benefits for miners** was adopted, which regulated the financing of three benefits [6]:

- **special benefit for miners**;
- **compensation for loss of earnings** (for both, the entitlement had to occur to 31 December 1992).
- **allotment of coal and wood** (if entitlement occurred to 16 January 1992).

The Act concerned only the benefits paid out by organizations of coal, uranium or ore mining and their payments were made in a form of subsidies from the state budget. The organizations had a legal claim for granting these subsidies. The extensive restructuring in the deep mining industry, implemented by the state in 1990-1992, was the reason for this new legal regulation. This period was marked by a sharp decline in mining and mass-firing of miners for organizational and health reasons. Other job placement of the released workers into the labour market was difficult; therefore the granting of these benefits provided them a partial compensation for health and social harm occurred as a result of work in dangerous workplaces.

The payment of social and health benefits is a long-term, although time-limited financial commitment of the state which is a result of adopting the strategic decision on the decline of coal, ore and uranium mining. The state as the owner of mining enterprises made a decision on them at the time when such benefits were granted to the affected miners, and therefore it must be guaranteed in the future as well.

However, the above legal regulation does not apply to the financing of these entitlements granted after 31 December 1992 for commercial companies, several of them occurred by the year 1994 (e.g. Ostravsko-karvinské doly, Mostecká uhelná, etc.). These companies must pay them out from their own funds.

The situation in state-owned enterprises is different – after the Act entered into force, the founder of state-owned enterprises (MIT) divided social and health benefits for miners funded from the state budget, as follows:

a) benefits financed pursuant to **Act No. 154/2002 Coll., on temporary financing of certain social and health benefits for miners** [6];
b) other benefits paid out according to Government Resolution No. 429/1993, then No. 911/1999 which was then replaced by **Government Resolution of the CR No. 1128/2003, on the programme for completion of the restructuring of coal mining**. [7]

Government Resolution No. 1128/2003 applies to ore and uranium mining and contains two appendices: the first set the update of the procedure of remediation of mining activities (deadlines for the completion of work at various sites covered by state-owned enterprises like Palivový kombinát Ústí nad Labem, Východočeské uhelné doly Trutnov, and DIAMO Stráž pod Ralskem). The second appendix defined costs that can be covered by funds from the state budget since 2004. These are both the obligatory social and health benefits resulting from Act No. 154/2002 Coll. (so-called **mandatory**

compensations), i.e. the entitlements occurred to 31 December 1992; and obligatory social and health **decline-based compensations**, the entitlements for which arose after 31 December 1992.

Government Resolution No. 1128 2003 further defined all declined sites till then incorporated in the DIAMO state-owned enterprise (in a form of transfer of remains of privatized enterprises or through sales of parts of enterprises).

The following tables 1, 2 provide an overview of the number of beneficiaries of social and health compensations and trends of total costs on their funding in the years 2001-2012. For DIAMO a big break came in 2004, when since 1 January, the enterprise has taken over the entire agenda of social and health reimbursements. As a result of this centralization, the biggest increase in the number of beneficiaries and thus the costs of benefits occurred in DIAMO. The state enterprise DIAMO today provides the payment of social and health reimbursements not only to own employees of the uranium mining (current and former ones), but the entity was charged to pay these benefits also to former employees of abolished, privatized or otherwise reorganized mining companies or their parts, for which DIAMO is a successor organization based on the founder decision. These include Rudné doly Příbram (Příbram Ore Mines) and Rudné doly Jeseník (Jeseník Ore Mines), Jihomoravské lignitové doly Hodonín (South Moravian Lignite Mines), Odra Mine and Barbora Mine and many other locations, including the Centre for payment of social benefits exempt from OKD a.s. Ostrava:

Tab. 1 Development of the total costs and the number of beneficiaries of social and health reimbursements (SHR) 2001-2006

SHR	2001	2002	2003	2004	2005	2006
Mandatory:						
Number of beneficiaries	44,767	41,749	39,765	36,970	33,884	31,948
Costs (in thousand CZK)	1,232,334	1,155,996	1,103,050	1,048,150	1,008,466	969,076
Decline-based:						
Number of beneficiaries	1,535	1,266	1,149	1,172	1,136	1,074
Costs (in thousand CZK)	33,201	25,928	22,268	23,617	22,424	19,524

Tab. 2 Development of the total costs and the number of beneficiaries of social and health reimbursements (SHR) 2007-2012

SHR	2007	2008	2009	2010	2011	2012
Mandatory:						
Number of beneficiaries	29,357	27,597	25,602	23,599	21,590	19,597
Costs (in thousand CZK)	943,505	874,045	815,960	737,188	673,428	620,500
Decline-based:						
Number of beneficiaries	978	876	833	743	662	578
Costs (in thousand CZK)	18,924	16,345	16,289	13,789	11,719	11,340

Due to the nature and length of time of payment of benefits it is assumed that their costs will decrease in the coming years, as evidenced by the following charts 1, 2, 3, 4 prepared on the basis of actual data on the various social and health reimbursements paid in the years 2001 to 2012.

3 ANALYSIS OF SOCIAL AND HEALTH REIMBURSEMENTS

As already mentioned, in 2002, the founder of state-owned enterprises (MIT) divided the social and health benefits for miners into two major groups.

3.1. Mandatory compensations

Also known as **major social and health reimbursements** whose structure is determined by the mentioned Act No. 154 2002 Coll., on temporary financing of certain social and health benefits for miners, the entitlement for which occurred to 31 December 1992, are financed from the state budget.

- **Compensation for loss of earnings (the so-called allowance) and the compensation for damage admitted to employees for occupational disease or work-related injury**

It is governed by Labour Code - Act No. 65 1965 Coll., as amended. [8] It is the largest volume of all paid benefits. The compensation under Article 195 is provided to employees in such amount so that, together with their earnings after an accident at work or occupational disease, plus any disability allowance granted for the same reason, it is equal to the average earnings before the damage occurred. Furthermore, this includes reimbursements for nutrition of survivors (Article 199), for which the employee who died due to an accident at work or occupational disease, provided nutrition, as well as reimbursements for expenses incurred from medical treatment (Articles 196-198 of Labour Code) which are not already paid by a health insurance company (e.g. curative rehabilitation stays, payments for medications, etc.). In case of death of the reported occupational disease a widow's allowance and expenses incurred for the funeral shall be paid. In addition, the consequences of accidents at work and occupational diseases shall be paid, thus damages for pain and suffering, and social impairment and damage to property to former political prisoners who worked in uranium mining. The allowance is provided monthly until granting the old age pension, up to 65 years maximum. In 2012, there were 2,420 beneficiaries of the allowance, the projected overall costs in this year amounted to CZK 570,000 thousand.

- **Special benefit for miners (health)**

Granted to employees fired on health grounds pursuant to Decree No. 19 1991 Coll., on the job placement and material welfare of workers in mining, incapable of previous work on a long-term basis [9], followed by Act No. 98 1987 Coll., on the special benefit for miners as amended by Act No. 160 1989 Coll. [10] The employee, who after the HPE or threat or occupational disease was transferred to another job or stopped working, but is employed, has the right to it. It is paid monthly, up to the age of 60, or granting a retirement or disability pension. In 2012, 317 beneficiaries drew health special benefit form miners (SBM); the projected overall costs were CZK 620,500 thousand.

- **Allotment of coal and wood**

For pensioners, widows and fired staff for health reasons, according to Decree of the Federal Ministry of Fuel and Energy of 13 January 1990 No. 1 1990, on allotment of coal and wood, reg. under Act No. 59 1990 Coll. [11], in the amount of unvalorized financial performance at a level of actual disbursements in 1992, the entitlement to which incurred on 16 January 1992. Allotment of coal and wood was not provided in the uranium mining. In the state enterprise DIAMO, this benefit had been paid until 2001, when there was a takeover of Jihomoravské lignitové doly Hodonín (South Moravian Lignite Mines Hodonín), taken over from the state enterprise Rudné doly (Ore Mines). Other substantial increase in the number of beneficiaries occurred after taking over the Barbora Social Centre of OKD Ostrava, a.s. as of 1 January 2004. In 2012, 16,860 beneficiaries drew the allotments; the projected overall costs were CZK 42,500 thousand.

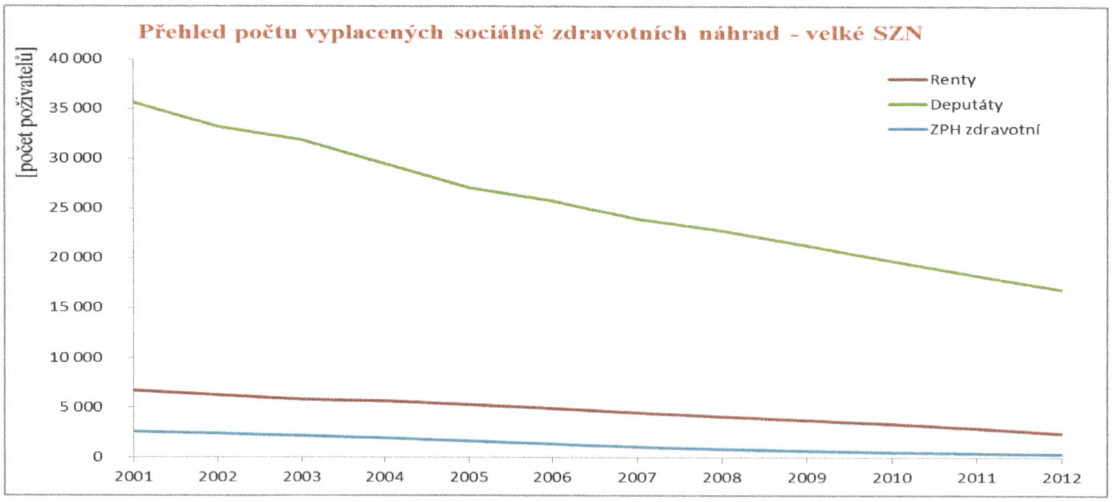

An overview of the number of paid out social and health reimbursements - major SHR
Number of beneficiaries
Allowances
Allotments
health SBM

Fig. 1 The number of beneficiaries of social and health reimbursements (mandatory) in the years 2001-2012

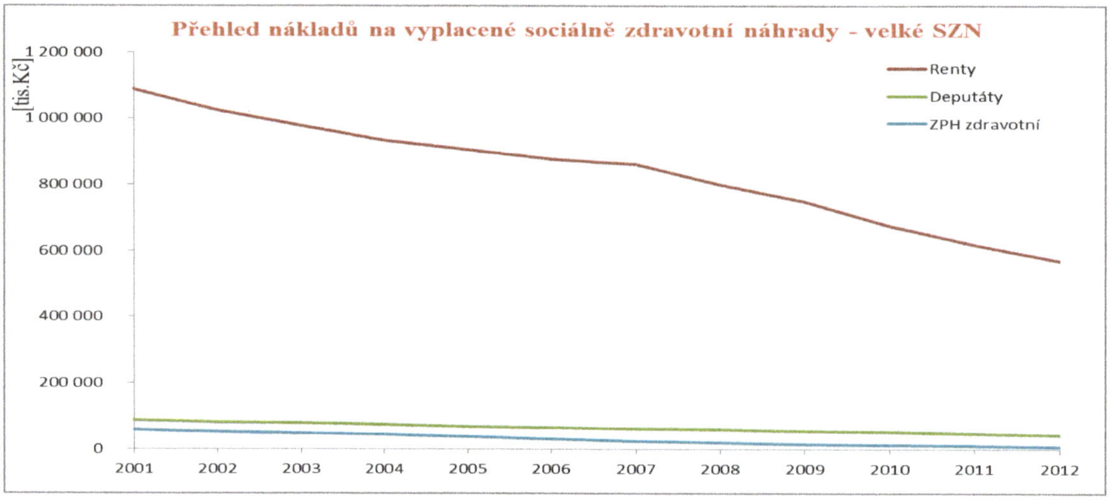

An overview of costs on paid out social and health reimbursements - major SHR
Thousand CZK
Allowances
Allotments
health SBM

Fig. 2 Total cost of social and health reimbursements (mandatory) in the years 2001-2012

Forecasts elaborated by the headquarters of the enterprise DIAMO s. p. based in Stráž pod Ralskem, according to the age of beneficiaries, type of work and other practice-verifiable data, suggest that a sharp decline in the number of beneficiaries of mandatory compensations will have occurred by about the year 2015, the complete termination of payment of the special health benefit for miners – by 2025, compensations of accidents at work, occupational diseases and allotments by the year 2035.

3.2. Decline-based compensations

Being denoted also as **minor social and health reimbursements**, they relate to the implementation of downturn and restructuring of coal, ore and uranium mining, paid according to Government Resolution No. 429 1993, further No. 9113 1999, which was then replaced by Government Resolution No. 1128 2003, on the programme for completion of the restructuring of coal mining. [7] The right to them arose after 1 January 1993. As shown in Tables 1 and 2, they are lower by volume (allowances are not included yet – since 1993 the beneficiaries have been being paid from Kooperativa). The source of funding is once again the state budget.

- **Severance pay (decline-based)**

According to Article 67, for reasons referred to in Article 52, letter a) to c) of Act No. 262 2006 Coll., Labour Code, or by agreement for the same reasons (i.e. organizational reasons), payable in the amount of three times the average monthly salary paid upon the termination of employment. [8] In 2012, four employees were fired, the cost of severance payments amounted to CZK 670 thousand.

- **Severance pay (health)**

According to Article 67, on the grounds referred to in Article 52, letter d), Labour Code (health reasons or achievement of HPE), payable in the amount of twelve times the average salary paid upon the termination of employment. [8] In 2012, two employees were fired; the cost amounted to CZK 800 thousand. In the period of 2008-2011, such an entitlement did not arise.

- **Special benefit for miners (decline-based)**

According to Article 2(4) of Act No. 98 1987 Coll., on the special benefit for miners provided in connection with the implementation of the decline programme [12]. The entitlement occurs to an employee who went from the work underground to the work outside the underground. It is a short-term benefit provided according to the length of service underground, for up to 30 months. From the year 2009 to present, such entitlement did not arise.

- **Special benefit for miners (health)**

Granted to employees pursuant to Article 2(1 to 3) of Act No. 98 1987 Coll., on the special benefit for miners, as amended by Act No. 160 1989 Coll., the entitlement for which arose after 1 January 1993. [12]. The conditions for its payment are the same as in the previous case. In 201, this benefit was paid to 363 beneficiaries, the cost of its payment amounted to CZK 9,000 thousand.

- **Wage compensation**, or **supplement to average earnings**

According to Decree No. 19 1991 Coll., on the job placement and material welfare of workers incapable of previous work on a long-term basis – entitlements occurred on 30 November 2003 in the meaning of Decree No. 405 2003 Coll., which repealed Decree No. 19 1991 Coll., including relevant transfer payments of social insurance and health insurance in compliance with Act No. 589 1992 Coll. and Act No. 592 1992 Coll., as amended. [13], [14], [15] It is paid monthly until the admission of old age pension. From the year 2006 to the present, it is paid only to one former employee (monthly – until the admission of old age pension).

- **Loyalty bonus for miners**

According to Act No. 62 1983 Coll., on loyalty bonus for miners and Government Regulation No. 67 1983 Coll., on implementing the act on loyalty bonus for miners. [16] It is paid annually on the "Day of miners" to those employees who perform or performed mining activities in mining operations. Most of the payments are in the so-called "stop state", they do not increase any more, and new entitlements to them arise only in a limited way. In 2012, 208 employees received the bonus, the total cost amounted to CZK 850 thousand.

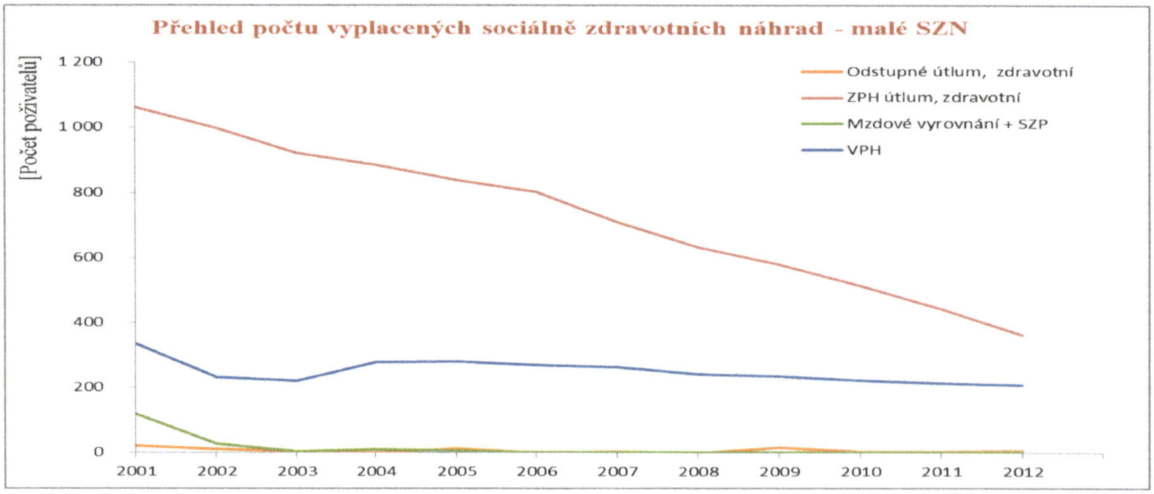

An overview of the number of paid out social and health reimbursements - minor SHR
Number of beneficiaries
SBM (decline, health)
Wage compensation - SHW
LBM

Fig. 3 The number of beneficiaries of social and health reimbursements (decline-based) in the years 2001-2012

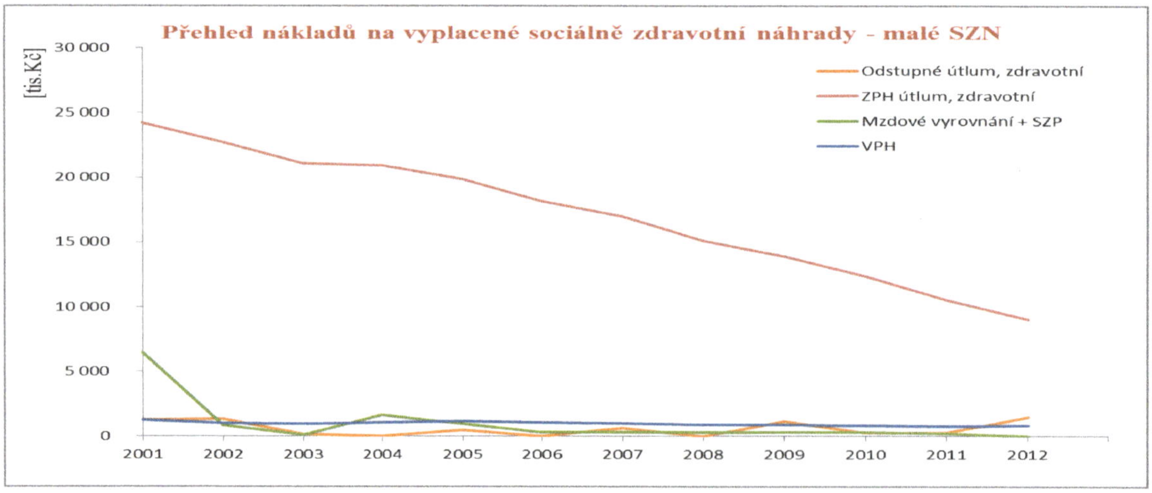

An overview of costs on paid out social and health reimbursements - minor SHR
Thousand CZK
SBM (decline, health)
Wage compensation - SHW
LBM

Fig. 4 Total cost of social and health reimbursements (decline-based, health) in the years 2001-2012

According to forecasts worked out by the headquarters of the enterprise DIAMO s.p., the payment of special benefit for miners, for which the entitlement occurred after 31 December 1992, will have been terminated by the year 2040.

4 CONCLUSION

The decline of mining induced by the change in economic conditions is not a problem just in the Czech Republic. Coal mining in Western Europe has begun to decrease since around 1960, and the decline continues to this day. Mainly German, French, English and Belgian mining, the most advanced in terms of capacity and used technology, along with smaller deposits in Spain and Italy, decreased their annual production outputs by tens to hundreds millions of tons. To prevent social and political impacts, the governments of the countries adopted the system of measures with significant financial subsidies on the part of the state.

In all these economically developed countries with well-developed coal industry, the governments organized and managed the process of decline and addressing the consequences, either directly or through state organizations, regardless of whether the coal industry was in public or private ownership.

The strategic goal of the decline programme for mining in CR was mainly the completion of mining operations in unprofitable locations, removal of the effects of mining and technical disposal of mine workings, i.e. putting the landscape into a similar condition as prior to mining, in compliance with all legal conditions (such as mining and environmental ones). From material, temporal, financial points of view, the decline procedure is developed in the "Technical and social projects of disposal of declined sites" being updated at least once every five years. Projects for the decline of a certain site are proposed by grant recipients and approved by the MIT CR.

Duration of the whole process and also its financial demands were at least initially affected by the fact that MIT CR had no clear and realistic concept of restructuring the coal industry, and therefore the original assumptions on its termination proved to be unrealistic. Only after 1998, the entire process was conceived as a long-term matter and individual procedures governing the participation of the national budget in the decline of mining were determined, specified and detailed. The completion of the entire process of mining decline is very expensive and will take many decades. Its anticipated end is still estimated by 2040 or 2050 and will cost at least another CZK 130 billion.

REFERENCES

[1] http:. www.mpo.cz dokument90131.html
[2] http:. www.edb.sk sk produkt utlum-tezby-a-upravy-uranu-p59115-dwC06039.html
[3] http:. www.psp.cz eknih 1993ps tisky t1733a00.htm
[4] http:. www.psp.cz eknih 1993ps tisky t0936_21.htm
[5] http:. slon.diamo.cz hpvt 2010 legislativa L_05.pdf
[6] http:. kormoran.vlada.cz usneseni usneseni_webtest.nsf 0 57579EC01A2AF7FEC12571B6006C02C2
[7] download.mpo.cz get 26820 27096 301998 priloha002.doc
[8] http:. www.psp.cz sqw sbirka.sqw cz 65&r 1965
[9] http:. www.pravnipredpisy.cz predpisy ZAKONY 1991 019991 Sb_019991_------_.php
[10] http:. www.pravnipredpisy.cz predpisy ZAKONY 1987 098987 Sb_098987_------_.php
[11] http:. slon.diamo.cz hpvt 2010 legislativa L_05.pdf
[12] http:. www.psp.cz sqw sbirka.sqw cz 98&r 1987
[13] http:. www.psp.cz sqw sbirka.sqw cz 405&r 2003
[14] http:. www.psp.cz sqw sbirka.sqw cz 589&r 1992
[15] http:. www.psp.cz sqw sbirka.sqw cz 592&r 1992
[16] http:. www.pravnipredpisy.cz predpisy ZAKONY 1987 098987 Sb_098987_------_.php

RESUME

Předložený článek se zabývá analýzou jednotlivých sociálně zdravotních náhrad pro bývalé a současné zaměstnance hornických provozů v souvislosti s útlumem uranového, rudného a uhelného hornictví v ČR. Nezbytnou součástí bylo provedení analýzy veškeré legislativy, upravující dané téma. Následují přehledné grafy, které byly zpracovány na základě reálných údajů, získaných z interních podkladů podniku DIAMO s. p. Grafy vypovídají o vývoji počtu poživatelů všech druhů poskytovaných mandatorních i útlumových sociálně zdravotních náhrad ve sledovaném období 2001-2012, včetně celkových nákladů, financovaných formou čerpání dotací ze státního rozpočtu. Je z nich dále patrná neustálá tendence klesání jednotlivých vyplácených dávek v čase. Prognóza vývoje sledovaných ukazatelů, zpracovaná na základě věku příjemců dávek, druhu práce a dalších praxí ověřitelných údajů potvrzuje, že k prudkému poklesu počtu příjemců mandatorních dávek dojde kolem roku 2015, k úplnému ukončení výplaty zvláštního příspěvku horníkům zdravotního do roku 2025, k odškodnění pracovních úrazů, nemoci z povolání a deputátů do roku 2035. Výplata zvláštního příspěvku horníkům, na který vznikl nárok po 31. 12. 1992, skončí do roku 2040. Dokončení celého procesu útlumu hornictví je finančně velice náročné a bude trvat ještě řadu desítek let. Jeho předpokládaný konec je zatím odhadován do roku 2040, resp. 2050 a bude stát nejméně dalších 130 mld. Kč, z toho 37,2 mld. Kč by měly činit náklady na sociálně zdravotní zabezpečení horníků.

23

REDUCING CARBON DIOXIDE EMISSIONS BY UNDERGROUND STORAGE IN AN ABANDONED COAL MINE - AN INITIAL STUDY

Josef CHOVANEC [1], Jaroslav ZÁVADA [2], Tomáš BOUCHAL[3], Vladislav BLAŽEK[4], Kamil OŽANA[5]

[1] *Ing., Ph.D., Institute of Mining Engineering and Safety, Faculty of Mining and Geology, VSB-TU Ostrava, 17. listopadu 15/2172, 708 33 Ostrava-Poruba, Czech Republic*

E-mail: josef.chovanec@vsb.cz

[2] *Ing., Ph.D., Department of Environmental Engineering, Faculty of Mining and Geology, VSB-TU Ostrava, 17. listopadu 15/2172, 708 33 Ostrava-Poruba, Czech Republic*

E-mail: jaroslav.zavada@vsb.cz

[3] *Ing., Ph.D., Department of Environmental Engineering, Faculty of Mining and Geology, VSB-TU Ostrava, 17. listopadu 15/2172, 708 33 Ostrava-Poruba, Czech Republic*

E-mail: tomas.bouchal@vsb.cz

[4] *Ing., Department of Environmental Engineering, Faculty of Mining and Geology, VSB-TU Ostrava, 17. listopadu 15/2172, 708 33 Ostrava-Poruba, Czech Republic*

E-mail: vladislav.blazek.st@vsb.cz

[1] *Ing., Institute of Mining Engineering and Safety, Faculty of Mining and Geology, VSB-TU Ostrava, 17. listopadu 15/2172, 708 33 Ostrava-Poruba, Czech Republic*

E-mail: kamil.ozana@vsb.cz

Abstract

This article is a basic study dealing with the issues of underground storage of carbon dioxide generated from different kinds of activities. Carbon dioxide can be stored underground as a free gas; gas dissolved in water, or can be adsorbed in the rock mass and in the remaining seams. The technology for processing and storage of carbon dioxide is known as Carbon Capture & Storage (CCS). The article focuses on the possibility to store CO2 underground at the Paskov Mine in the Czech Republic.

Abstract

Tento článek je základní studií zabývající se problematikou ukládání oxidu uhličitého, vznikajícího z různých druhů činností, do podzemí. Oxid uhličitý je možno ukládat do podzemí jako volný plyn, plyn rozpuštěný ve vodě nebo může být adsorbován v horninovém masívu a ve zbylých slojích. Technologie zpracování a ukládání oxidu uhličitého je označována jako Carbon Capture & Storage (CCS). Článek se zaměřuje na možnost ukládání CO_2 do podzemí v Dole Paskov v České republice.

Key words: carbon dioxide, Carbon Capture & Storage (CCS), geological storage, abandoned coal mine

1 INTRODUCTION

A number of evidences and studies suggest that one of the causes of global warming may be an increasing concentration of carbon dioxide in the atmosphere. Worldwide CO2 creates more than 60 % of the enhanced greenhouse effect. In industrialized countries, CO2 represents more than 80 % of greenhouse gas emissions. [1] For these reasons, it is desirable to properly dispose of CO2 (so that it does not damage individual components of the environment). One option being verified is to store CO2 back into the ground, from which its increased content originates actually. Also, the project of CO2 storage in deep mines appears and is experimentally verified as one of the many ways to store carbon dioxide into deeper layers of the earth's crust. However, it was demonstrated as well that this method is quite promising for a number of problems that need to be resolved. Significant efforts in this direction is manifested by the Environmental Protection Agency (EPA), which attempts to obtain information about the values of emissions and also initiate programs that would eliminate this negative phenomenon. EPA also applies a uniform classification of proportions of individual gases in the greenhouse effect and expresses it by an CO2 equivalent ($MtCO_2eq$), which means one million metric tons of the CO2 equivalent. For example, in this classification, methane has 20 times higher effect on the greenhouse effect than CO2. [2]

In June 2009, Directive 2009/31/EC of the European Parliament and Council of 23 April 2009 on the geological storage of carbon dioxide entered into force. This new directive, in order to contribute to the fight against climate change, establishes a legal framework for the environmentally safe geological storage of carbon dioxide (CO2) for the purpose of permanent capture. The essence is a relatively new technology consisting in the capture of CO2 from industrial installations, its transport and subsequent intrusion (injection) into a suitable underground geological structure (storage) for the purpose of permanent capture. This method is called Carbon Capture & Storage (CCS). The most difficult part of the process seems to be the very capturing of CO2 from flue gas. The use of CCS is extremely energy-intensive process. The technology consumes about 10-40 % of the energy produced by the plant itself. [4, 15]

In our country, taking into account the reserves of fossil fuels, it is expected to use CCS, particularly in brown-coal burning plants. The local energy company CEZ has already determined in advance a suitable plant for this principal project. The best location seems to be a territory owned by the mining company Moravské naftové doly (MND). In its vicinity, there is a power plant standing in a town of Hodonín, as well as the area of Ledvice in the North Bohemian region where new brown-coal blocks are to be built. The construction of – in this sense – a pioneering power plant is expected in 2015. It is assumed that the costs of construction of this plant will amount to CZK 10 billion. [4] Emissions of gases in CR in the year 2010 are shown in Tab. 1 [5].

Tab. 1 Carbon dioxide emissions from individual sources in the Czech Republic in 2010 [5]

Source of CO_2	**$MtCO_2eq$**
Production of electricity and heat	62
Other emissions from the energy sector	1
Industry	18
Households	10
Services	5
Transport	19
Total emissions in the Czech Republic	115

$MtCO_2eq$... million metric tons of CO_2 equivalent

This paper examines the possibilities of CO_2 storage in abandoned underground mines. This issue, particularly from the area of the Upper Silesian Basin, has already been dealt with by many authors, and laboratory experiments were performed as well. [18, 19, 20] Generally, with an increasing storage depth, pressure and temperature increase as well. In underground mines of the Ostrava - Karviná District, we expect a confining pressure of 12 MPa at a depth of 500 m below the surface, and 25 MPa at a depth of 1000 m. The temperature at a depth of around 1000 m can be more than 30 °C. Such conditions significantly determine the state of CO_2. Fig. 1 illustrates the phase diagram of CO_2 [6]. The diagram shows that when storing CO_2 to greater depths, where the pressure may be more than 10 MPa and the temperature higher than 30 °C, CO_2 enters the phase of fluid, possibly even supercritical fluid.

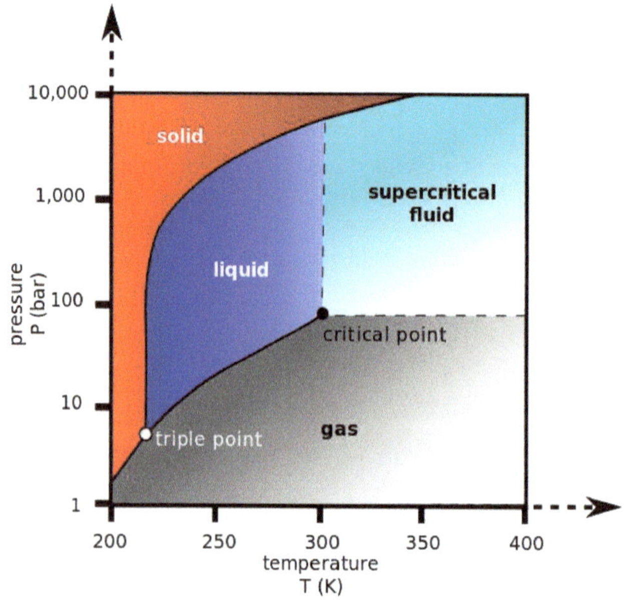

Fig. 1 The phase diagram of carbon dioxide (1 bar = 10^5 Pa) [6]

2 PRINCIPLE OF CARBON CAPTURE & STORAGE TECHNOLOGIES

The Carbon Capture & Storage (CCS) technology itself is not so complex as could be think. In the first step, the carbon dioxide is separated from the flue gas and captured with a special separation unit, which is built near the plant. Subsequently, the separated gas is transported (through pipeline) to the location, which was determined for its permanent storage. As suitable locations, we consider mined deposits of oil or natural gas and underground mines. In these locations, the gas is injected into the ground (of an abandoned working). The disadvantage, however, is that the company producing CO_2 must be built near the appropriate location (excavated sites). Already today, it is possible to permanently store up to 90 % of carbon dioxide emissions to the ground in such a way. [3, 7]

After determining a suitable site, carbon dioxide is injected into an underground reservoir; in the rock mass, it is accumulated in cracks or comes between grains and thus displaces present substances (e.g. water, oil, natural gas). For this reason, appropriate locations are the sites where rocks with high porosity and permeability are present. These requirements meet most sedimentary rocks of the so-called sedimentary basin. In some locations, permeable rocks alternate the impermeable ones. These rocks then act as an impermeable seal. Sedimentary basins are known by a frequent occurrence of oil and natural gas, and possibly natural CO_2. The long-term sealing ability of such rocks is confirmed by that these rocks were able to hold gaseous and liquid hydrocarbons for many years.

While the figures, which illustrate storing CO_2 underground, depict the rock composition in a simply way (as homogeneous cake layered structure), in fact these are unevenly distributed rock formations, reservoirs and seal rocks that are affected by local faults which together form complex heterogeneous structures. [8]

The European Union has issued a directive on storage of carbon dioxide into the ground and set up a consultation centre at the Technical University of Athens. In this centre, various technical problems are dealt with, associated with environmental risks, issues of management and geological research. The most serious problem in the use of this method, however, still remains its price, which is then reflected in increasing prices for produced electricity, which can grow up to 3 cents per kilowatt hour. The solution to rising prices could be the utilization of the deposited carbon dioxide to displace the oil into the borehole, which would increase the yield of oil fields.

In Europe only, 4 billion tonnes of carbon dioxide per year are discharged into the atmosphere. If commitments to reduce greenhouse gases set by the Kyoto Protocol are to be met by 2012, it will be necessary to realize this project for CO_2 storage in the ground as soon as possible. It is clear that the combined effect of improving energy efficiency and use of renewable energy sources will not be sufficient. [7]

Carbon dioxide can be stored into underground workings as a free gas; gas dissolved in water, or can be adsorbed in the rock mass and in the remaining seams.

The injection of carbon dioxide into the abandoned mine workings, however, is accompanied by many risks. In many areas, there is methane and other toxic volatile gases. The injection of carbon dioxide into the abandoned mine workings may result in the release of dangerous methane into the atmosphere and the consequent threat to the population in the area. [16, 17]

Another risk is the release of heavy metals and other elementary particles which may be present in coal seams and after the release may contaminate underground water, soil, but also air. Since the underground water supplies are strategic sources, the foregoing facts may represent a considerable problem for the storage of CO_2 in the ground. [9] The right selection of a suitable location for a long-term storage of CO_2 underground is a complex process that requires detailed pieces of knowledge on the geological structure of the site and considerable geological experience.

The most important prerequisites for the establishment of underground storage of CO_2 are as follows:

- appropriate porosity, permeability and storage capacity;
- presence of overlying impermeable rock – so-called "sealing rock" (e.g. clay, claystone, marlite, salt), which prevents CO_2 from the migration towards the surface;
- the existence of so-called "structural traps" - phenomena, such as the arched construction of a sealing rock, which can control and regulate the extent of migration of CO_2 in the storage formation;
- storing in a depth of 800 m, where there is a sufficiently high pressure and high temperature to ensure the storage of CO_2 in the compressed liquid phase, so as to ensure maximization of the stored quantities;
- absence of drinking water (CO_2 will not be injected into water intended for consumption and other use by population). [8]

3 POSSIBILITY OF STORING CARBON DIOXIDE AT A SITE IN THE CZECH REPUBLIC

As a default model site, the Paskov Mine in the southern part of the Ostrava-Karvina District was selected. This once original mine, now a plant, still produces a significant amount of residual methane. After the depletion of reserves it could be used as storage for CO_2. During this time period, however, it can only serve as a model example herewith that similar regularities must be examined for each additional convenient location.

The Paskov Plant is not connected with other mines, which this location along with the fact that in 2001 it was finally closed at all accesses with enclosures, predetermines for permanent storage of CO_2 underground. The location and geological profile of that area around the mine are shown in Fig. 2.

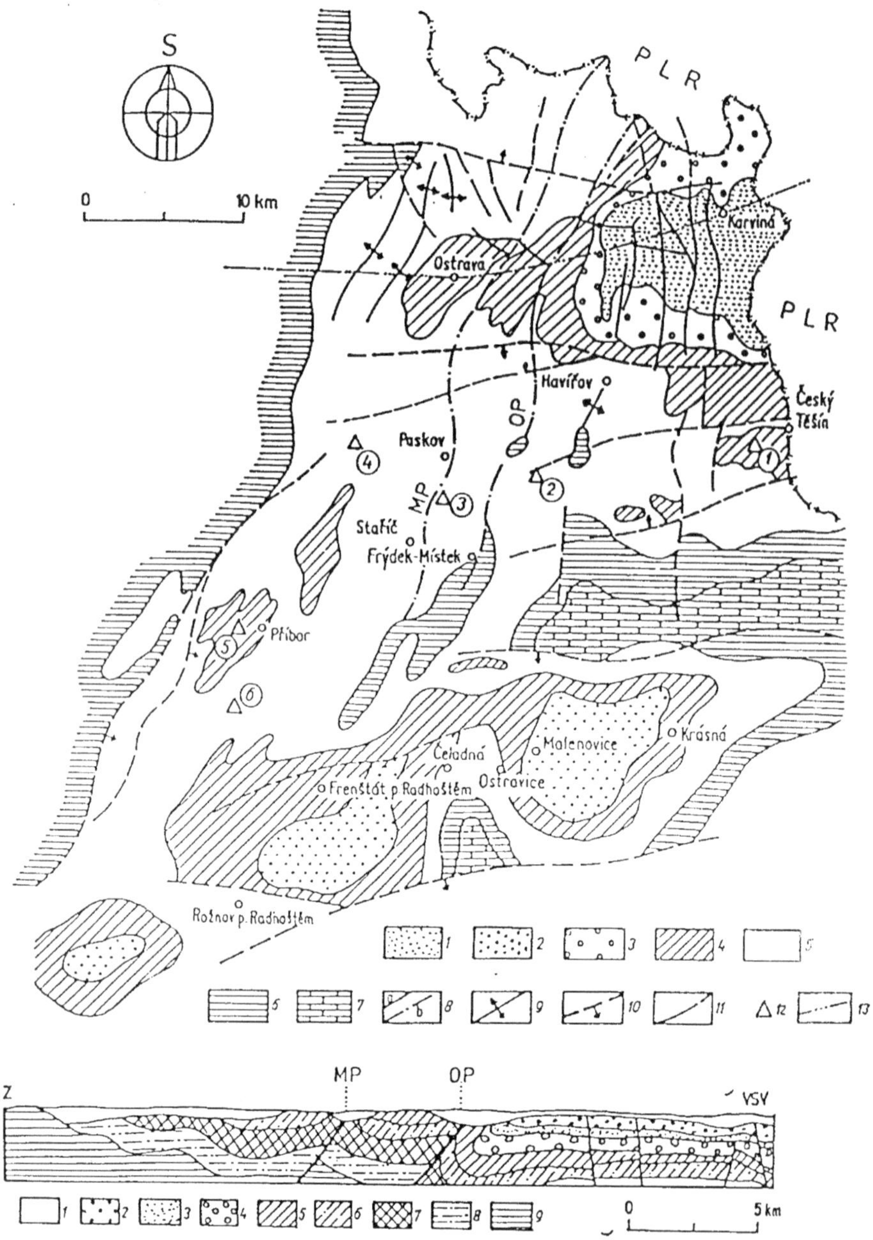

Fig. 2 Exposed geological map of the Carboniferous system in the Czech part of the Upper Silesian Basin and a geological section through the Ostrava ridge of OKR [12, 14]

Explanatory notes to the map: 1-3 - Karviná Formation - layers: 1 - Doubrava and Suchá, 2 - Suchá, Saddle including Prokop Seam, 3 - Saddle including Prokop Seam, 4-5 - Ostrava Formation - layers: 4 - Poruba and Jaklovec, 5 - Hrušov and Petřkovice, 6 - dinant (kulen), 7 - devon and dinant (carbonates), 8 - Michálkovice (MP) and Orlová (OP) fault: and mines-verified course, b - expected course, 9 - major anticline axes, 10 - breaks of fundamental importance, 11 - other faults, 12 - significant gas deposits, 13 - geological cut lines. Numbers (circled) next to gas deposits: 1 - Žukov, 2 -. Bruzovice, 3 - Staříč, 4 -Krmelín, 5 - Příbor - North, 6 - underground gas reservoir Příbor - South.

Explanatory notes to the cut: 1 - cover, 2-8 - Upper Carboniferous: 2-4 - Karviná Formation - layers: 2 - Doubrava, 3 - Suchá, 4 - Saddle, 5-8 - Ostrava Formation - layers: 5 - Prokop and Poruba seams, 6 - Jaklovec, 7 - Hrušov, 8 - Petřkovice, 9 – dinant.

3.1 Mining and geological situation of the Paskov Plant

Fig. 3 shows the mining area. It illustrates geological strata and coal seams. The mining area was formed by the Carboniferous cluster with a total mining area of 19.6 km^2. The dimension of the inlet of pits (mining and upcast) at the surface is +250.8 m B. p.v. (Balt after settlement). From the deepest level at an elevation of -630 m, dip workings were headed up to the elevation of -761 m. The mine was made available from the surface by 4 pits, from a total of which 2 are in the central part of the mining field and 2 in the marginal eastern and western parts. The highest first level was set at an elevation of -150 m.

Rocks of sets of superincumbent beds of carbon reach a thickness from 300 m up to 900 m and are nearly impermeable for the release of CH_4 from the mine area. This impermeableness was not violated either by undertaken excavations. In the period since the mine's establishment, mining areas of the order of 9.2 million m^3 were released by mining activities. With regard to their subsequent compression, collapse and consolidation over a certain time horizon, the remaining free volume can be deduced from the regularities of compression of the rock material. The compressibility for this case (in the long term horizon, i.e. years and over) from 25 % to 40 % can be a priori determined. This is a free volume of mining area from 2.3 million m^3 to 3.6 million m^3.

Technological factors participate in the decisions on possibilities of filling this area with the stored gas, as well. It is also necessary to take into account the fact that a number of existing mine workings (cross-cuts, corridors) was filled with various kinds of dams from simple dam counters to explosion-proof dams. These counters are designed for a maximum pressure of 1 MPa. Due to the distortion of rocks in their vicinity and the expected pressure in the reservoir (15 to 22 MPa), sealing dams thus do not affect the communication of gas throughout the space of the mine being damped down.

The mine operation in the period of its existence was marked by the emergence of coal and gas outbursts. This was associated with a high gas bearing property of the location and low permeability of any its parts. In the past period, theoretical and practical pieces of knowledge were obtained that can be used in the project and the assessment of the suitability of the proposed CO_2 storage. These include especially gathering a large set of values of permeability of the rock layers and the gas flow under these conditions.

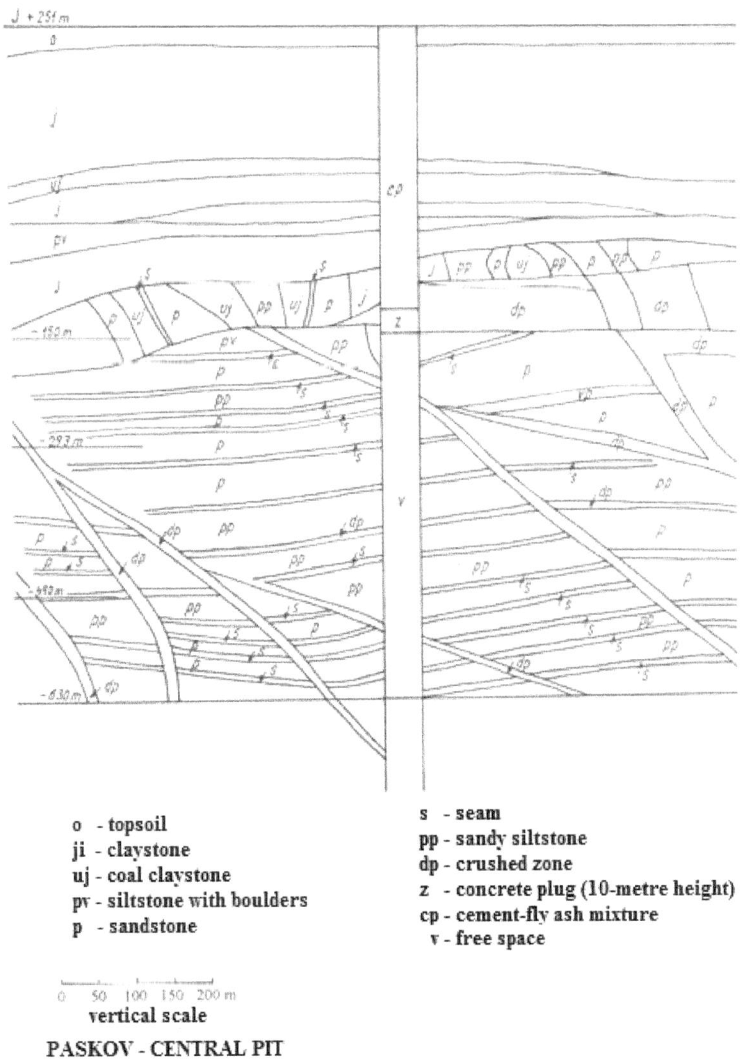

Fig. 3 The adjusted geological profile of the Paskov Mine with rendering the layers between the surface and the -150 m dimension [12, 14]

The average thickness of the mined seam reached 0.9 m. Accompanying rocks are composed of sandstone (about 40 %), siltstone (40 %) and clays (20 %). The firmest rocks reached the maximum compressive strength of 70 MPa (sandstone). Their thickness does not exceed 20 mA and due to tectonic segmentation no tremors of rocks have ever occurred in the Paskov locality. In terms of stability of the mine it is a quiet area without seismic activity. Activity induced by tremor phenomena can be excluded as well.

An important condition for the storage of CO_2 is an estimate to calculate the time period of flooding of individual levels of the mine. After the completion of mining activity at the mine the inflow of mine water is determined for about 259 m^3 per day at the tendency of flooding to a dimension of -30 to -50 m of the absolute depth, which is the position of the piezometric level of the detrital aquifer. If during a period of 23.7 years the mine area would be flooded to the third level (-630 m), there is still a free volume of 5.3 million m^3 for 56.7 years, and the additional volume of 1.5 million m^3 for 17.1 years. Because the consolidation of rocks will be gradual, the actual time period of flooding may be further increased (indicatively up to 40 years).

3.2 Permeability between the underground and the surface

To determine the critical point of the project, i.e. the possibility of gas leak from underground to the surface, the calculation of linear filtering according to the Darcy's theory was used (Equation 1).

$$Q = \frac{S \cdot (p_S - p_O) \cdot k_p}{2 \cdot \eta \cdot h} \qquad (1)$$

Where:

S calculation was made for the area S = 400 m^2 which corresponds to the area of the subsidence trough above excavations

p_s anticipated pressure of 15 MPa in the tank

p_0 atmospheric pressure at the surface of 0.1 MPa

k_p permeability coefficient for topsoil 10^{-8}, claystone $5 \cdot 10^{-18}$, claystone coal 10^{-14}, coal seam 10^{-10}, siltstone with boulders 10^{-14}, geologically very diverse layer just above the level – 150 m 10^{-16}, excavations 10^{-10} (m^2). Using the arithmetic mean depending on the thickness of each layer, „k_l" between the level of -150 m and the surface $k_l = 10^{-16}$ m^2 was found out.

h critical distance between the top level and the surface of 400 m

η_{CH4} dynamic viscosity of CH_4 10,26$\cdot 10^{-6}$ Pa·s (at 0°C)
η_{CH4} dynamic viscosity of CO_2 13.9$\cdot 10^{-6}$ Pa·s (at 0°C)
γ specific mass for CO_2 1,9 kg/m^3 (at 0°C)

The permeability of layers was investigated in-situ for the subject area in the earlier works like [10, 11].

The result of the equation is the total volumetric flow rate Q of gas that can escape from the mine to the surface with a surface area of 400 m^2. The volume flow rate of carbon dioxide amounts according to the equation to 5.35·10-5 $m^3 \cdot s^{-1}$, which is a slight flow rate indeed, but in a year this leak could achieve a volume of 1,687 m3. It is difficult to see how this theoretical leakage of CO2, which is 4.22 cubic meters per square meter of the area per year, could affect vegetation, as shown in [9].

Taking into account that the free space in the mine, the premises of which occupy 2.3 to 3.6·106 m^3, can be filled with the amount of carbon dioxide proportional to this volume, then the ratio between the stored and leaked gas is significantly positive in respect to the protection of the atmosphere.

To verify the value calculated according to Equation 1, the Fluent calculation program of the Institute of Economics and Management Systems, VSB-TU Ostrava, was used. The input data for this model was deliberately limited to the rock formation between -150 mA and the surface. The resulting values were then adjusted as different areas of the mine may have different values of permeability.

3.3 Verification of the opportunities and risks when storing CO_2 in the location of the Paskov Mine using the Fluent program

The permeability of individual layers occurring in the area has been selected as follows:

- Topsoil $1 \cdot 10^{-6}$ m^2
- Claystone $1 \cdot 10^{-14}$ m^2

- Coal claystone $1 \cdot 10^{-12}$ m^2
- Siltstone with boulders $1 \cdot 10^{-13}$ m^2
- Sandstone $1 \cdot 10^{-11}$ m^2
- Siltstone $1 \cdot 10^{-10}$ m^2
- Seam $1 \cdot 10^{-8}$ m^2
- Crushed zone $1 \cdot 10^{-7}$ m^2
- Cement-fly ash mixture in the borehole $1 \cdot 10^{-5}$ m^2

Pressure acting underground is 10 MPa. It corresponds to a geostatic pressure at a given depth below the surface. The shoed depth is 450 metres. The gas permeability of cement - fly ash mixture in the borehole is irrelevant as it is closed on all sides. We can, therefore, assume that there is an airtight mixture in the borehole. The model of geological layers is shown in Fig. 4.

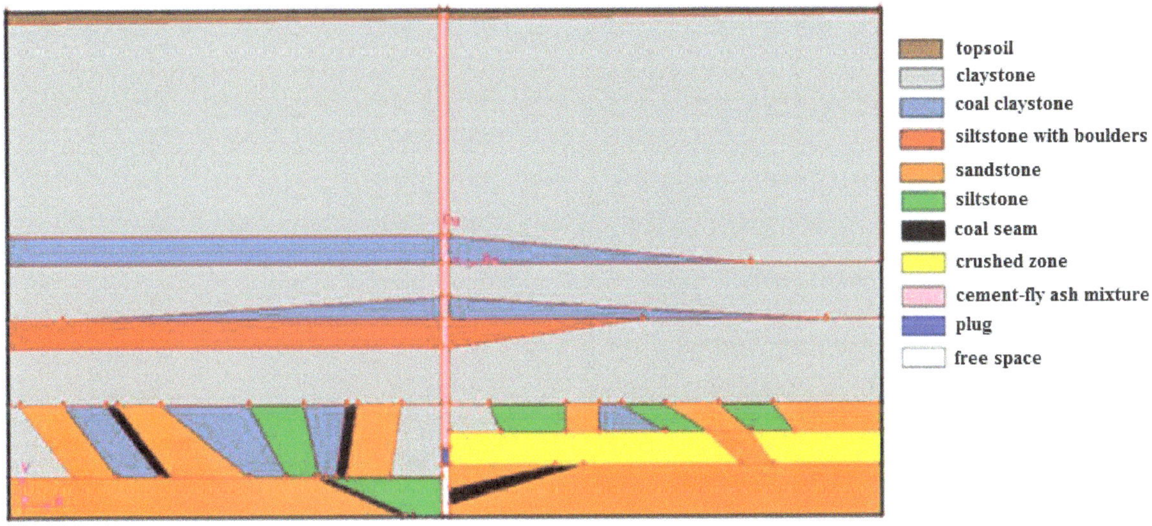

Fig. 4 Model of geological layers for the Fluent program [13]

The resulting values of static pressure calculated in the Fluent program from the specified input parameters are shown in Fig. 5. The results show that in the environment modelled in Fig. 5, the static pressure is divided in the left and right parts of the model, depending on the configuration of geological layers. Just below the surface it reaches a value of 0.5 MPa which causes that the mass flow rate of CO_2 reaches 0.026 kg/s above ground. When compared with the result in accordance with the equation (1), the value is 3 orders higher, and it would probably cause significant problems. However, it must be counted with such contingency for the final draft. The negative value in the model at the surface (in the layer of topsoil) means that there is a pressure by 134 Pa lower than the atmospheric one, so that the surface pressure should be 101,200 Pa. [14]

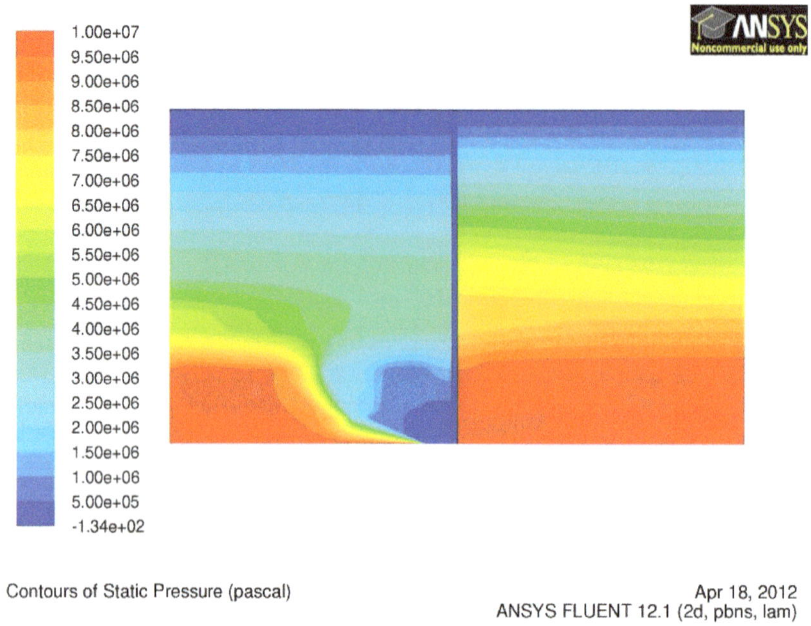

Fig. 5 Results of the distribution calculation of static pressure [13]

4 CONCLUSIONS

Even although there is no consensus of the scientific community, whether it is provable global warming is caused by increased carbon dioxide emissions, yet many individuals and organizations efforts are developed, this phenomenon not to be underestimated and timely implementation of measures to be realized to reduce the concentrations of CO2 in the atmosphere. As part of this activity, a number of scientific studies were performed and also some practical projects were implemented. Most suggestions look for a way to reduce the CO_2 content in a way of storing it underground. The procedure is quite logical, because most human activity of produced carbon dioxide comes from fossil fuels, which occurred in underground environments of the Earth. Also there are certain reservations to how it is stored underground. In any case, it would be advisable to invest not only into the CCS technology, but also into the technologies for generating energy from renewable resources. The article dealt with the possibility of storing CO_2 underground at the abandoned Paskov Mine. This is only a conceptual idea. To implement such a storage method, it is necessary to acquire a large range of input data, in particular the data on the geological survey of the site position and situation at the mine. It is also necessary to determine the properties of the rock mass, strength characteristics and in particular the data on gas permeability of layers, seams and the environment that was affected by mining activities.

REFERENCES

[1] Skleníkové plyny-výklad. *(Greenhouse gases - interpretation.)* In: *Change: Jak můžete vy sami odvrátit změnu klimatu? (In: Change: How can you yourselves prevent the climate change?)* [online]. European Commission, 2011, 15.6.2012 [cit. 2012-06-29]. Available at: http://ec.europa.eu/clima/sites/campaign/pdf/gases_cs.pdf

[2] Česká republika. Závěrečná zpráva z hodnocení dopadů návrhu zákona 82/2012 Sb o ukládání oxidu uhličitého do přírodních horninových struktur a o změně některých zákonů. In: 2011. *(Czech Republic. The final report from the evaluation of the impact of the Act draft 82/2012 Coll on storage of carbon oxide in natural rock structures and on amending certain acts. In: 2011.)*

[3] BLUMA, Aleš. EU chce dostat skleníkový plyn pod zem: Evropská unie chce s globálním oteplováním bojovat i prostřednictvím uskladňování oxidu uhličitého. *(EU wants to get the greenhouse gas under the ground: The European Union wants to fight global warming through storage of carbon dioxide.)* In: *IHNED.cz*: *Hospodářské noviny (Economic News)* [online]. Prague: Economia, 2008, 31.10.2008 [cit. 2012-06-29]. Available at: http://tech.ihned.cz/veda/c1-29709220-sklenikovy-plyn-pod-zem

[4] Klimaticko-energetický balíček. *(The climate and energy package.)* In: *Ministerstvo životního prostředí (Ministry of the Environment)* [online]. Prague: Ministerstvo životního prostředí *(Ministry of the Environment)* [cit. 2012-06-29]. Available at: http://www.mzp.cz/cz/klimaticko_energeticky_balicek

[5] Klíčové indikátory ŽP ČR. *(Key indicators of the environment of the Czech Republic.)* CENIA. *EMISE SKLENÍKOVÝCH PLYNŮ – vyhodnocení indikátoru (GREENHOUSE GAS EMISSIONS - indicator evaluation)* [online]. 4.10.2011 [cit. 2012-09-07]. Available at: http://issar.cenia.cz/issar/page.php?id=1508

[6] Oxid uhličitý. *(Carbon dioxide.)* In: *Wikipedie: Otevřená ncyklopedie (Open Encyclopedia)* [online]. 2010 [cit. 2012-06-29]. Available at: http://cs.wikipedia.org/wiki/Oxid_uhli%C4%8Dit%C3%BD

[7] EU chce dostat skleníkový plyn pod zem. *(EU wants to get the greenhouse gas under the ground.)*: In: *All for power: Informační portál a časopis (Information portal and magazine)* [online]. All for power, 2008, 18.6.2009 [cit. 2012-07-01]. Available at: http://www.allforpower.cz/clanek/eu-chce-dostat-sklenikovy-plyn-pod-zem/

[8] CO2GeoNet. In: *The European Network of Excellence: Co to vlastně je geologické ukládání CO_2? (What exactly is the geological storage of CO_2?* [online]. 2011 [cit. 2012-06-29]. ISBN: 978-80-7075-767-3 Available at: http://www.co2geonet.com/UserFiles/file/Rowena/czech%20brochure_protected.pdf

[9] Carbon Dioxide: from Necessity for Life to Greenhouse Gas. In: *Bright hub: The Hub for Bright Minds* [online]. New York, 2007 [cit. 2012-07-01]. Available at: http://www.brighthub.com/guides/carbon-dioxide.aspx

[10] DOPITA, M., V. HAVLENA and J. PEŠEK. *Ložiska fosilních paliv. (Deposits of fossil fuels.)* Prague: SNTL, 1985.

[11] URBAN, Petr. *Řešení plynových poměrů při ukončení hornické činnosti s aplikací na Důl Paskov-Závod Paskov, o.z., OKD, a.s. (The solution of gas conditions at the end of mining operations with the application to the Paskov Mine-Paskov Plant, o.z., OKD, a.s.)* Ostrava, 2000. Disertační práce. (Doctoral thesis.) VŠB-TU Ostrava. Vedoucí práce Pavel Prokop. *(Thesis Supervisor Pavel Pokop.)*

[12] LÁT, J. *Dobývání nízkých, ploše uložených slojí ve složitých důlně-geologických podmínkách. (Mining of low, flat stored seams under complex mining and geological conditions.)* Prague: MŠMT, 1988. *(Ministry of Education, Youth and Sports)*

[13] STAŠA, Pavel. *Zjišťování plynopropustnosti vrstev pro uložení CO_2 v lokalitě dolu Paskov v programu Fluent. (Determination of gas permeability of layers for CO_2 storage in the locality of the Paskov Mine in the Fluent program.)* Ostrava: VŠB-TU Ostrava

[14] CHOVANCOVÁ, Kamila. *Snižování emisí oxidu uhličitého a jeho využití k ukládání do podzemí. (Reducing emissions of carbon dioxide and its use for storage underground.)* VŠB-TU Ostrava, Faculty of Mining and Geology, 2012. Bakalářská práce. *(Bachelor thesis.)*

[15] POLANECKÝ, K. at al., 2010: *Chytrá energie: konkrétní plán ekologických organizací, jak zelené inovace a nová odvětví mohou postupně proměnit energetický metabolismus české ekonomiky - a srazit znečištění, dovoz paliv i účty za energii. (Smart energy: a specific plan of environmental organizations how green innovations and new industries can gradually transform the energy metabolism of the Czech economy - and reduce pollution, fuel imports and energy bills.)* Hnutí Duha, Brno. Available at: http://www.chytraenergie.info/images/stories/chytra_energie.pdf.

[16] LUTYŃSKI, M. (2010). A concept of enhanced methane recovery by high pressure CO_2 storage in abandoned coal mine. *Gospodarka Surowcami Mineralnymi / Mineral Resources Management, 26*(1), 93-104.

[17] PIESSENS, K., DUSAR, M.: Feasibility of CO_2 sequestration in abandoned coal mines. (2000) Geologica Belgica Volume, 7 (3-4). Brussels

[18] TARKOWSKI, R., ULIASZ-MISIAK, B. Struktury geologiczne perspektywiczne do składowania CO_2. (2005) Polityka Energetyczna, 8, pp. 551-560.

[19] LABUS, K., BUJOK, P. (2011). CO_2 mineral sequestration mechanisms and capacity of saline aquifers of the upper silesian coal basin (Central Europe) - modeling and experimental verification. Energy, 36(8), 4974-4982.

[20] LABUS, K., & BUJOK, P. (2011). Hydrogeochemistry of CO_2 aquifer sequestration - preliminary modeling and experimental examinations. [Hydrogeochemia sekwestracji CO_2 w poziomach wó donośnych - Wstepne badania modelowe i eksperymentalne] Biuletyn - Panstwowego Instytutu Geologicznego, (445 PART 1), 355-362.

RESUMÉ

Tento článek se zabýval pouze základní studií možnosti uložení odpadního CO_2 do podzemních prostor již uzavřeného Dolu Paskov. Studie ukázala, že je nutno získat velkou škálu vstupních údajů. Jsou to především geologická pozice lokality a to ze širšího pohledu oblasti i užší situace na předmětném dole. Dále je nutno zjistit vlastnosti horninového masivu, pevnostní charakteristiku a zejména údaje o plynopropustnosti (permeabilitě) vrstev, slojí i prostředí, které bylo ovlivněno hornickou činností. Stávajícími matematicko – fyzikálními zákonitostmi bylo teoreticky řešeno, zda pod tlakem uložený oxid uhličitý bude nebo nebude mít možnost unikat na povrch.

24

STUDY AND RESEARCH ON CLEANING PROCEDURES OF ANAEROBIC DIGESTION PRODUCTS

Silvie Heviánková [1], *Miroslav Kyncl* [2], *Jana Kodymová* [3]

[1] *assoc. prof., Environmental engineering department, Faculty of Mining and Geology VSB-Technical University of Ostrava, 17. listopadu 15, 708 00, Ostrava- Poruba, Czech Rebublic*

e-mail silvie.heviankova@vsb.cz

[2] *prof., Environmental engineering department, Faculty of Mining and Geology VSB- Technical University of Ostrava, 17. listopadu 15, 708 00, Ostrava- Poruba, Czech Rebublic*

e-mail miroslav.kyncl@vsb.cz

[3] *Dr., Environmental engineering department, Faculty of Mining and Geology VSB- Technical University of Ostrava,17. listopadu 15, 708 00, Ostrava- Poruba, Czech Rebublic*

e-mail jana.kodymova @vsb.cz

Abstract

This paper focuses on increasing quantities of digestate, a final product of anaerobic digestion, in biogas stations being built as alternative sources of energy. The potential use of digestate is limited due to its rather specific physical properties. This paper presents current approaches to digestate and digestate management. The objective is to compare the properties of digestate and the products of its separation via centrifuging, i.e. the liquid phase known as digestate liquor, and the solid phase referred to as digestate fibre. Its focus is mainly laid on techniques for dewatering digestate in low-solid anaerobic processes only, which have been tested for the effectiveness of the basic physical and chemical methods including their combinations. The measured results show that the use of coagulants and flocculants for this purpose would be very problematic in practice with regard to their high consumption as well as the need to input other elements into the process.

Abstrakt

Předložený článek se zabývá problematikou narůstajícího množství digestátu jako konečného produktu anaerobní digesce v bioplynových stanicích, budovaných jako alternativní zdroje energie. Digestát má v současné době omezené využití, které je dáno jeho specifickými fyzikálními vlastnostmi, zejména extrémně problematickou separací jednotlivých složek, ale také chemickým a mikrobiologickým složením. Jeho vlastnosti jsou z největší části ovlivněny druhem zpracovávané biomasy. V úvodní části tato studie předkládá současné možnosti nakládání s digestátem. Předmětem experimentální části je popis odběru a analýzy vzorků digestátu včetně jeho zpracování v laboratorních podmínkách použitím odstředění a následné posouzení vlastností digestátu a produktů separace, tj. fugátu a separátu a dále testování koagulačních a flokulačních činidel pro účinnější separaci tuhé fáze z digestátu, resp. fugátu. Použití koagulantů a flokulantů pro tyto účely v praxi by bylo značně problematické s ohledem na jejich vysokou spotřebu a vnos dalších prvků do procesu.

Key words: biogas plant, biogas, whole digestate, digestate liquor, digestate fibre, flocculation and coagulation agents

1 INTRODUCTION

Digestate is produced by biogas plants (BGPs) whose number significantly rises. The growth is caused by the increasing interest in alternative, or renewable energy sources, and the requirements for reducing biodegradable waste. The use of alternative energy sources is also an integral part of the energy policy in the Czech Republic, similarly to the European Union, where biogas shows a growing trend.

The process used in biogas plants is a process of anaerobic digestion (also called fermentation). It is a complex biochemical, chemical and physicochemical process of organic matter conversion in the absence of air (oxygen) to a gas mixture and a fermented undecomposed organic residue. This process consists of four consecutive stages: hydrolysis, acidogenesis, acetogenesis and methanogenesis. The product of one phase serves as a substrate for the second phase, and failure or disruption of one of the activities can break the dynamics of the system. These process steps are subsequent, but in continuous operation they occur simultaneously. The degree of lability and stability within the hydrolysis of organic substances (e.g., the difference between carbohydrate raw materials and sawdust with high lignin content) is an important factor as well. (Dohányos at al., 1998; Schulz, 2004; Straka et al., 2006; Kára et al., 2007)

The process of anaerobic fermentation affects the inaccessibility of air, pH (optimum 6.3 to 7.8), substrate composition (which requires a correct ratio of organic matter, nitrogen and phosphorus, in addition to the presence of micronutrients and vice versa the absence of heavy metals, oxidants, etc.), residence time and applied technology. (Dohányos et al., 1998; Schulz, 2004; Kasali & Senior, 2007; Dohányos, 2014)

Biogas, which is a mixture primarily of methane and carbon dioxide, is the main product of anaerobic digestion besides several other minor compounds, and is used mainly for the production of electricity and heat. In addition to biogas, biogas plants produce a fermentation residue called digestate. The amount of the produced digestate increases proportionally with the increasing number of biogas plants. The growth in the number of BGPs are supported by grants from the EU.

The anaerobic digestion process can be theoretically performed with almost any organic material with high organic content and solid content of less than 50%. Feedstock for BPSs may be biodegradable municipal waste (BMW), sewage sludge, agricultural crops and wastes from livestock production.

It is possible to divide the anaerobic digestion processes into two types according to the solid contained in biomass – dry and wet, which is associated with the way of transporting biomass to a digester. Although the dry running method is cheaper (less heat loss during water heating), the wet anaerobic digestion process is used more frequently as the biogas production thus achieves higher stability. (Mužík and Slejška, 2014)

1.1 Digestate

Digestate is a stabilized material containing material undigested during digestion and necrotic microorganisms. Its properties are mostly affected by the kind of processed biomass and process parameters.

Digestate may be classified according to various aspects, e.g. dry matter, feedstock or methods of its application. Most often, we encounter digestate which is the feedstock from agriculture and is most widely used for fertilization. It should be noted that from a view of farmers, there is a problem consisting in its large amount and the fact that it is too fluid (low dry matter content usually not exceeding 8 wt. %).

Digestate may be perceived from several points of view. First, it is a by-product in biogas generation, applicable as a mineral fertilizer. Second, it is waste which may cause penetration of hazardous or even toxic substances in the soil, which are not degradable by soil microorganisms. Therefore, they accumulate there that is unsustainable in the long run.

On the basis of the previously acquired information and experience, it can be concluded that the composition of digestate is suitable for fertilization, because up to 50% of organic nitrogen, such as ammonium ion, is released in fermentors by plants in the form of mineral uptake. If we assess the amount of nutrient in fertilizer from the whole digestate, especially a mineral and nitrogen fertilizer (in absolute values), it might seem that it contains high amounts of phosphorus and potassium, but in terms of agrochemicals the amounts are relatively low. (Kolář et al., 2010). On the other hand, it contains large amounts of relatively stabilized indigestible organic matter, because almost all biodegradable organic substances were converted during the long process of anaerobic digestion to biogas, and it is not clear whether the organic matter present in the digestate will be well spent by microorganisms in the soil. There is very problematic high water content, however, mainly in that part of the digestate, to which nitrogen in an unstable form is bound, which is well applicable for plants (crops). Dry matter content of the typical digestate is between 2-8%, usually not exceeding 4%.

The current research and available literature deal with the fermentation process itself, its intensification and various types of substrates, i.e. agricultural crops specially grown for the purposes and waste from agriculture. For more information, see Kolář et al., 2010; Hansges el al., 2011; Bühle et al., 2011; Braguglia et al., 2012; Martín et al., 2002; Dewil et al., 2007. The research has been carried out on the properties of digestate improving its applicability as fertilizer to enhance the plant growth or enhancement of soil structural properties with respect to aeration. For more information, see Kolář et al., 2011; Mareček et al., 2010; Kolář et al., 2010. If the research is carried out by agricultural organizations, they often aim to increase the proportion of dry matter in the digestate or decrease the quantity of the liquid phase, which is more advantageous in terms of the economy of fertilizing. However, the issue of digestate has not been studied in its complexity. On the one hand, useful components are to be used to maximum, on the other hand, the environment should be protected.

1.2 Possibilities of whole digestate management

The following diagram shows theoretical possibilities of disposing the whole digestate or cleaning the digestate, and was generated on the basis of available literature sources. (Blumerg, 2014; Chambers, 2011; Chauzy et al., 2010; Fuschs et al., 2010; Nielsen et al., 2005; Degremont. Heliantis™ Solar Sludge Drying, 2012)

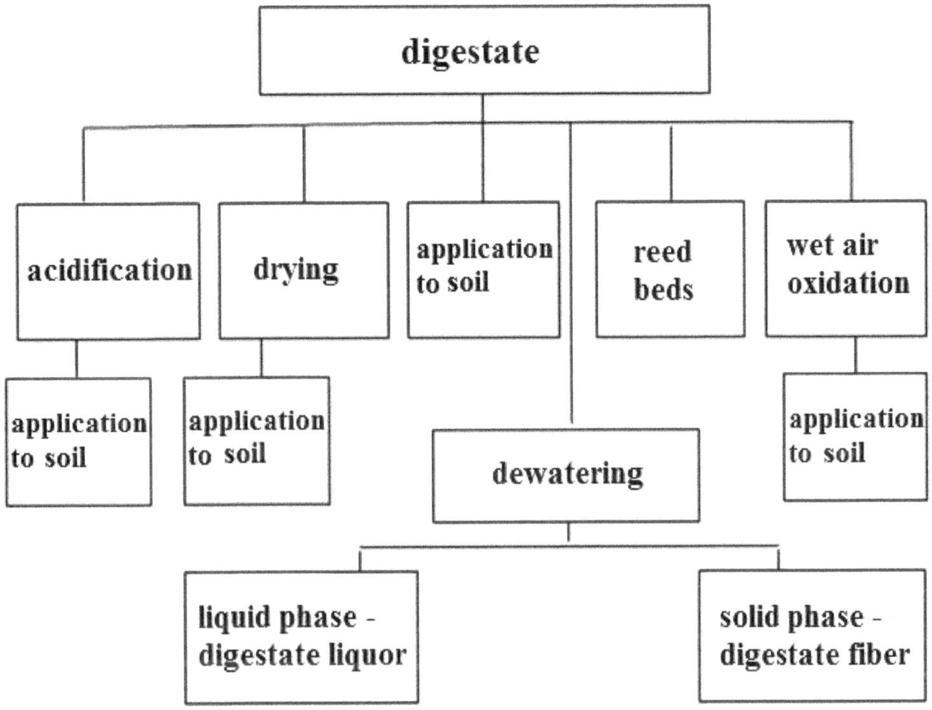

Fig. 1 Variants of digestate management

Digestate can be applied directly to the soil, which is a commonly used procedure. To prevent the release of ammonia, the acidification process can be used; however, it is necessary to take into account that this digestate management method will be connected with soil acidification, which is undesirable for the most types of soils.

By drying, if it is dried, the digestate has got rid of excess water. The dried digestate can be applied to soil as a fertilizer and is much easier to handle. However, in general, the thermal methods are very expensive. Biogas releases from cogeneration units a large amount of heat, but the amount is not able to ensure the drying of the digestate so it is often used only for drying wood biomass. Although the digestate volume is relatively large in a BGP, the drying of digestate is not frequently used in practice. It is also possible to use solar drying. In this case, the digestate is fed into ventilated greenhouses. Nevertheless, there is a disadvantage due to the requirement for a large area, which cannot be ensured for following volumes of digestate, despite the possibility to provide the waste heat from cogeneration units. Each method has its drawbacks, whether the high energy demand or the needed amount of space.

The evaporation, which is able to ensure about 20% dryness of digestate, is a less economically demanding process than drying.

Wet air oxidation is the oxidation of liquid sludge (digestate), under air conditions of increased pressure (4-6 MPa) and temperatures (200-300°C), which represents up to 90% of organic matter transferred in the sludge into the liquid phase in a form of mixtures of short-chain fatty acids and methanol. Even in this case, there are complicated and economically demanding technology requirements, such as high temperature and pressure, and final treatment of resulting gas and waste water.

Reed beds are a good solution for the production of low volume digestate for sites with low volume production of digestate. These are areas below ground level where reeds are planted and the digestate is being cleaned in the root system. Simultaneously, the digestate is thickened by evaporation. Water flowing from such system is useful for example for irrigation. Mineralization of digestate takes up to 10 years. Then it is possible to extract the sludge and apply it as an inorganic fertilizer. This technology is highly dependent on the size of required area.

2 THICKENING AND DEWATERING OF DIGESTATE

As already mentioned, there are problematic properties of digestate, such as low solid content, so thickening and dewatering is the only option and a separate chapter of the digestate management. Requirements for these processes are as follows: gravitational forces, centrifugal forces or pressure.

When thickening the digestate, 5-10 % solid content of sludge can be reached. A horizontal belt filter or centrifuge can be used. The process can be intensified by adding a polymer, which causes flocculation of colloidal particles. However, the required amount of polymers is extremely high (see results below).

When dewatering the digestate, 18% solid content of slurry and more can be achieved. For these purposes, we can use sieve belt presses or centrifuges as well as the process can be intensified in this case by dosing a polymer. The separated solid phase is known as digestate fibre and the separated liquid phase as digestate liquor. The digestate slurry or the solid digestate from high solids "dry" anaerobic digestion often has 18-25% total solid content and some technologies produce even drier digestate. But there are only tens of "dry" biogas plants in Europe.

2.1 Possibilities of digestate fibre management

The following diagram (Fig. 2) summarizes theoretical possibilities of disposing the digestate fibre, and was generated on the basis of available literature sources. (Chambers, 2011; Fuschs et al., 2010; Pyreg, 2011; Pell Frichmann consultans Ltd., 2012; Degremont. Heliantis™ Solar Sludge Drying, 2012)

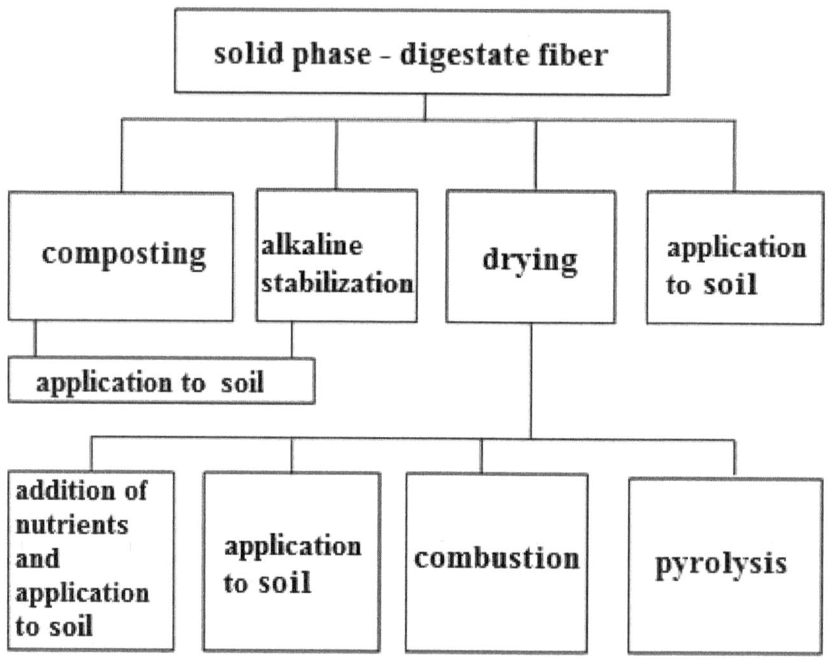

Fig. 2 Selected options for digestate fibre management

The options include the composting of digestate fibre, lime stabilization, which results in the destruction of pathogens, however, an increase in pH, which almost always causes the release of ammonia also included in the digestate fibre (but several years ago, fertilization with ammonia was common so it should not be a problem for farmers). The direct application to the soil is the most frequently used option. Finally, it is also possible to dry the fibre fraction and to enrich it with nutritional elements, and thus, improve its quality as a fertilizer, and apply it into the soil. The fibre fraction of digestate is a good solution to lighten and aerate the soil. The improved access of air to the root system and soil improvement hydrolimits can have a good yield effect, which unfortunately is often mistakenly ascribed only to nutrients. (Vaněk et al., 1998)

The dried digestate fibre may be used as fuel for combustion or fuel production such as pyrolysis and gasification. Adding suitable inert soil to the digestate fibre (or wash it with water), and then use it as a substrate for pot culture is another theoretical option.

2.2 Possibilities of digestate liquor management

The following diagram (Fig. 3) summarizes the options for the management of digestate liquor. (De la Rubia, 2010; Botheju and Bakke, 2010; Driver, 1998; Gao, 2010; Guštin, 2011; Iyovo and Du, 2010; Maurer et al., 2001; Tchobanoglous et al., 2004; Evans, 2009)

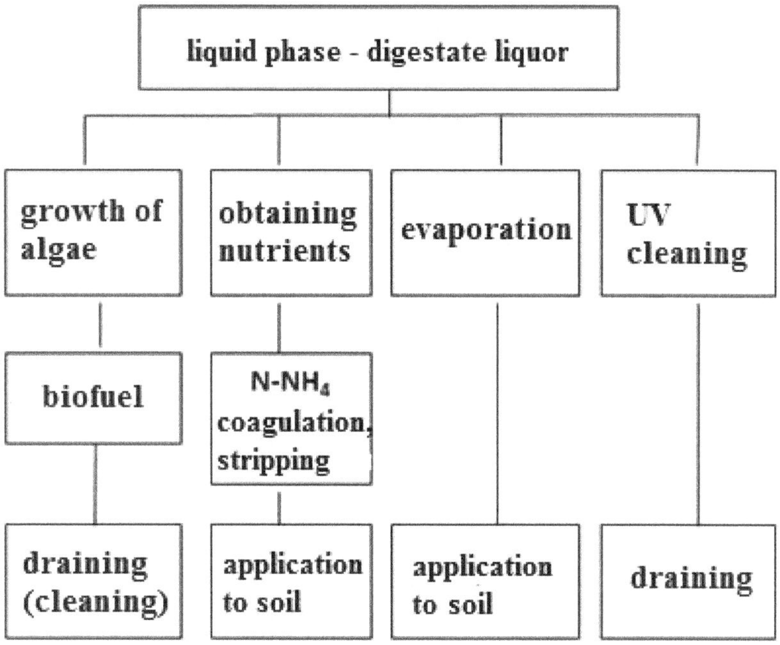

Fig. 3 Selected options of digestate liquor management

The liquid phase can be used for the production of algae as a source for fuel production. In Netherland, digestate liquor is used as a raw material for the growth of algae which is further used for fuel production.

The digestate liquor contains valuable elements. It is especially ammonium nitrogen that is most appropriately directly ploughed into the soil; however, the struvite precipitation is another theoretical option.

Evaporation (wet surface heat exchangers), typically utilised for whole digestate treatment, can be realized with equipment operating on the principle of a heat exchanger. After the evaporation, the material can be re-mixed with digestate fibre and used for fertilization. This system was launched in Scotland and is able to reduce the volume of digestate liquor to 15% of the initial volume, while the waste heat comes from a cogeneration unit.

Cleaning the digestate liquor using ultrafiltration and reverse osmosis – authors in foreign publications refer to the laboratory results of these methods, but their practical application in biogas plants in available resources have not been found. That is why their use is highly questionable with regard to the fouling of membranes, optionally filters, serving as protection against microscopic colloidal particles.

3 EXPERIMENTAL PART

Within the experimental part, all digestate samples were collected from agricultural biogas plants in the Moravian-Silesian and Olomouc Regions; the samples were subjected to tests of separation of the solid phase, and digestate products (digestate liquor and digestate fibre) were analysed and coagulation and flocculation agents were tested for the intensification of the process of separation of the solid phase of whole digestate.

3.1 Sampling and analysis of whole digestate, digestate liquor and digestate fibre

Digestate samples were collected from agriculture biogas plants in two-month intervals. The monitored BGP processed organic waste from pigs, silage from well silage agricultural crops (corn and beet tails), and at the end of summer months, fresh grass or sorghum, and during winter and spring months, silage with high solid content (more than 50% of dry matter) were added. After centrifuging the sample of digestate on laboratory conditions, the digestate liquor and digestate fibre fractions were subjected to a chemical analysis. They were monitored for CODCr, BOD5, total phosphorus and ammonia contents.

The aim was to compare the properties of the whole digestate and the separated products by centrifugation, (liquid phase – digestate liquor and solid phase – digestate fibre). The obtained results were compared. Laboratory tests and cleaning of the whole digestate and digestate liquor were carried out in the laboratory at the Technology and Water Management Department of the Faculty of Mining and Geology at the VSB-Technical University of Ostrava. The analyses of the samples were performed also in an accredited laboratory.

The results that have been obtained from both laboratories are the same (even if every time different units were used). The digestate and digestate liquor density is about 1000 kgm^{-3} so the results can be mutually compared even if they are expressed in $mg.kg^{-1}$ of the original weight or in $mg.l^{-1}$ of the sample. The analyses of the samples were carried out in the company Laboratoř MORAVA, s.r.o., an accredited laboratory. In the samples of the whole digestate, digestate liquor and digestate fibre, solids were determined according to EN 12879: total phosphorus by JPP ÚKZÚZ (a uniform workflow of the Central Institute for Supervising and Testing in Agriculture), ammonia nitrogen by JPP ÚKZÚZ, $CHSK_{Cr}$ according to ISO 6060, and BOD_5 according to EN 1899-1.

Further analyses of digestate and fugate were conducted in the university laboratory. The samples density was determined using the pycnometer method, total phosphorus according to EN ISO 6878, ammonia nitrogen according to ISO 7150-1, COD_{Cr} according to TNV 757520, and BOD_5 according to EN 1899-1.

4 RESULTS AND DISCUSSION

According to a subjective assessment of all digestate samples, it can be concluded that the digestate is a heterogeneous viscous mixture of dark brown-green colour with visible unaltered remains (e.g. corn silage) and with the typical putrefactive and ammonia odour.

The measured properties of digestate from various agricultural sources have comparable values. Due to the high concentrations and the need for high dilution of samples, the measurements have a degree of uncertainty; however, understanding the issue is sufficient.

The average value of total phosphorus was P_{total} 800 $mg.kg^{-1}$ of the initial sample weight. The average value of ammonia was $N\text{-}NH_4$ 2.500 $mg.kg^{-1}$ of the initial sample weight. The average value of organic substances, expressed as COD_{Cr} in the whole digestate was 30.000 $mg.kg^{-1}$ of the original sample weight. The average value of biodegradable substances expressed as BOD_5 in the whole digestate was about 3200 $mg.kg^{-1}$ of the original sample weight. BOD_5 values are approximately 10 times lower than the values of $CHSK_{Cr}$.

4.1 Separation of digestate fibre

First the digestate samples were filtered through the KA1-M filter paper, but the digestate was not filterable under atmospheric pressure. The digestate sample was not filterable either in vacuum. For the solid phase separation (digestate fibre), the laboratory centrifuges of the Merci company were used. The digestate was dispensed into plastic tubes of 200 ml and subjected to centrifugation at 4100 rpm for 5 min.

Fig. 4 shows photographs of products separated by centrifugation of the solid digestate which is in the liquid phase – digestate liquor representing around 70-80% of the initial volume of the whole digestate. The digestate liquor shown on the left, can be characterized as a viscous brownish-green liquid lighter than the starting mixture (whole digestate is shown on the right) with a lower content of fine sludge. There is a noticeable residue of silage no longer. Also sensory properties were significantly better than those of the whole digestate.

The further products obtained by the centrifugation were in the solid phase – digestate fibre constituting about 20-30% of the initial volume of the whole digestate. The digestate fibre can be characterized as a solid engravable mixture containing the remains of the undecomposed input substrate and fine sludge with a typical smell.

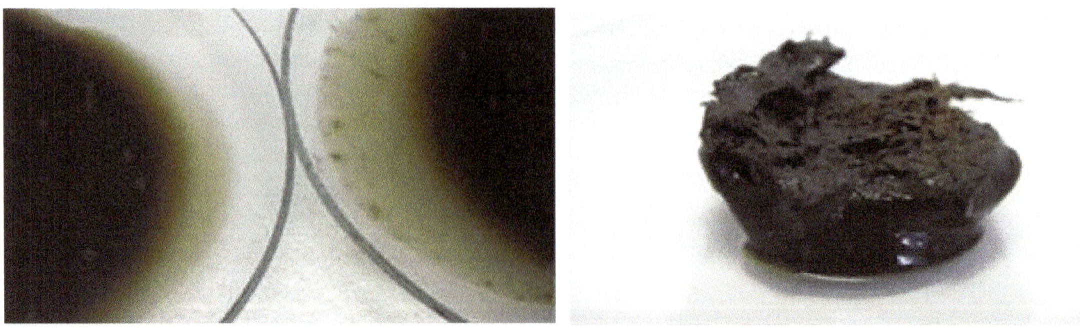

A B C

Fig. 4 Samples of digestate liquor (A); whole digestate (B); digestate fibre (C)

The results of compositions of digestate fibre and digestate liquor from individual agriculture biogas plants were also compared.

The average value of total phosphorus was Ptotal 800 $mg.kg^{-1}$ of the initial sample weight. The average value of ammonia was $N\text{-}NH_4$ 2.500 $mg.kg^{-1}$ of the initial sample weight. The average value of organic substances expressed as COD_{Cr} in the whole digestate was 30.000 $mg.kg^{-1}$ of the initial sample weight. The average value of biodegradable substances expressed as BOD_5 in the whole digestate was about 3200 $mg.kg^{-1}$ of the initial sample weight. The BOD_5 values were approximately 10 times lower than the values of $CHSK_{Cr}$.

Using the ratio of BOD_5:COD_{Cr}, it can be concluded that it is a heavily biologically degradable substance. Digestate fibre is mainly applicable in agriculture for soil aeration, although it contains a relatively high proportion of total phosphorus and ammonia; but it is attached to the stabilized organic material and it is almost inaccessible for plants. It had been expected as the instable organic matter in the feed material was decomposed by anaerobic fermentation for about 40 days or more.

With regard to the concentrations of ammonia nitrogen and phosphorus, it is apparent that phosphorus is mainly in the separator, while the distribution of ammonia in the two phases is approximately the same.

Tab. 1 Average values of indicators in samples of whole digestate, digestate liquor and digestate fibre

Indicator	Unit	Whole digestate	Digestate liquor	Digestate fibre
Dry mass	%	4.32	1.52	16.3
P_{celk}	$mg.kg^{-1}$ of orig. weight	800	200	1 600
$N\text{-}NH_4$	$mg.kg^{-1}$ of orig. weight	2 500	2 500	2 500
$CHSK_{Cr}$	$mg.kg^{-1}$ of orig. weight	30 000	26 000	97 000
BSK_5	$mg.kg^{-1}$ of orig. weight	2 800	3 200	8 000

Digestate liquor (liquid phase) is heavily polluted, so cleaning procedures will be tested and analysed. It is important to realize that the digestate liquor contains valuable elements for farmers, therefore the aim, besides the cleaning, will also be to get the solid phase which is also useful for fertilization. It is necessary to consider the effectiveness of digestate liquor cleaning, because in agricultural practice, the application of liquid-suspension fertilizers is quite common.

4.2 Testing coagulating agents

Coagulation and separation is a set of processes that may be used for removing colloidal and macromolecular organic compounds from wastewater. Ferric or aluminium salts are used as coagulating agents.

During the laboratory testing, hydrates of ferric chloride and ferric sulphate were used, containing a comparable amount of ferric ions. If water is too dirty, it is essential to choose an appropriate dose based on the detected contamination. Therefore, it is appropriate to use the basic benefit formula $DZ = 8 * c$, where $c = COD$ (Pitter et al., 1983) for the calculation.

As it is usual in the coagulation tests, series of tests were conducted with multiple basic doses from 0.1 to 1.2 times the calculated base dose. The experiments were conducted with both the digestate liquor and the whole digestate.

The used coagulants caused a decrease in pH, therefore it was necessary to correct the course of coagulation and adjust the pH value to about 5, which was carried out by dosing calcium hydroxide. It was also required to add a certain amount of dilution water that was added in the process of dispensing the solutions of coagulants. The experiments were carried out through the use of 20% solutions of $FeCl_3$ or $Fe(SO_4)_3$. During the experiments, the suspension with lime milk was mixed using the LAVAL magnetic stirrer for 1 min. After that a dose of coagulant was added and the sample was mixed with the magnetic stirrer for 2 minutes at maximum speed and additional 10 minutes at low speed. Subsequently, the pH value was measured and the sample was filtered through the filter paper for quantitative analyses KA-4 at medium speed. The monitored indicators in the filtrate were determined.

4.3 Coagulant $FeCl_3$

It is necessary to add coagulant into the digestate (digestate liquor) to reduce the pH value to about pH 2. The value of pH is modified by the dose of $Ca(OH)_2$ as a 10% suspension. In order to achieve efficient coagulation, the doses of 50 ml of 10% slurry of $Ca(OH)_2$ and 160 ml of 20% $FeCl_3$ solution to 1 litre of digestate (digestate liquor) were used.

After the use of the effective doses of agents, it can be stated that the value of COD_{Cr} was reduced to about 600 $mg \cdot l^{-1}$ (η 99%); $N\text{-}NH_4$ 350 $mg.l^{-1}$ (η 85%); P_{total} 3 $mg.l^{-1}$ (η 99%). Conversely, there were found out increased concentrations of Cl^- 9300 mg.l^{-1}; Ca^{2+} 1200 mg.l^{-1}, pH was 4.8. In the analysis of the sludge, N, P, C, Fe and chlorides prevailed. (Vidlářová, 2013)

4.4 Coagulant $Fe_2(SO_4)_3$

It is necessary to add coagulant into the digestate (digestate liquor) to reduce the pH value to about pH 3. The value of pH is modified by the dose of $Ca(OH)_2$ – 10% suspension. To achieve efficient coagulation, the doses of 80 ml of 10% slurry of $Ca(OH)_2$ and 160 ml of 20% solution of $Fe_2(SO_4)_3$ in 1 l of digestate (digestate liquor) were used.

After the use of the effective doses of agents, it can be stated that the value of COD_{Cr} was reduced to about 500 mg · l^{-1} (η 98%); N-NH_4 1000 mg.l^{-1} (η 60%); P_{total} 2 mg.l^{-1} (η 99%); and Ca^{2+} and SO_4^{2-} concentrations increased to 600 mg.l^{-1}; 8000 mg.l $^{-1}$ respectively. The pH value was 5.8. In the analysis of the sludge, N, P, C, Fe and sulphates prevailed. (Vidlářová, 2013)

4.5 Evaluation of the use of coagulants

In order to achieve coagulation, extremely high doses of agents were used. The composition of the filtrate corresponded with the reagent. We achieved a significant reduction of organic compounds in the liquid phase, but there was also a significant increase in sulphate, or chloride and calcium ions. Likewise, in the filter cake, enormously high concentrations of nitrogen, phosphorus, organic matter iron and sulphates, or chlorides were found.

Regarding the usability of coagulants in the separated solid phase, which should primarily be used for fertilization of the soil, the content of nitrogen, phosphorus, calcium and organic substances is generally beneficial for fertilization. Also an iron concentration is beneficial, but the dose of coagulant must be high enough to carry out successful coagulation. Furthermore, the chloride content in fertilizers leads to the acidification of soil so it is preferable to use an organic material containing sulphates which can be commonly found in fertilizers (e.g. in fertilizers used for the cultivation of rape plant).

Coagulant tests use the same procedure for both the whole digestate and the digestate liquor (Fig. 5). It has been discovered that the samples of the digestate are filterable after coagulation when using the same volumes as for the coagulants in testing the digestate liquor, despite the fact that the undiluted digestate has a higher concentration than the undiluted digestate liquor COD_{Cr}, i.e. 30,000 mg.l^{-1} compared to 26,000 mg .l^{-1}. In several cases, foam formation was found. This can be eliminated by adding skimmers; or the flotation process can be tested in the whole digestate management, however, this will be the subject of further research. (Vidlářová, 2013)

A B

Fig. 5 Sample of whole digestate after using coagulation agents (A); filtrate (B) ($Fe_2(SO_4)_3$)

4.6 Testing flocculation agents

Flocculation with an organic polymeric flocculant is a process of destabilization of polymer colloids (flocculants) and there are two mechanisms. In the first case, it acts as a coagulant neutralizing the negative surface charge of sludge particles to permit subsequent aggregation, and in the other, it serves as a row bridging mechanism to activate the formation of flakes.

Digestate, or digestate liquor, is a system of macroscopic and colloidal particles which consists mostly of negatively charged particles.

Flocculants were tested as 0.1 or 0.05% solutions. For individual experiments, after dispensing the selected dose of flocculants, the suspension was always stirred with a glass rod and left for sedimentation. If the macro flakes were created, the pH value was measured and the suspension was filtered through the filter paper for the quantitative analysis KA-4 at medium speed. The monitored indicators were determined in the filtrate.

Anionic flocculants (SOKOFLOC 12 and 26) were the first test agents, which, as expected, did not cause the system destabilization, and, therefore, were excluded from further experiments. We also tested various types of cationic flocculants (SOKOFLOC 56, SOKOFLOC 61, SOKOFLOC 65, SOKOFLOC 68, SOKOFLOC 109, Superfloc C475 and C496 Superfloc). Good results were achieved, i.e. the destabilization and the formation of macro flakes, when using e.g. SOKOFLOC 61 and 65.

To achieve smooth flocculation when using the flocculants, an additional and large volume of water was required in comparison with the experiments with coagulants. As for the composition of the filtrate, after the application of effective doses of agents, we claim that the $CHSK_{Cr}$ value decreased to about 2000 mg $\cdot$ l^{-1} (η 93%); NH_4-N 360 mg.l^{-1} (η 85%); P_{total} 10 mg.l^{-1} (η 99%). The pH value was around 7. (Souček, 2014)

Fig. 6: Sample of whole digestate after flocculation (Sokoflok 61)

5 CONCLUSIONS

Digestate is one of the major environmental problems regarding the growing number and characteristics, especially its low dry matter content. The paper presents various theoretical possibilities for the treatment of whole digestate, digestate liquor and digestate fibre that were found in available literature. It also presents the results of a basic chemical analysis of individual components (whole digestate, digestate fibre and digestate liquor) which clearly shows that the digestate liquor is a very stable material containing elements which are difficult to use for plants. Then the presentation of used methods for the separation and purification of the whole digestate follows. The digestate fibre obtained by centrifugation is available directly to fertilization, and the digestate liquor is the liquid phase of the whole digestate. The digestate liquor is heavily polluted, but also contains important elements useful for the purpose of fertilization. In the research of the purification of the whole digestate and digestate liquor, first the properties of samples from five agriculture BGPs were evaluated. It was determined that the properties are comparable so the cleaning procedures generally applicable to the digestate can be applied. The methods for separation of the solid phase of the digestate – digestate fibre or digestate liquor were tested as well. If flocculants and coagulants are not used, centrifugation may be applied as a separation process. When adding coagulation and flocculation agents to intensify the process, their consumption for effective coagulation and flocculation was enormously high; the solid phase was well-separable by sedimentation and filtration. The use of coagulants and flocculants in practice would be highly questionable and problematic due to their high consumption and the need to input further elements into the process and hence to the product separation.

ACKNOWLEDGEMENTS

The paper was elaborated with the support of the Ministry of Agriculture within the Project QJ1320159 "Research in processing, use and disposal of waste products from biogas stations".

REFERENCES

[1] Blumerg I. *Reed Bed Sludge Treatment.* [online] Available: http://www.blumberg-engineers.com/Gallery/Reed-bed-sludge-treatment.html (2014-09-12)

[2] Botheju D. S., Bakke R. Digestate Nitrification for Nutrient Recovery. *The Open Waste Management Journal,* 2010. No.3, 1 – 12. ISSN 1876-4002

[3] Braguglia C.M., Gianico A., Mininni G.: Comparison between ozone and ultrasound disintegration on sludge anaerobic digestion. *Journal of Environmental Management.* 2012. Vol. 96. pp. 139-143. ISSN: 0301-4797

[4] Bühle L., Stűlpnagel R., Wachendorf M.: Comparative life cycle assessment of the integrated generation of solid fuel and biogas from biomass (IFBB) and whole crop digestion (WCD) in Germany. *Biomass and Bioenergy.* 2011. Vol. 35. pp. 363-37. ISSN: 0961-9534

[5] Chambers B. Digestate Utilisation on Agricultural Land. 2011. [online] Available: http://warrr.org/1226/ (2014-11-2)

[6] Chauzy J; Martin J. C.; Cretenot D.; Roisere J. P. Wet Air Oxidation of Municipal Sludge: Return Experience of the North Brussels Waste Water Treatment Plant. *Water Practice & Technology*, 2010. Vol. 5. ISSN 1751-231X

[7] De la Rubia Á. M.; Walker M.; Heaven S.; Banks C., B.; Borja R. Preliminary trials of in situ ammonia stripping from source segregated domestic food waste digestate using biogas: Effect of temperature and flow rate. *Biosource Technology*, 2010. No.101, 9486 – 9492. ISSN 0960-8524.

[8] Degremont. Heliantis™ Solar Sludge Drying. 2012

[9] Dewil R, Appels L., Baeyens J., Degreve J.: Peroxidation enhances the biogas production in the anaerobic digestion of biosolid. *Journal of Hazardous Material.* 2007. Vol. 146. pp. 577-581.

[10] Dohányos M., Zábranská J., Jeníček P., Fialka P., Kajan M. *Anaerobní čistírenské technologie.* 1. vyd. Brno: NOEL 2000 s.r.o., 1998.345 s. ISBN 80-860020-19-3.

[11] Dohányos M.: *Anaerobní reaktor není černou skřínkou - teoretické základy anaerobní fermentace.* [online] Biom.cz: <http://biom.cz/cz/odborne-clanky/anaerobni-reaktor-neni-cernou-skrinkou-teoreticke-zaklady-anaerobni-fermentace>. (2014-09-08)

[12] Driver J. Phosphates recovery for recyling from sewage and animal wastes. *Phosphorus & Potassium*, 1998. Vol. 216, pp. 17 – 21. ISSN 0031-8426

[13] Evans T. D. Recovering ammonium and struvite fertilisers from digested sludge dewatering liquors. In: *Resource Recovery Not Wastewater Treatment Conference*. London: Aqua-Enviro, 2009. ISBN 1 84339 001 9

[14] Fuschs W.; Wager F.; Kirchmayr R.; Braun R.; Dorsg B. DigestateTreatment: Comparison and Assessment of Existing Technologies. In: *Third International Symposium on Energy from Biomass and Waste.* Venice: CISA, 2010. ISBN 978-88-6265-008-3

[15] Gao T. L. Using thermophilic anaerobic digestate effluent to replace freshwater for bioethanol production. *Bioresource Technology*, 2010. Vol. 102, pp. 2126 – 2129. ISSN 0960-8524

[16] Guštin S. M.-L. Effect of pH, temperature and air flow rate on the continuous ammonia stripping of the anaerobic digestion effluent. *Process Safety and Environmental Protection*, 2011. Vol. 89, pp 61 – 66. ISSN 0957-5820

[17] Hansges L.,Richter F., Wachendorf M.: Integrated generation of solid fuel and biogas from green cut material from landscape conservation and private household. *Bioresource Technology.* 2011. Vol. 102. pp. 10441-10450. ISSN: 0960-8524

[18] Iyovo G. D.; Du G. Chen J. Sustainable Bioenergy Bioprocessing: Biomethane Production, Digestate as Biofertilizer and as Supplemental Feed in Algae Cultivation to Promote Algae Biofuel Commercialization. *Journal of Microbial and Biochemical Technology.* 2010. Vol. 2, issue 7. pp.100 ISSN 1948-5948

[19] Kára J., Pastor, Z; Přibyl, E.; Hanzlíková I.; Andert D.; Gerndtová I.; Hutla P.; Měkovotá P.; Abrham Z.; Mužík O.; Hlinka J.; Bradna J. *Výroba a využití bioplynu v zemědělství.* Praha: VÚZT, v.v.i., 2007. 117 pp. ISBN 978-80-86884-28-8.

[20] Kasali G. B; Senior E. Effects of temperature and moisture on the anaerobicdigestion of refuse. *Journal of Chemical Technology and Biotechnology.* 2007. Vol. 44, no. 1, 31-41. ISSN 1097-4660

[21] Kolař, L., Kužel, S., Peterka, J., and Borova-Batt, J., Agrochemical value of the liquid phase of wastes from fermenters during biogas production. *Plant, Soil and Environment,* 2010. Vol.56: pp. 23-27.ISSN 1214-1178

[22] Mareček J., Vostoupal B..: Influencing the quality of biogas station digestate with respect to its subsequent use as fertilizing watering in plant production [online]. Brno 2010. Available at

http://eagri.cz/public/web/file/93920/Ovlivneni_kvality_digestatu_bioplynovych_stanic_vzhledem_k_jeho_naslednemu_vyuziti_jako_hnojive_zalivky_v_rostlinne_vyrobe.pdf>. (in Czech)

[23] Martín M. A., Raposo F., Borja R., Martín A.: Kinetic study of the anaerobic digestion of vinasse pretreated with ozone, ozone plus ultraviolet light, and ozone plus ultraviolet light in the presence of titanium dioxide. *Process Biochemistry.* 2002. Vol 37. pp. 699-706.

[24] Maurer M.; Mucke J.; Larsen T. A. *Nitrogen Revovery and Reuse.* [online] Available: agrienvarchive.ca/bioenergy/download/maurer_IWA_2001.pdf. (2014–09-12)

[25] Mužík O.; Slejška A. *Možnosti využití anaerobní fermentace pro zpracování zbytkové biomasy.* [online] Biom.cz: < http://biom.cz/cz/odborne-clanky/moznosti-vyuziti-anaerobni-fermentace-pro-zpracovani-zbytkove-biomasy >. (2014-09-10)

[26] Nielsen S.; Willoughby N. Sludge Treatment in Reed Bed Systems and Recycling of Sludge and Environmental Impact. In: *10th European Biosolids and Biowaste Conference.* UK, Aqua Enviro Technology Transfer, 2005. ISBN 1 84339 001 9

[27] Pell Frichmann consultans Ltd. Enhancement and treatment of digestates from anaerobic digestion. Project code OMK006-002. 2012

[28] Pitter P. F.; Tuček J.; Chudoba L. *Laboratorní metody v technologii vody.* Praha: SNTL, 1983. 304 pp.

[29] Schulz H., Eder B. *Bioplyn v praxi.* Ostrava: HEL, 2004. 168 s. ISBN 80-86167-21-6.

[30] Souček R. Testování intenzifikace oddělení tuhé fáze z digestátu při odstřeďování. *Diplomová práce,* Ostrava: VŠB-TUO, 2014. 62 pp.

[31] Straka F.; Dědek, J.; Dohanyos, M.; Kunčarová, M.; Malijevský, A.; Novák P. J.; Zábranská, J. *Bioplyn.* Praha : GAS s.r.o., 2006. 706 s. ISBN 80-7328-090-6.

[32] Tchobanoglous G.; Burton F. L.; Stensel D. H. *Waste Water Engineering Treatment and Reuse.*4th Edition, New York: McGraw Hill, 2004. ISBN 978-0070418783

[33] Vaněk, V.; Kolár, L.; Pavlíková ,D.; Tlustoš, P.(eds.) "Reasonable Use of Fertilizers", In: *Proceedings of the International Conference*, Czech University of Agriculture, Prague, 1998.

[34] Vidlářová P. Testování koagulačních činidel při separaci tuhé fáze v odpadním produktu z anaerobní digesce. *Bakalářská práce*, Ostrava: VŠB – TUO, 2013. 45 pp.

[35] Kolář, L.; Kuže, S.; Peterka, J. and Borová-Batt, J.(2011). Utilisation of Waste from Digesters for Biogas Production, Biofuel's Engineering Process Technology, Dr. Marco Aurelio Dos Santos Bernardes (Ed.). Available from: http://www.intechopen.com/books/biofuel-s-engineering-process-technology/utilisation-of-waste-from-digesters-for-biogas-production. ISBN: 978-953-307-480-1

[36] PYREG. 2011. Biomass to Biochar using Pyreg's Slow Pyrolysis Technology.

RESUME

Předložený článek se zabývá problematikou narůstajícího množství digestátu jako konečného produktu anaerobní digesce v bioplynových stanicích, budovaných jednak jako alternativní zdroje energie a dále jako zařízení schopné fermentovat, a tím snižovat množství ukládaného biodegradabilního odpadu na skládky.

Digestát je stabilizovaný materiál, který vzniká při anaerobní digesci v bioplynové stanici vedle hlavního produktu bioplynu. Producenty digestátu jsou bioplynové stanice, jejichž počet výrazně roste, a to s ohledem na zvyšující se zájem o alternativní resp. obnovitelné zdroje energie spolu s požadavky na redukci biodegradabilních odpadů.

Digestát má v současné době omezené využití, které je dáno jeho specifickými fyzikálními vlastnostmi, zejména extrémně problematickou separací jednotlivých složek, ale také chemickým a mikrobiologickým složením. Jeho vlastnosti jsou z největší části ovlivněny druhem zpracovávané biomasy. Digestát představuje environmentální problém s ohledem na jeho rostoucí množství a vlastnosti, především nízký obsah sušiny.

Odstředěním digestátu získáme separát, a fugát. Mezi možnosti nakládání se separátem se řadí kompostování separátu. Nejvyužívanější možností je přímá aplikace přímo na půdu. V neposlední řadě je také možné separát sušit a dále s ním nakládat, a to obohacením o nutriční prvky a tímto zvýšit jeho kvalitu jako hnojiva a aplikovat na půdu. Vysušený separát je možné také využívat jako palivo při spalování nebo ho využít pro výrobu paliva, a to uplatněním procesů jako jsou pyrolýza a zplyňování.

Kapalná fáze digestátu - fugát, která je silně znečištěná, nicméně obsahuje důležité prvky rovněž využitelné pro účely hnojení. Kapalnou fázi je možné využít pro produkci řas jako zdroje pro výrobu paliva. Z fugátu mohou být získány cenné prvky (týká se to především amoniakálního dusík).

V rámci výzkumu čištění digestátu a fugátu byly nejprve zhodnoceny vlastnosti vzorků digestátu z 5 zemědělských bioplynových stanic. Byly sledovány ukazatele $CHSK_{Cr}$, BSK_5, celkový fosfor a amoniakální dusík. Cílem bylo srovnat vlastnosti digestátu a produktů jeho separace odstředěním, tj. kapalné fáze tzv. fugátu a tuhé fáze, tzv. separátu. Na základě srovnání bylo zjištěno, že vlastnosti digestátů z různých zemědělských zdrojů mají srovnatelné vlastnosti Na základě tohoto zjištění můžeme konstatovat, že lze uplatňovat postupy čištění, které by byly aplikovatelné na digestát obecně.

Dále byly testovány způsoby separace tuhé fáze z digestátu, resp. fugátu. Bez použití flokulačních a koagulačních činidel lze pro separaci uplatnit proces odstřeďování. Co se týče posouzení vhodnosti použití koagulačních činidel, v odseparované tuhé fázi, která by měla být použita především ke hnojení půd, je obsah dusíku, fosforu, vápníku a organických látek za účelem hnojení obecně přínosný. Rovněž určitá koncentrace železa je přínosná, nicméně v dávce koagulantu potřebné k úspěšné koagulaci, je jeho obsah vysoký. V případě použití flokulantů byl pro dobrý průběh flokulace, oproti experimentům při použití koagulantů, nutný přídavek většího objemu ředící vody.

Pro intenzifikaci procesu odstřeďování použitím koagulačních a flokulačních činidel účinné, ale jejich spotřeba pro účinnou koagulaci a flokulaci byla enormně vysoká. Tuhá faze je dobře oddělitelná i sedimentací a filtrací. Použití koagulantů a flokulantů pro tyto účely v praxi by bylo značně diskutabilní a problematické i s ohledem je na jejich vysokou spotřebu a vnos dalších prvků do procesu a tedy do produktů separace.

SEWERAGE PUMPING STATION OPTIMIZATION UNDER REAL CONDITIONS

Lukas Chabal [1], Stefan Stanko [2]

[1] *Ing. PhD., Department of Sanitary and Environmental Engineering, Faculty of Civil Engineering, Slovak University of Technology in Bratislava*
Radlinského street 11, Bratislava
lukas.chabal@stuba.sk

[2] *doc. Ing. PhD., Department of Sanitary and Environmental Engineering, Faculty of Civil Engineering, Slovak University of Technology in Bratislava*
Radlinského street 11, Bratislava
stefan.stanko@stuba.sk

Abstract

Construction of a sewer system on a flat area is characterized by a large number of pumping stations (PS), which implies the probability of occurrence of technical problems and increases operational costs. The article focuses on drainage methods for municipalities situated on a flat area where it is necessary to build up a large number of pumping stations. Problems occur in case of multiple serial, parallel or combined connections of PSs. Energy costs can outweigh other costs, especially if the PS runs more than 2000 hours per year (Wilson et al., 2010). It has been shown that there is a large technical and economic potential for energy savings in sewage pumping. The pumping of waste waters in Slovakia is mostly based on the START-STOP method. This means that the pumps operate at all times at full power. In practice, we can also meet with oversizing of pumps. These and other facts lead to increased power consumption.

Also rainwater infiltration and "black" stormwater connections belong to significant present problems in sewer system operation. Large amounts of storm water in a sewerage PS lead to increasing intensity of waste-water pumping, which is reflected in increasing operational costs.

Optimization of such sewer system is based on mathematical modelling and was implemented in the "Ivanka pri Dunaji" municipality, close to Bratislava – the capital of Slovakia where the above mentioned problems in wastewater discharge cause considerable operating costs and inefficient performance of the system as a whole for a long time. Due to the enormous houses development, the system is inadequately loaded by rain waters as well. Also poor discipline of property owners contributes to the inauspicious situation to a great extent. Despite a ban of connections, paved areas are drained into the sewage system very often. The identification of such connections is very problematic with regard to ownership rights. Rain waters in sewages can degrade the quality of wastewaters, increase operating costs of pumping and reduce system life cycle. It was demonstrated by mathematical modelling that with the use of information technologies, it is possible to make the existing sewerage systems more effective or propose a new system of pumping and discharging waste waters.

Abstrakt

Pre výstavbu stokovej siete v rovinatom území je typické značné množstvo čerpacích staníc (ČS), čo implikuje pravdepodobnosť vzniku technických problémov a zvyšuje náklady na prevádzku. Príspevok je zameraný na spôsoby odkanalizovania obcí situovaných v rovinatom území, kde je potrebné budovať veľký počet čerpacích staníc (ČS). Problémy nastávajú pri niekoľkonásobnom sériovom, paralelnom alebo kombinovanom zapojení ČS. Spotreba energie tvorí často najväčšiu nákladovú položkou celkových nákladov generovaných počas životného cyklu čerpadla. Náklady na energiu môžu dominovať ostatným nákladom, hlavne ak je čerpadlo v chode viac ako 2000 hodín za rok (Wilson et al., 2010).

Bolo dokázané, že existuje veľký technický a ekonomický potenciál úspor energie pri čerpaní splaškových vôd. V praxi sa môžeme veľmi často stretnúť s predimenzovaním čerpadiel. Tieto a ďalšie skutočnosti vedú k zvýšeniu spotreby elektrickej energie.

Významným problémom pri prevádzke stokových sieti je aj infiltrácia dažďových vôd a nedovolené pripojenie dažďových zvodov do splaškovej kanalizácie. Množstvo dažďových vôd v ČS v značnej miere ovplyvňuje čerpanie OV, čo sa prejavuje zvýšením nákladov na prevádzku ČS.

Optimalizácia takýchto sietí je v značnej miere založená na matematickom modelovaní a bola použitá na kanalizácií obce Ivanka pri Dunaji, kde vyššie uvedené problémy v odvádzaní odpadových vôd už dlhší čas spôsobujú nemalé prevádzkové problémy - zvýšenie prevádzkových nákladov a neefektívny výkon systému ako celku. Systém je v dôsledku enormnej výstavby rodinných domov neprimerane zaťažovaný aj vodami, ktoré zaťažujú systém v čase dažďa. Nemalou mierou k tomu prispieva zlá disciplína majiteľov nehnuteľností v zmysle napojenia spevnených plôch na kanalizáciu. Tieto plochy sú aj napriek zákazu pripojenia mnohokrát zvedené práve do systému splaškovej kanalizácie. Identifikácia takýchto napojení je problematická vzhľadom na vlastnícke práva majiteľov nehnuteľností a tieto napojenia nielenže zhoršujú kvalitu odpadovej vody, ale implikujú zvýšené prevádzkové náklady na čerpanie a znižujú životnosť systému.

Matematickým modelovaním bolo preukázané, že pri využití informačných technológii môžeme zefektívniť prevádzku už jestvujúcich sieti alebo navrhnúť nový systém odvádzania a prečerpávania odpadových vôd.

Key words: waste water, sewage pumping stations, rainwater infiltration

1 INTRODUCTION

Within Eastern Europe, it is very difficult to encounter any publications describing a solution how to optimize existing sanitary sewer networks with a combination of several pumping stations (PSs). Most of the literature with similar issues focuses on optimization only. Due to the fact, it was necessary to combine several studies and expert articles dedicated to the optimization such as the "Optimization Guidance Manual for Sewage Works" (Nutt, Ross, 2010) dealing with energy savings and efficiency, saying also that pumping is one of the processes, for which it has been shown that it has a great technical and economic potential in energy savings. To achieve energy savings, it is necessary to develop new energy-saving devices and systems.

The article shows that by linking of individual water infrastructure systems, such as CTD, GIS and mathematical modelling a powerful tool can be obtained to reduce operating costs and predict negative events which could cause significant financial losses.

This article also shows the possibility of using hydroinformatics systems to optimize existing sewage systems. Hydroinformatics, as a relatively new discipline, combines elements of latest information technologies and hydraulics. Created hydroinformatic tools, such as Mike Urban, have a great capability to realistically view operational and emergency situations in water and wastewater related systems. They enable to answer not only basic questions about hydraulic system behaviour, but also questions related to forecasts of quantitative and sometimes also qualitative parameters and their simulations for specific events as in the article "Utilization of Hydroinformatic Tools in Slovakia" (Kučera, Gibala, 2006).

At present, we can meet with designs of sewerage and collecting wastewater into one central waste water treatment plant for large conurbations. In case of talking about gravitation drainage, we can talk about a preferred solution, but if it is necessary to pump wastewaters with a greater number of PSs, we can encounter many problems in design and operation (Sousa et al., 2002). Several unusual operating conditions were simulated there that were discussed with operating staff of Bratislava Water Company. It is also aimed at using hydroinformatics methods in safe sewerage disposal and demonstrating potential benefits of using them. On the basis of supporting documents of operating order and GIS and their proper calibration with supporting data from SCADA, it is possible to assess the current state. This enables to calculate basic hydraulic parameters such as flow capacity of each pipe, speed of sewages, and optimum diameter of sewer pipes and synergies of PSs.

Also infiltration of rainwater represents an enormous problem in operation of a sewer network. When developing wastewater planning strategies, water utilities face one of the greatest challenges that is to find such solutions that are not only cost effective in the short term, but also provide a desired level of service over the life of the infrastructure. Uncertainties in predictions of future population size, climate change impacts, rates of system deterioration and accuracy of hydraulic model calibration can have a significant impact on planning solutions. (Wilson et al., 2010). In the course of nonlinear climate processes in nature where extreme events occur increasingly, solving the problem is indispensable. Rainwater can infiltrate through pipe joints and leak through holes in manhole shafts. Leaks can occur due to the use of low-quality materials in the network construction. Large amounts of storm water in a drainage system lead to increasing intensity of pumping, which is reflected in increasing operating costs.

Nowadays, the operation of pumping stations needs to use computing technologies. To better understand a pumping system, relevant data must be processed depending on their importance and used to solve the problem. Proper system design, installation and possibly operation are the first important step. The achieved goals of simulations will be incorporated into the security plan of the sewerage system, which can also be applied to other similar systems. The article may serve as an analysis of current situation and also help predict and accurately identify weak points of the system and optimize effectiveness of operation and performance. The contribution of this work consists in an innovative approach to understanding a sewerage network and can serve as a guide for water companies in their reconstruction.

2 DESCRIPTION OF THEORETICAL METHOD AND CHOSEN AREA

The village of Ivanka pri Dunaji is located near the Danube River eastward from the capital city of Slovakia – Bratislava. The total population amounts to 5531. At present, it has 2910 inhabitants. In the simulation, connections with all residents were considered. The territory of the municipality is flat terrain with moderate waves. The terrain slightly declines to the Sursky channel. The municipal sewerage system is designed as a separate sewer. Rain waters and other waters are disposed at the source and they are excluded from the sewer system.

The Sursky channel as a man-made stream with the function of a trapping channel is the main recipient of the village.

There is no large industrial plant running in the village of Ivanka pri Dunaji. In this region, there are only minor operations, restaurants, cafes, further shops, schools and medical facilities. Waste water produced in these facilities has a municipal wastewater character.

The technological process of the village sewer consists of the following operations:

- collecting wastewaters discharged through household connections in sewer system
- wastewater transport through gravity pipeline to pumping stations,
- pumping wastewater to WWTP in Bratislava-Vrakuna

The sewer system consists of 14 PS. The wastewaters are several-fold pumped to the biggest and main PS5 which is located at the end of the village.

Profiles of house connections are DN 150 or DN 200 PVC, main sewer network from DN 300 to DN 600. The bulk of the gravity sewer system in the village is DN 300 with minimum slope 3.5 ‰.

On the main collector system, PSs are built. On the collector A, there are 3 PSs. On the collector B, 1 PS, on the collector D, there are 2 PSs. The collector E includes 4 PSs and the sewerage system F then 1 PS.

On the collector system A, PS5 is placed, from which sewages are pumped through a pressure pipe DN 600, 4673 m in length to the WWTP in Vrakuna (Fig.1).

3 MATHEMATICAL MODELLING OF SEWER NETWORK

In times of rising energy prices in transport of fluids, effectiveness should be considered. On the foreign market, software applications are available that help us reconcile the system utilization in order to avoid excessive spending, whether in the procurement, construction, operation or maintenance of the network.

This chapter describes the formation of a strategic mathematical model of shared sewage drains stated in the report of Bratislava Water Company (BVS) using the simulation software Mike Urban followed by computing a building simulation model of the sewerage network comprising pipelines retrieved from GIS and sewer PSs. The resulting model is then calibrated according to the data obtained from the CTD (central technology dispatching) such as information about course levels in PSs, inflows and consumption.

When compiling the mathematical model, the following was made:

- Identification of problems, or questions to be solved by the model,
- Defining real elements of the sewage network in terms of the mathematical model,
- Operating system characterization during the examined time periods,
- Model calibration based on the data from CTD,
- Launch of calculating the mathematical model and solving the issues mentioned above.

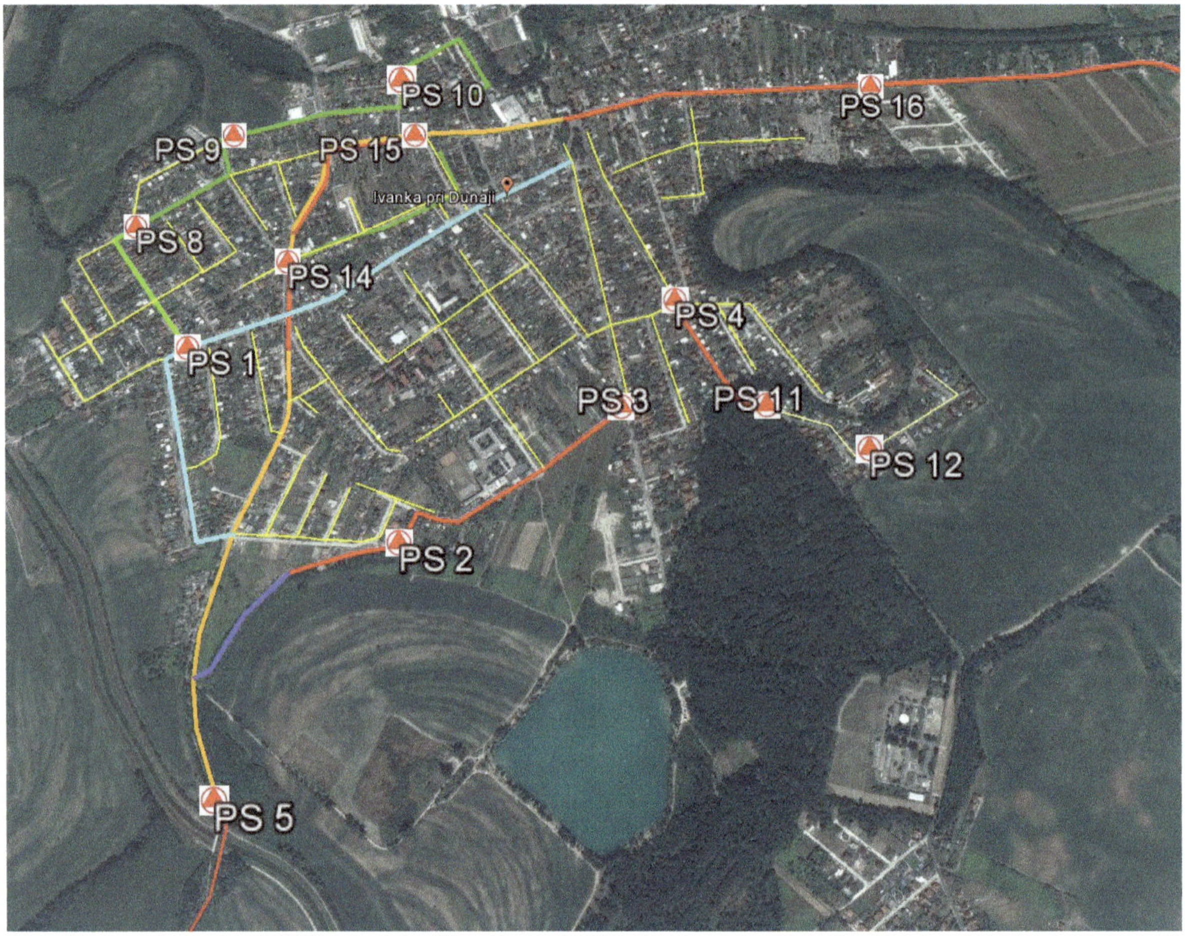

Fig. 1 Site layout of sewer system in Ivanka pri Dunaji

3.1 Unusual cases in network modelling

This chapter is focused on the assessment of non-standard conditions, for example during a power failure. The question was how long it would take before all PSs in the village stop pumping and the system collapses. The information is very important for the attended operation in order to act quickly in similar situations to prevent damage of property.

In this simulation, we considered only sewage waters without rain waters. The simulation time was set to 7 hours per day during 5 working days and 2 weekend days. For each type, separate curves of daily irregularities during the day were used.

It should be said that when the simulation was set to the existing conditions, no defects of PSs and electrical outages occurred; the current system works properly. The collecting system appeared to be slightly oversized because it was designed for prospective conditions that are not currently met.

The simulation started at 00.00 a.m. on 05/19/2014. After running the simulation, flooding at the edge inspection chamber occurred. It should be noted that these conditions became evident after 8 hours of simulation in PS16. The simulation shows other problematic parts upstream PS7, PS10, PS11 and PS16. The sewer part before PS16 was flooded first. The discharge of wastewaters to landscape occurred on 19/05/2014 at 8:21 p.m. followed by PS7 at 19:17 p.m. PS11 was flooded at 19:38 p.m. that day (Fig. 2). PS10 was flooded after the last 4 days of simulation time on 23/05/2014 at 01:38 p.m. It was due to the low population density in the area.

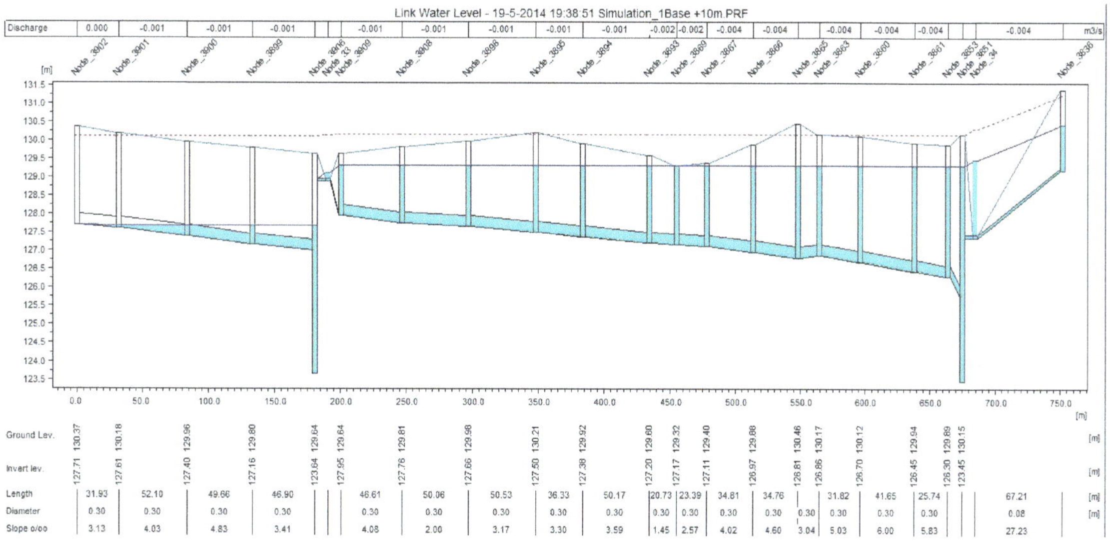

Fig. 2 Flooded sewerage pipes and pumping station PS11

3.2 Illegal connections of rain waters from roofs to sewage system

In the next step of simulation, we focused on illegal connections from roofs to the sewage system. Due to the insufficient solution of rainwater drainage in the village, property owners are often forced to discharge the water into the separate sewerage system. Residents do so in contradiction to the contract on connection to the public sewer system. There was a number of cases where the people have been connected from roofs into the sewer house connections already during the house construction, either directly or through an overflow edge of the rain water basin. Residents, however, do not realize that the inflow of rainwater from few roofs could exceed the designed capacity of a PS.

The operator can defend against such cases as follows:

- Do not allow any new inhabitancy without resolving and testing the rainwater drainage system, despite the fact that the newly built sanitary sewer will be finally approved.
- Increase the awareness of the population of possible financial penalties. Smoke tests together with camera monitoring should demonstrate who has a "black" connection. It should be in the interests of the village council and the sewer operator so the confirmation of such suspicions should be the priority.
- An expensive, but permanent solution consists in building storm sewers, or constructing an accumulation "rain" tank near a PS where the mixed water can be accumulated during the rainfall and then, after the rainfall end, pumped back to the system.

To simulate these conditions, 15 minute rainfall with a frequency of 0.5 was used for the territory of Bratislava subsequently divided into the simulated zones through each catchment area of PSs. The simulation was carried out in three variants with a share of 10%, 15% and 30% connection from roofs of the total interested area. Within each simulation of illegal connections, there were significant operational states.

For each alternative, we focused on the rate of flooding each particular boundary part of the sewer network. In these places, along the longitudinal profiles, the sewer lays in a smaller depth. Right here, there was the assumption that flooding inspection chambers would happen sooner and gravitational flow be pressurized. We considered only the PSs that demonstrated the greatest impact of rain runoff from roofs as PS16, PS11, PS1, PS7 and PS5. In the individual situations, we can see the places highlighted in blue colour which were loaded by the runoff from roofs (Fig.3, Fig.4, and Fig.5). The points, or the shafts which occurred during the simulation discharge of mixed wastewaters over the landscape are highlighted in red.

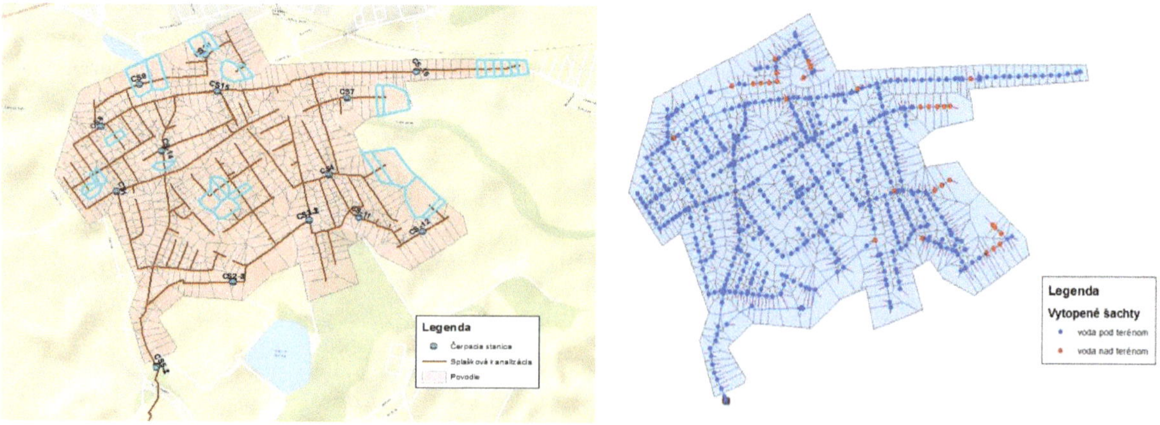

Fig. 3 Alternative with 10% connection of rainwaters from roofs

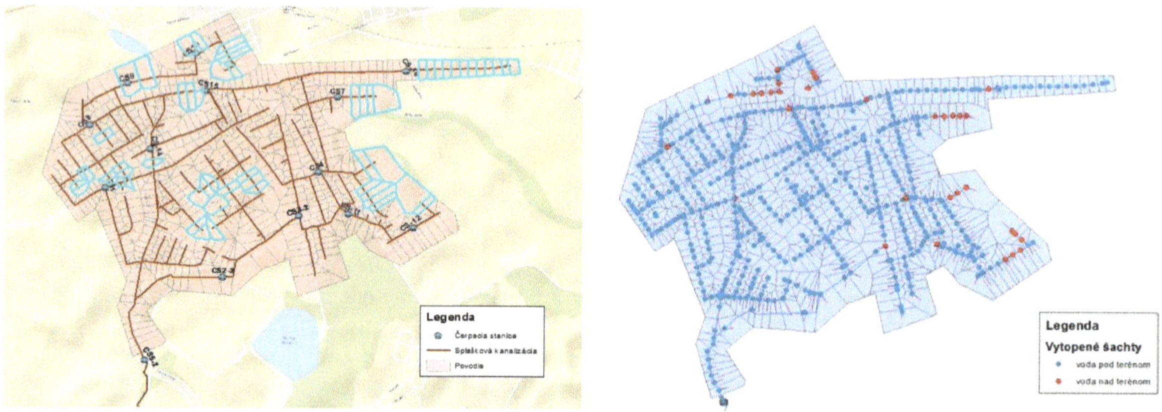

Fig. 4 Alternative with 15% connection of rainwaters from roofs

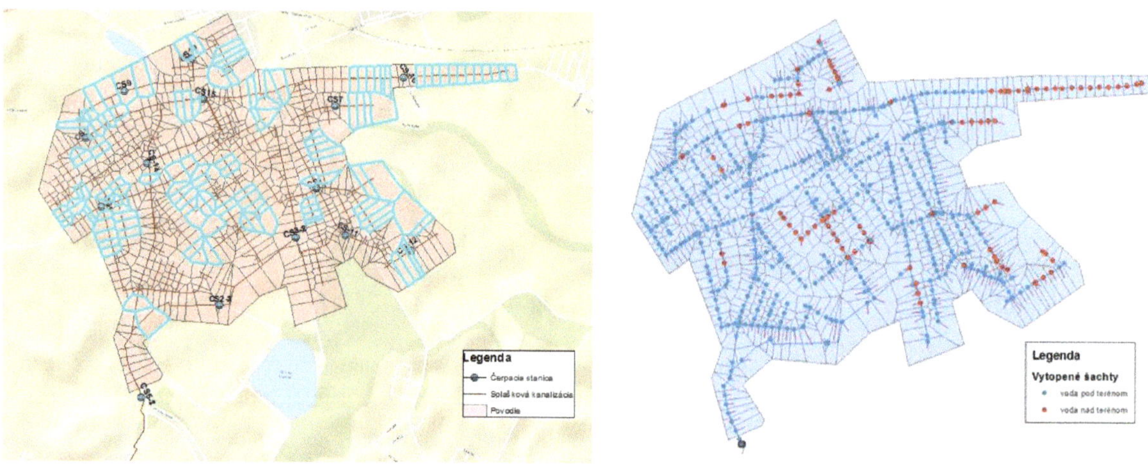

Fig. 5 Alternative with 30% connection of rainwaters from roofs

All three alternatives showed the danger of connecting rain waters to the separate sewage system. Mostly edge areas were flooded, and it was confirmed that the larger proportion of the connections has a significant impact such as the size of the flooded area as well as the speed of flooding. Already in the first alternative, there was a pressure flow instead of gravity flow in places with a pure gravity sewer, which has an adverse effect on the operation as well as shortens operating lifetime of pipes and joints. The worst results showed the option of 30%. The simulation leads to the pressure flow and backflow water nearly at 95% of the network.

3.3 Use of variable frequency drive

The simulation of using a variable frequency drive can demonstrate potential savings in operating costs, whether electrical energy savings in running pumps, or savings due to a prolonged period of repairs and their life. The simulation required to change the characteristic curves of various pumps offered by manufacturer. It is important to note that it is not recommended to regulate the rotation speed in the whole range from 0% to 100%. Manufacturers recommend the pump regulation ranging from 70% to 100%. Therefore, the simulation was carried out with 100%, 85% and 70% of the characteristic efficiency. The sewerage system in the village of Ivanka pri Dunaji uses four types of pumps: Sigma 80 GFHU; ABS AFP 1035; ABS Piranha M26/2D; ABS Piranha M55-2D pumps. We can mention the pump type ABS AFP 1035 which is installed at the last pump station PS5. The simulation was running in one particular day without rainfalls. The characteristic curve had to be transferred to the simulator program MIKE URBAN (Fig. 6).

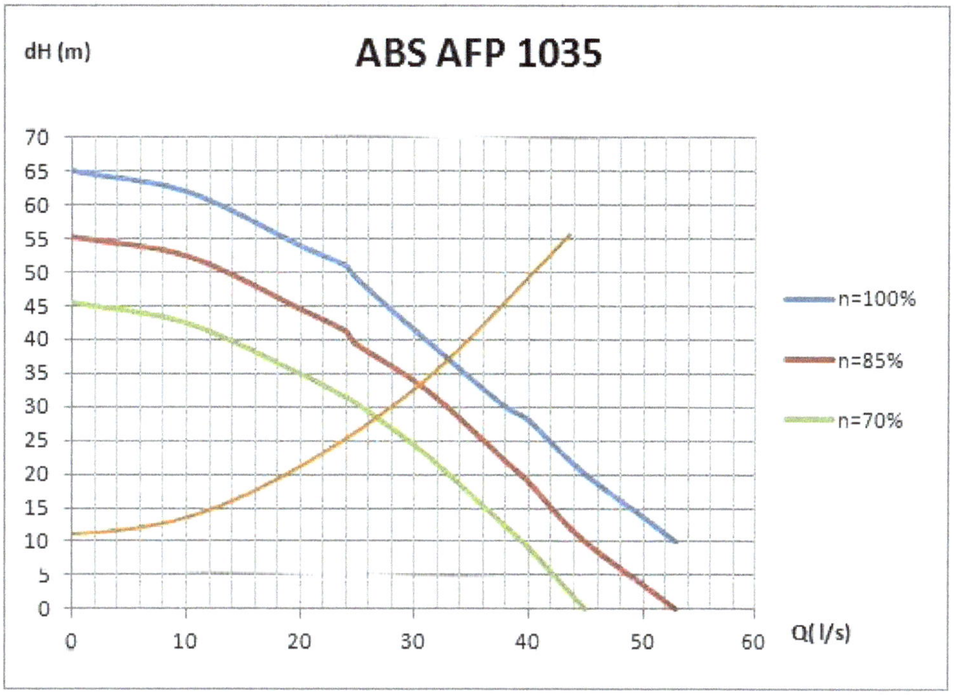

Fig.6 Characteristic performance curve of ABS AFP 1035 pump

The graphics drawing of the ABS AFP 1035 pump (PS5) at 100% efficiency shows that the pump started pumping 32 times a day, while the longest pumping time was 40 min from 14:45 to 15:25. The maximum flow rate of the pump was 0.032 $m^3.s^{-1}$, which is 32.0 $l.s^{-1}$ (Fig. 7).

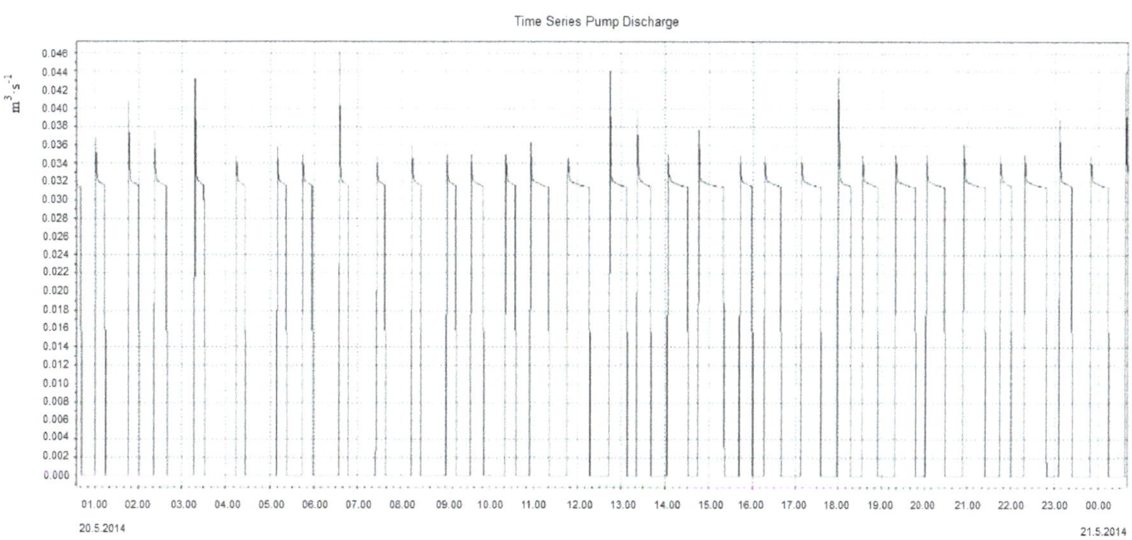

Fig.7 Graphics pumping performance of ABS AFP 1035 pump at 100%

The graphics performance shows pumping at 85% efficiency where it is apparent that the pump started pumping 29 times a day, while the longest pumping time was 43 min from 21:05 to 24:48. The maximum flow rate of the pump was 0.03 m3.s-1, which is 30.0 l.s-1 (Fig. 8).

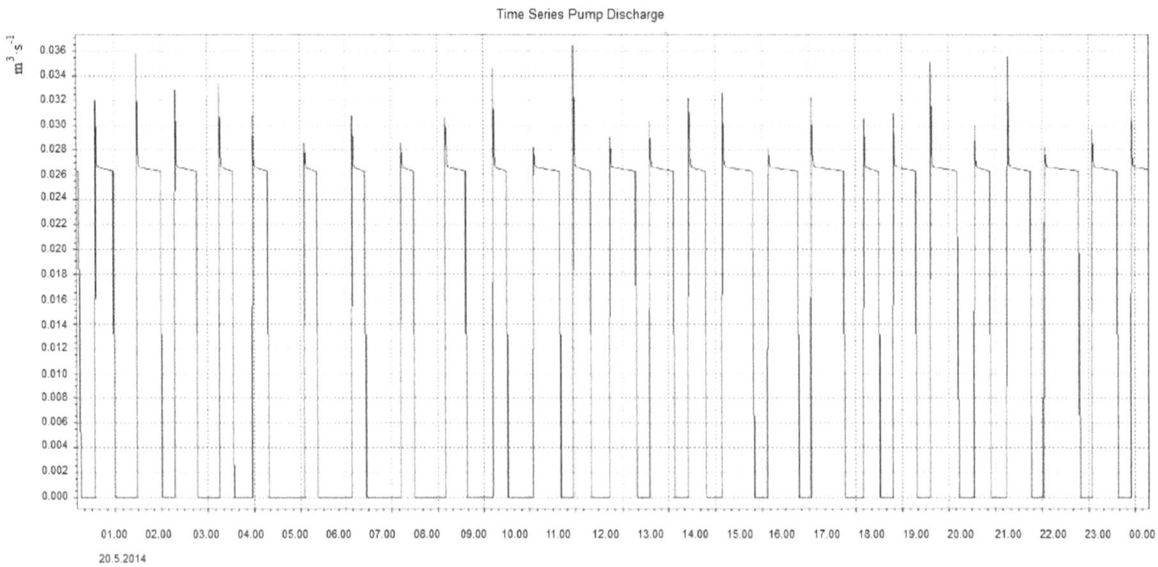

Fig. 8 Graphics pumping performance of ABS AFP 1035 pump at 85%

The graphics drawing at 70% efficiency shows that the pump started pumping 26 times a day, while the longest pumping time was 50 min from 21:00 to 21:50. The maximum flow rate of the pump was 0.026 m3.s-1, which is 26.0 l.s-1 (Fig. 9).

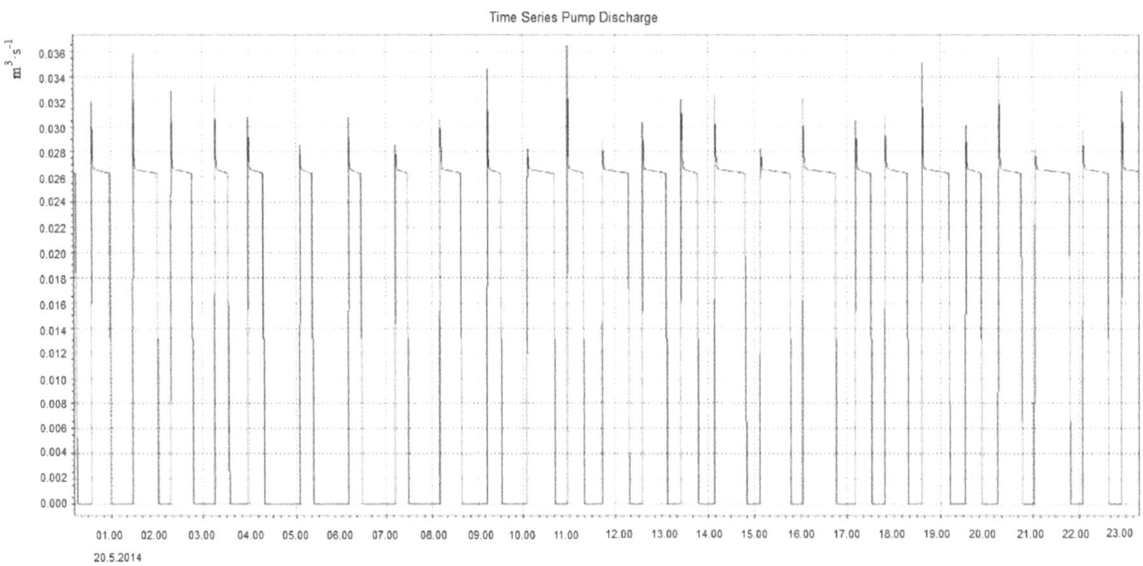

Fig. 9 Graphics pumping performance of ABS AFP 1035 pump at 70%

In this case, the simulation showed considerable savings in operating costs. They are induced by the ideal pump design and by that the efficiency points are in the range of pump efficiency, which corresponds to the flow rates for individual efficiency levels. At 70% efficiency, the saving potential is 703 euros/year. When the price of VSD amounts to around 500 Euros, the return on investment is 14 months of operation (Tab. 1).

Tab. 1 Calculation of electric energy consumption for the ABS AFP 1035 pump at various efficiencies

PS5 - ABS AFP 1035	100%	85%	70%
Pumping time (hr.)	11,0	11,5	12,6
Pump power (kW/h)	26,0	24,0	21,0
EUR/day	25,7	24,8	23,8
EUR/year	9395,0	9067,0	8692,0

4 CONCLUSIONS

The article shows the possibility of using hydroinformatics systems to optimize existing sewage systems. Several unusual operating conditions were simulated that were discussed with the attended operation of Bratislava Water Company.

The first simulation was focused on power failure during pumping in PSs and the ability of the sewer system to accumulate sewages. The simulation showed that it took about eight and a half hours, until sewages sprained above the surface. The water company operations need to ensure the removal of faeces car first from PS16, PS11 and PS7. Due to the low built up area density in the peripheral parts of the sewer network, the sewers can be accumulated for few days.

The next step of simulations included illegal connections from roofs to the sewage system. Three variants of the simulation were carried out with a 10%, 15% and 30% share of connections from roofs of the total area in question. On peripheral edges of sewer, where pipes of smaller diameters in smaller depths were designed, the 10% simulation showed a problem with flooding as first. Also PS16 was the most problematic as it was flooded in 15 minutes before the start of the simulation. In the alternative of 15% connection, sewages sprained above the surface in the 14-th minute and in the case of 30% connection already in the 10-th minute. Flooding of PS7, PS12 and PS10 occurred with less time differences. The strategic PS5 drained sewages reliably in all three alternatives.

The application of a variable frequency drive was the third part of the simulations. It aimed at demonstrating potential savings in operating costs and due to a prolonged period of repairs and life cycle. The simulations were conducted with 100%, 85% and 70% characteristic efficiency. The possibility of savings in operating costs was demonstrated in two cases: for PS1 and PS5. For PS5, the return of investments was within 14 months of operation and for PS1 up to 15 months. It is important to note that the reduction in efficiency also reduces the pump flow, which may negatively affect the function of macerator pumps. Result may be the clogging of the pump by stiff particles.

The achieved goals of simulations will be incorporated into a security plan of the sewerage system which can also be applied to other similar systems. The article may be used as an analysis of current situation and also help predict and accurately identify weak points of the system and can optimize effectiveness of operation and performance. It has been shown that by linking of individual water infrastructure systems, such as CTD, GIS and mathematical modelling, a powerful tool can be obtained to reduce operating costs and possibly predict negative events which could cause significant financial losses.

ACKNOWLEDGMENTS

This work was supported by the Slovak Research and Development Agency under the contract No. APVV-0372-12 held at the Department of Sanitary and Environmental Engineering, Faculty of Civil Engineering, Slovak University of Technology.

REFERENCES

[1] Nutt, S.G. and D. Ross 2010: Optimization Guidance Manual for Sewage Works, presented at the 24th Annual Water Environment Association of Ontario Conference, Toronto, Ontario.

[2] Kučera M., Gibala T. 2006: Využívanie prostriedkov hydroinformatiky na Slovensku, Vodní hospodářství 6, ISSN: 1211-0760.

[3] Moe 2008: Design Guidelines for Sewage Works. ISBN 978-1-4249-8438-1.

[4] Urcikán, P., Rusnák, D. 2004: Stokovanie a čistenie odpadových vôd, Stokovanie I, Navrhovanie stokových sietí, STU Bratislava, ISBN 80-227-2136-0.

[5] Bedford, M. 2004: GIS for Water Management in Europe, ESRI, United Kingdom, ISBN: 9781589480766 .

[6] Garr, J., Sanks, R. 2008:Pumping Station Design, Revised 3rd Edition, ISBN 978-1-85617-513-5

[7] Sousa, J., Ribeiro, A., Cunha, M.,Antunes, A. 2002: An Optimization Approach to Wastewater Systems Planning at Regional Level, Journal of Hydroinformatics 04.2, 115-123.

[8] Keyser, W., Amerlimck, Y., Urchegui, G., Harding, T. 2014: Detailed Dynamic Pumping Energy Models for Optimization and Control of Wastewater Applications, Journal of Water and Cimate Change, Vol. 5, No.3, pp 299-314.

[9] Wilson, J., Smith, P., Witt, A., Canning, M. 2010: Wastewater Collection System Optimization for Cost-Effective and Sustainable Capital Improvement Planning, Water Practice and Technology, Vol.5, No.4.

[10] Brandstetter, A., Engel, R., Cearlock, D. 2007: A Mathematical Model for Optimum Design and Control of Metropolitan Wastewater Management Systems, Jawra Journal of the American Water Resources Association, Vol.9, No.6, pp 1188-1200.

RESUMÉ

Cieľom príspevku bolo vytvoriť metodiku využitia prostriedkov hydroinformatiky v oblasti bezpečného odvádzania odpadových vôd a preukázanie možných výhod ich využitia. V rámci východnej Európy sa len veľmi ťažko stretneme s publikáciami, ktoré by opisovali riešenie optimalizácie jestvujúcej splaškovej kanalizačnej siete s kombináciou niekoľkých čerpacích staníc. S realizovanými pokusmi optimalizovať systémy odvádzania OV sa môžeme stretnúť hlavne v zahraničí. Z tohto dôvodu bolo množstvo informácií čerpaných zo zahraničnej literatúry a vedeckých štúdií.

Pre splnenie zadefinovaných cieľov bola spravená analýza implementácie informačných technológií v podobe matematického modelu v prepojení na dátové zdroje BVS. Dôležitou súčasťou príprav a stavby matematického modelu sústavy bol transfer dát. Pre správne fungovanie bolo nutné model nakalibrovať s využitím dát s centrálneho technického dispečingu. Nakalibrovaný model charakterizoval súčasné prevádzkové pomery na sieti, vďaka čomu mohol byť aplikovaný pri riešení predpokladaných reálnych neštandardných stavov. Zároveň sa preukázalo jeho možné využitie na iné sústavy.

Dosiahnuté výsledky bude možné zapracovať do plánu bezpečnosti odvádzania OV, ktorý sa bude môcť aplikovať aj na iné, podobné sústavy. Práca môže slúžiť, ako analýza súčasného stavu a tiež pomôže predpovedať a presnejšie identifikovať slabé miesta systému, môže optimalizovať z cieľom zefektívnenia prevádzky a hospodárnosti systému. Bolo preukázané, že prepojením jednotlivých sústav vodárenskej infraštruktúry, ako CTD, ZIS, GIS a matematického modelovania získame silný nástroj pre zníženie prevádzkových nákladov a možnú predikciu nepriaznivých udalostí, ktoré môžu spôsobiť značné finančné škody na majetkoch.

26

CREATING MODEL FOR TECHNOLOGICAL PROCESS OF INERTISATION

Lukáš OTTE [1], Vladislav VANČURA [2], Roman DANEL [3], Michal ŘEPKA [4]

[1] *Ing, Ph.D, Institute of Economics and Control Systems, Faculty of Mining and Geology, VŠB – Technical University of Ostrava, 17. listopadu 15, Ostrava*
e-mail: lukas.otte@vsb.cz

[2] *Ing, Ph.D, Institute of Economics and Control Systems, Faculty of Mining and Geology, VŠB – Technical University of Ostrava, 17. listopadu 15, Ostrava*
e-mail: vladislav.vancura@vsb.cz

[3] *Ing, Ph.D, Institute of Economics and Control Systems, Faculty of Mining and Geology, VŠB – Technical University of Ostrava, 17. listopadu 15, Ostrava*
e-mail: roman.danel@vsb.cz

[4] *Ing, Ph.D, Institute of Economics and Control Systems, Faculty of Mining and Geology, VŠB – Technical University of Ostrava, 17. listopadu 15, Ostrava*
e-mail: michal.repka@vsb.cz

Abstract

Underground coal mine environment is an environment with dynamic manifestations of methane. In this environment it is quite difficult to control various technological processes occurring therein. The technological process of collapse preventing inertisation is carried out by supply of compressed nitrogen to areas at risk of spontaneous combustion of coal. Petri nets allow the modelling of parallel dynamic systems and systems with discrete time. In conjunction with the software HPSim or WinPeSim, the individual processes can be modelled and then the results processed using a spreadsheet (Excel). Based on the results of the performed simulations, it is then much easier to determine the optimal solution or decision.

Abstrakt

Prostředí uhelného hlubinného dolu je prostředí s dynamickými projevy výskytu metanu. Takovéto prostředí je poměrně složité pro řízení jednotlivých technologických procesů, které se v něm vyskytují. Technologický proces preventivní inertizace závalu se provádí přívodem stlačeného dusíku do oblastí s rizikem vzniku samovznícení uhlí. Petriho sítě umožňují modelování paralelních dynamických systémů a systémů s diskrétním časem. Ve spojení se softwarem HPSim, popř. WinPeSim lze jednotlivé procesy modelovat a následně výsledky zpracovat pomocí tabulkového procesoru. Na základě výsledků provedených simulací lze pak mnohem snáze určit optimální řešení či rozhodnutí.

Key words: Inertisation, Petri Nets, Underground Mine, Modelling

1 INTRODUCTION

Inertisation of a gob (goaf) area in a longwall coal mine with nitrogen gas is an effective measure against spontaneous heating of coal. With regards to spontaneous heating indicators (CO concentration and coal temperature) the use of nitrogen inertisation can be applied in three different cases:

- preventive inertisation,
- inertisation to suppress spontaneous combustion,
- inertisation to extinguish fire.

The preventive inertisation aims at reaching safe levels of oxygen concentrations alongside the gob (goaf) area. In terms of the Ostrava-Karvina Coalfield, this means an O2 concentration of 10%. In case of inertisation used to suppress spontaneous combustion, when the temperature of the coal is higher than the temperature of the surrounding rock, injection of nitrogen is used to reduce oxygen concentrations alongside the gob (goaf) area below 10% (i.e. 3 - 5 % O2).

It is good to use the CFD modelling of spontaneous combustion zones for a longwall face using Y type ventilation. Modelling results show that the most likely areas (7~18% oxygen) for spontaneous combustion to occur spatially in the goaf are: the goaf edge on the belt road side some 200 m behind the face and 60 m into the goaf; within 50 m behind the chocks, and the unconsolidated areas about 60 m inside the retaining wall along the bleeder (return) road up to the start-up line. Air leakage is most excessive in the unconsolidated goaf boundary inside the retaining wall along the bleeder road. Both goaf gas monitoring and proactive inertisation are recommended to minimise the occurrence of spontaneous goaf heating. Optimum goaf inertisation can be achieved by pumping inert gas (nitrogen at a rate of 0.25 $m^3.s^{-1}$) at least 100 m behind the face on the belt road side, or ideally, if surface access permitting, via surface goaf hole(s). [10], [11]

For the purpose of preventive inertisation, Nitrogen Generator Units (hereinafter referred to as NGUs) are located on the surface, which provide nitrogen flow rate of 8,500 m^3h^{-1}. If need be, for short periods of time, the NGUs can increase the flow rate up to 18,000 m^3h^{-1} in order to extinguish mine fires. Measuring the flow rate of nitrogen gas at various points of the underground piping system is a standard part of the system used to adjust and optimize the inertisation process in a particular longwall panel in a coal mine. [1], [2]

The coal mine in question presently does not have any control system that would allow to control the flow rate of nitrogen based on an assessment of the current situation. However, the MTA Ostrava company intends to develop an autonomous control system that will be able to evaluate the situation and control the inertisation process based on current flow rate measurements in different parts of the mine and other metrics. In order to explore all possibilities of controlling this technological process (hereafter TP), appropriate models of the situation need to be used.

The aim of this paper is to present an initial study on the possibilities of using Petri nets to model the process of gob (goaf) inertisation in longwall coal mines. The HPSim v1.1 modelling software is used for modelling the situation; the data is then imported into the Microsoft Excel spreadsheet application and evaluated.

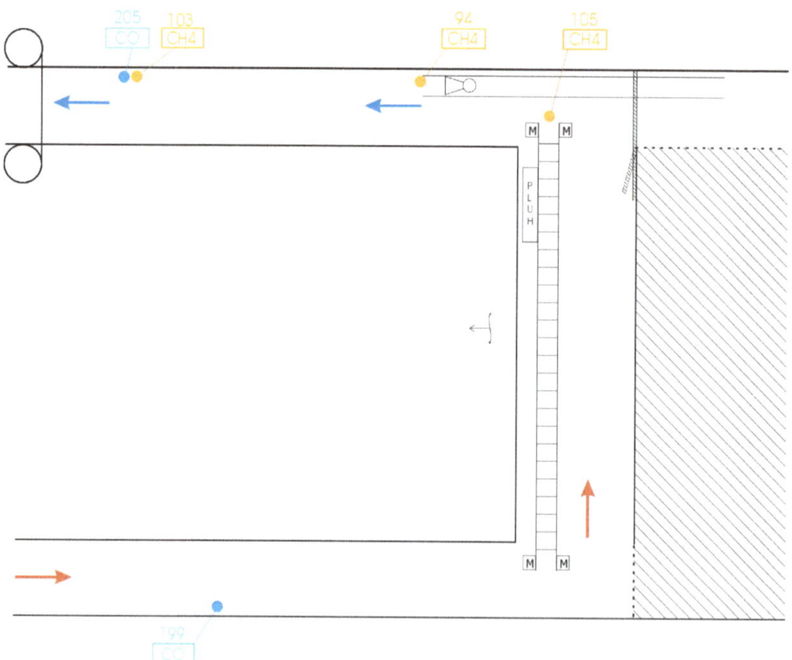

Fig. 1 Scheme of a longwall panel fitted with the CO and CH_4 concentration sensors.

2 MODEL

The overall model for the technological process of gob (goaf) inertisation will ultimately consist of several sub-models linked together. It will be a model for the NGU (nitrogen generator unit), which supplies gas to the piping network and is capable to temporarily increase the nitrogen production multiple times. For the purpose of the initial study, a stable source of nitrogen shall be considered, which will be delivered to the network and then used up in full volume in individual longwall panel operations. The network output, i.e. the throttle valve located in the gob (goaf) area, comprises a sub-model in itself. It is also necessary to model the CO sensors, which function as primary indicators of fire. The model has to incorporate elements that will generate problem situations, i.e. elements that will change the levels of CO and will simulate the occurrence of spontaneous heating.

END POINT

A sub-model, called *End Point*, is one of the fundamental elements of the whole model. This End Point is situated in the longwall panel operation in the immediate vicinity of the gob (goaf) area. A throttle valve controls the flow of nitrogen supplied to the gob (goaf) area. This sub-model is shown in Fig. 2. The throttle valve is represented by two Petri net elements (in the upper left part), called Transitions. In order to simplify the model only two valve states are considered – valve fully open and valve throttled to 50 %. These transitions, together with evaluated edges, are responsible for transporting a given amount of nitrogen from a NGU to an element labeled *End Point P1*. Test (read) edges and inhibitor edges represent the conditions for collecting a certain amount of nitrogen from the NGU. The element located in the center of the image represents opening the valve to 100 % or 50 % and the subsequent delay of the inertisation effect in terms of the CO concentration measured by the sensor located in the tailgate. The valve throttling conditions are also represented by test and inhibitor edges leading from the models of the CO concentration sensors. This part of the End Point sub-model is very complex. It is necessary to consider and model all conditions that affect the opening of the throttle valve in response to the measured CO concentration.

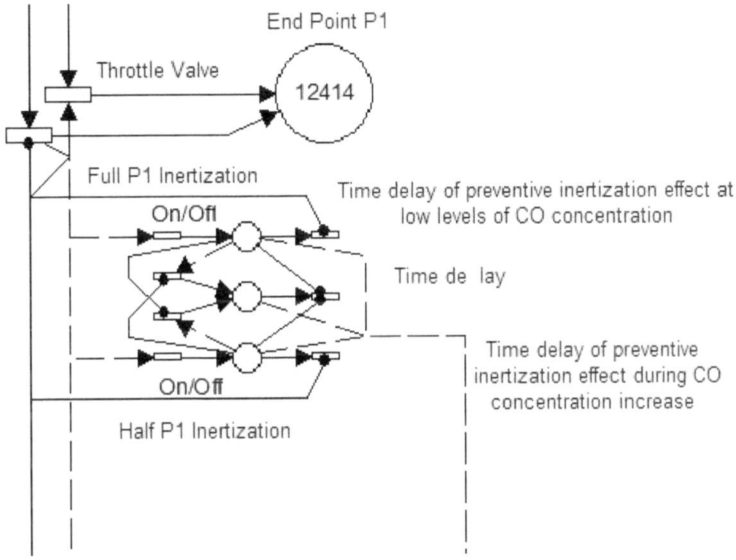

Fig. 2: Sub-model representing the End Point with the throttle valve

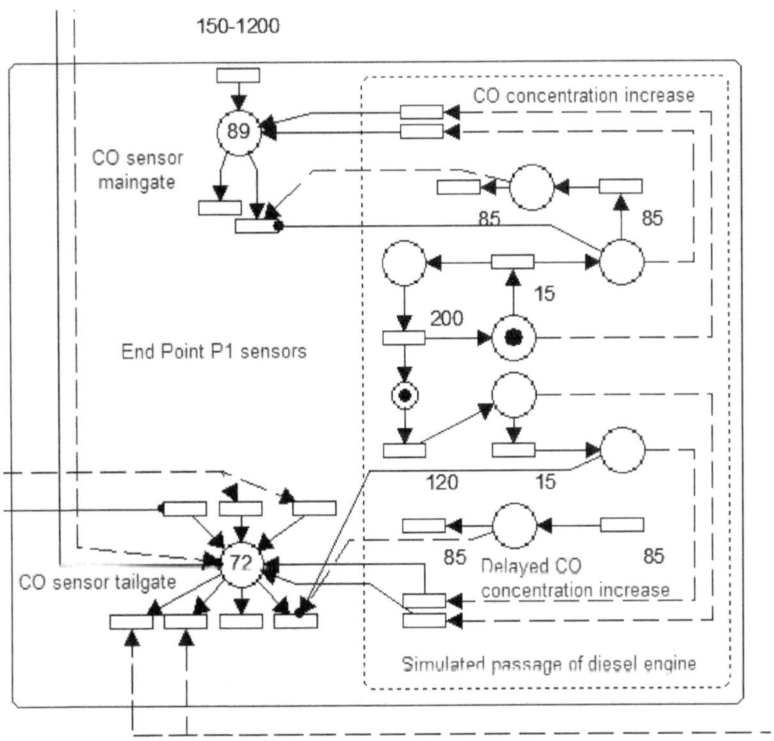

Fig. 3 Sub-model representing the CO sensors and passage of the diesel engine

CARBON MONOXIDE SENSORS

Another fundamental element of the model is a CO sensors sub-model, which can be seen in Fig. 3. Every longwall panel operation has two CO concentration sensors (in Fig. 1, they are labeled as 199 and 205). One sensor is located in the maingate, the other one in the tailgate. For the model, it is important that the first sensor is linked to the simulated passage of the diesel engine (right part of Fig. 3). Thus the control model will be able to evaluate the effect of the passing engine on the increasing concentration of CO measured by the second sensor. The simulated passage of the engine is represented by the area delimited by the dashed line.

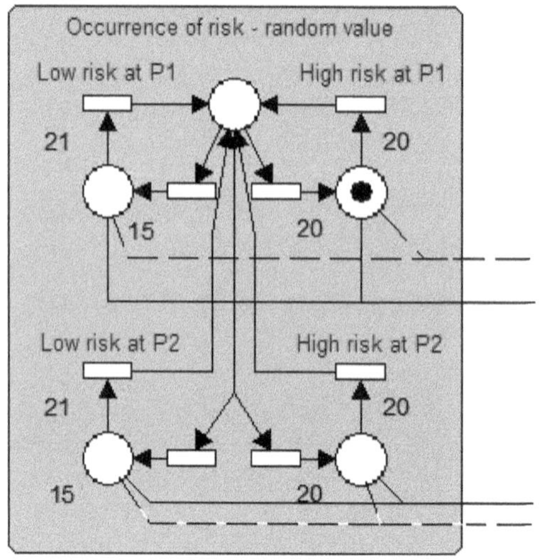

Fig. 4 Sub-model for a random increase in the concentration of CO

RANDOM SIMULATION OF SPONTANEOUS HEATING

Fig. 4 illustrates how the model incorporates the occurrence of "risk". It is the part of the model that invokes a CO concentration increase only at the sensor located in the tailgate which simulates the occurrence of spontaneous heating. Increasing the concentration of CO is simulated at two levels. The element at the top of the model represents the starting point with one initiation token. This will enable to generate different levels of risk in terms of various levels of CO concentrations to different points of delivery (i.e. individual longwall panel operations). However, it is necessary to set the transitions to generate random time delays within certain limits (i.e. Time Mode: Uniform Distribution).

After creating individual elements of an End Point, the overall model was created based on a model situation. This model works with defined elementary conditions. The model (Appendix No.1 - Figure 7) thus consists of two End Points. At these end points, a random passage of a diesel engine through the maingate and random emergence of risk, i.e. increased CO concentration levels at the sensor located in the tailgate, are being simulated. The modelled control system is capable to respond to increased concentrations of CO in one particular longwall panel by throttling the valve in another panel where no risk is being indicated. While maintaining a constant volumetric flow rate entering the network, the flow rate in the area at risk increases. If the CO concentration at the sensor drops below a certain level, the modelled control system will set the volumetric flow rate of nitrogen in both panels to the original settings.

The HPSim software allows exporting the data to a CSV file. The modelling results stored in the CSV file were imported into Microsoft Excel, graphically visualized and evaluated. Fig. 5 shows the data obtained from the software HPSim after a time simulation. The software HPSim allows to use the output data in a form of integer values only. Within the design of the models, it was therefore necessary to count with the option to convert the decimal numbers. It is visible especially in the concentration of the CO sensors. The real output values of these sensors are in a form of decimal numbers. In the model, this is implemented by integer values of concentration, which are ultimately converted to decimal numbers. In this figure, you can see this in the field for the CO sensor maintage – Mod. It is a modified field, where the initial value is obtained from the model (CO sensor maintage) and then it is divided by 100.

It is possible to compare the result of the model with measured values of concentration after the correct setting. The first graph in Fig. 5 shows the graphical representation of the CO concentration measured by sensors for the point P1. The figure shows that the concentration of CO increased in the first phase at the input of the mine working (labelled as P1) as well as at the output. This corresponds to a passing locomotive. In the second simulated phase, the concentration increased only at the output of mine, suggesting a formation of heating.

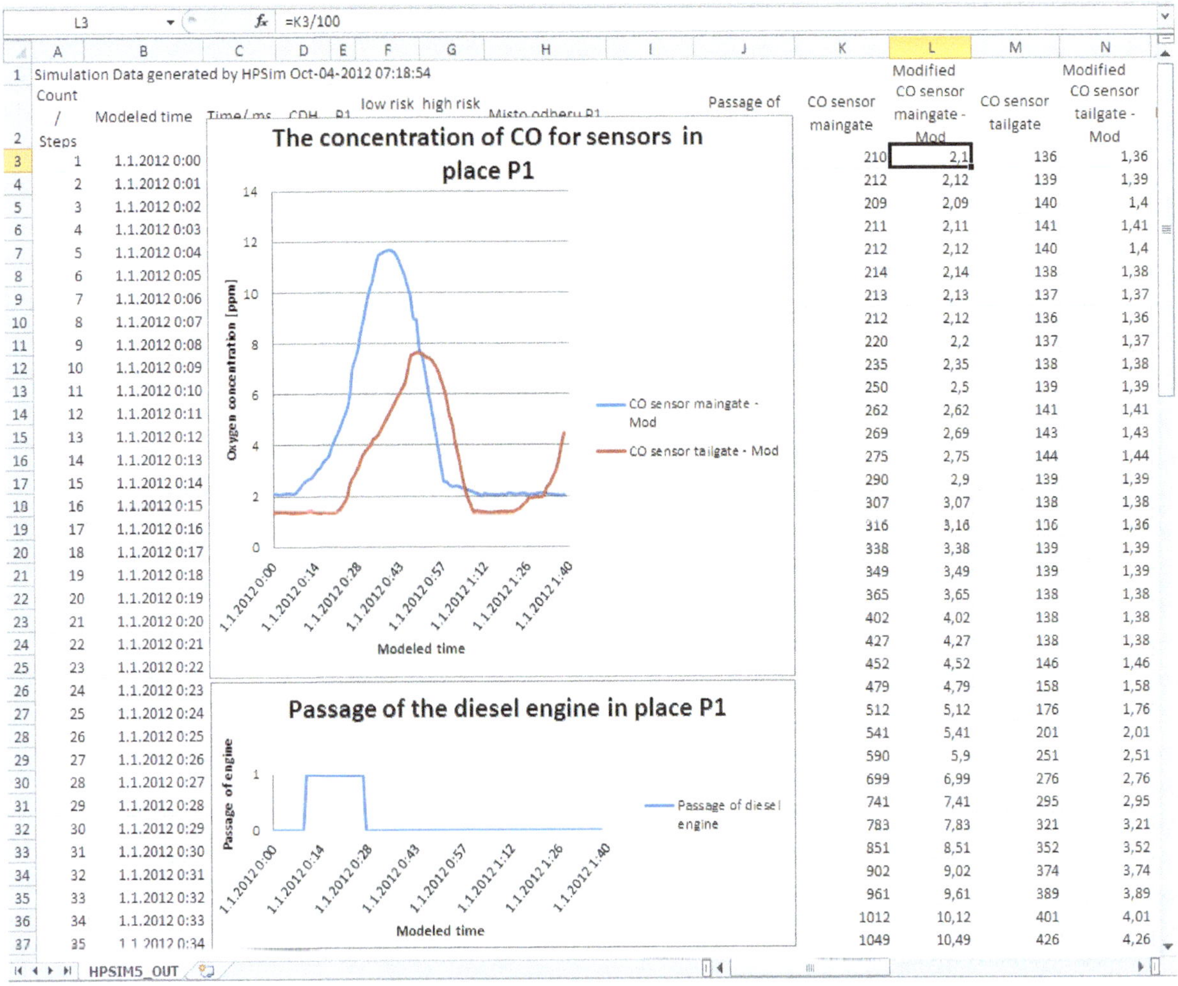

Fig. 5 Output simulation of the HPSim software

After the verification of the results, further efforts were made to create individual files for sub-models, which would enable a simple modular application for future use in other modelled situations. WinPeSim was used as a complementary software tool that enables to create, save and insert other sub-models into the overall model. Fig. 6 shows such a sub-model, which includes another sub-model labeled as Sensor I/O P1.

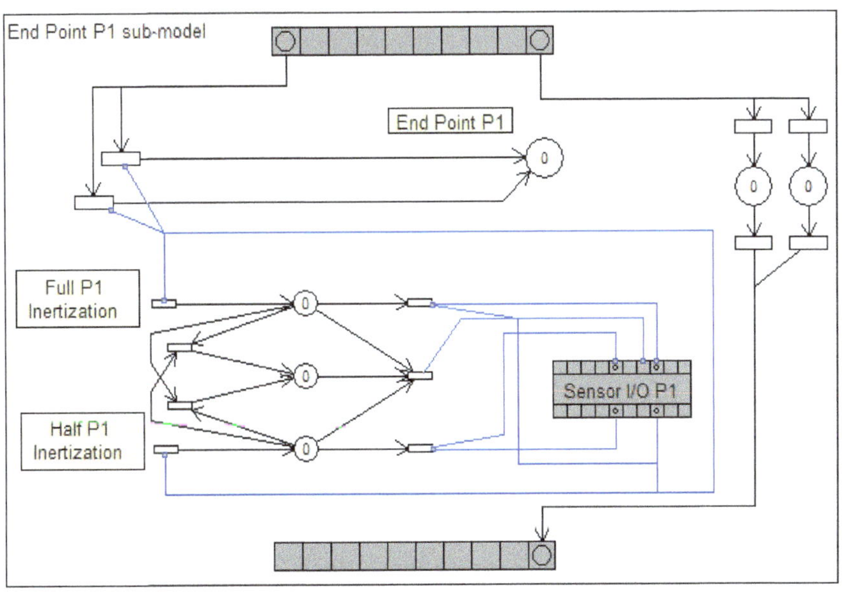

Fig. 6 End Point sub-model

3 CONCLUSIONS

By modelling the technological process in the HPSim software and by subsequent simulations, it was verified that this application is viable, but the created models are not simple and modular. The WinPeSim software was tested as a tool that enables to create, save and re-use sub-models. But it was just a not fully functional Beta Preview version of the software and errors occurred due to interconnected inputs and outputs. Therefore, we concluded that Petri nets do not comprise an optimal solution for the needs defined by the MTA Ostrava Company.

REFERENCES

[1] Adamus A., Hanák Z. & Molnári, P. *Listovka ZÁCHRANÁŘ*, Czech Republic, vol. 3, pp 6-9, 2003.

[2] Adamus A. *Měření objemového průtoku plynného dusíku v potrubí*, Uhlí-Rudy Geologický průzkum, Czech Republic, vol. 10, pp 14-20, 2002.

[3] Škorpík J. *Škrcení plynů a par*, Transformační technologie [online]. 2006[cit. 2012-04-13]. ISSN 1804-8293. Dostupné z: http://www.transformacni-technologie.cz/skrceni-plynu-a-par.html

[4] Vančura V. *Analýza informačních obsahů signálů automaticky snímaných veličin na plynujícím dole*. Ostrava, 2005. Disertační práce. VŠB - TU Ostrava. Vedoucí práce BURÝ A.

[5] Strakoš V. & Vančura V. *Intelligent model of safety management on gassy mines*, 20th World Mining Congress & Eexpo2005, Tehran. Iran, November 7, 2005.

[6] Vančura V. & Otte L. *Model of Safety Control Included into Monitoring Systém.* International Carpathian Control Conference ICCC`2006, Rožnov pod Radhoštěm, Czech Republic, May 29-31, 2006, p. 589-591, ISBN 80-248-1066-2.

[7] Vančura V., Otte L. & Kodym, O. *News Working Information on Gassy Mines*. Mine Planning and Equipment Selection – MPES 2006, September 20-22, 2006, Torino, Italy, p. 208-21, ISBN 88-901342-4-0.

[8] Wesley, C R , Wynne, T M, Urosek,, J E and Diederich, K S, 2006. *The successful recovery of the Dotiki Mine after a major mine fire. Proceedings*, Eleventh US Mine Ventilation Symposium, State College, Pennsylvania, Balkema, The Netherlands, 337-343, June2006.

[9] Otte L. *Využití Petriho sítí pro tvorbu simulačních modelů horizontální dopravy na uhelných hlubinných dolech*. Ostrava, 2009. Disertační práce. VŠB - TU Ostrava. Vedoucí práce BURÝ A.

[10] Ren, T & Balusu, R, 2009. *Proactive goaf inertisation for controlling longwall goaf heatings*, Procedia Earth and Planetary Science Volume 1, Issue 1, September 2009, pp 309-315.

[11] Ren, T., Wang, Z., Nemcik, J., Aziz, N. & Wu, J. *Investigation of spontaneous heating zones and proactive inertisation of longwall goaf in Fenguangshan Mine*, 12th Coal Operators' Conference, University of Wollongong & the Australasian Institute of Mining and Metallurgy, 2012, 212-220.

RESUMÉ

K tomu, aby případný model mohl fungovat správně, je zapotřebí prostudovat základní teoretická východiska a analyzovat danou situaci. Je tedy nutné zjistit informace o centrálním dusíkovém hospodářství, potrubní síti, kterou je dusík rozváděn do důlních prostor a podmínkách řízení TP v současné době. Potrubní síť není v tomto případě přímočará a je tvořena odbočkami různých tvarů, oblouky zúžením, filtry, měřidly apod. V těchto částech potrubní sítě vzniká tlaková ztráta, která je mnohem intenzivnější než na rovném úseku potrubí. Z pohledu tlakové ztráty se tyto prvky nazývají místní odpory. Za speciální případ místního odporu, lze považovat i výstupy dané sítě, v tomto případě škrtící ventily. Škrcením ventilu se sníží průtok a zvýší se tlaková ztráta, respektive sníží se tlak za ventilem, což je způsobeno nehomogenním prouděním v oblasti nejužšího průřezu a vířením.

V naší problematice vycházíme z naměřených hodnot oxidu uhelnatého. Oxid uhelnatý vzniká při hoření za vyšších teplot při nedostatku kyslíku, redukcí oxidu uhličitého v žáru, rozkladem vody na žhavém uhlí a také redukcí oxidu uhličitého vodíkem. Nejčastějším zdrojem výskytu oxidu uhelnatého jsou trhací práce a z havarijních případů je to vznik požáru nebo zápar.

Požadavkem společnosti MTA Ostrava, s.r.o. je najít vhodný nástroj k modelování technologického procesu inertizace závalu stěnového porubu uhelného hlubinného dolu. Za určitých okolností lze technologický proces inertizace považovat za systém, který je ovlivněn stochastickými, tedy náhodnými procesy. Petriho sítě jsou moderní prostředek pro modelování stochastických systémů a cílem bylo prověřit možnosti použití Pteriho sítí pro účely modelování již zmíněného technologického procesu.

Modelováním technologického procesu v softwaru HPSim a následnými simulacemi bylo ověřeno, že tento prostředek lze použít, ale vytvořené modely nejsou jednoduché a modulární. Následně byla ověřena i možnost použití softwaru WinPeSim, který udává právě možnost tvorby, ukládání a znovupoužití dílčích modelů. Jde však o tzv. WinPeSim Beta Preview verzi, která není zcela funkční a dochází k chybám vlivem uložení propojených vstupů a výstupů. Proto jsme dospěli k závěru, že Petriho sítě nejsou pro definované potřeby společnosti MTA Ostrava, s.r.o. příliš vhodné a doporučujeme využít dalších možností.

Appendix No. 1

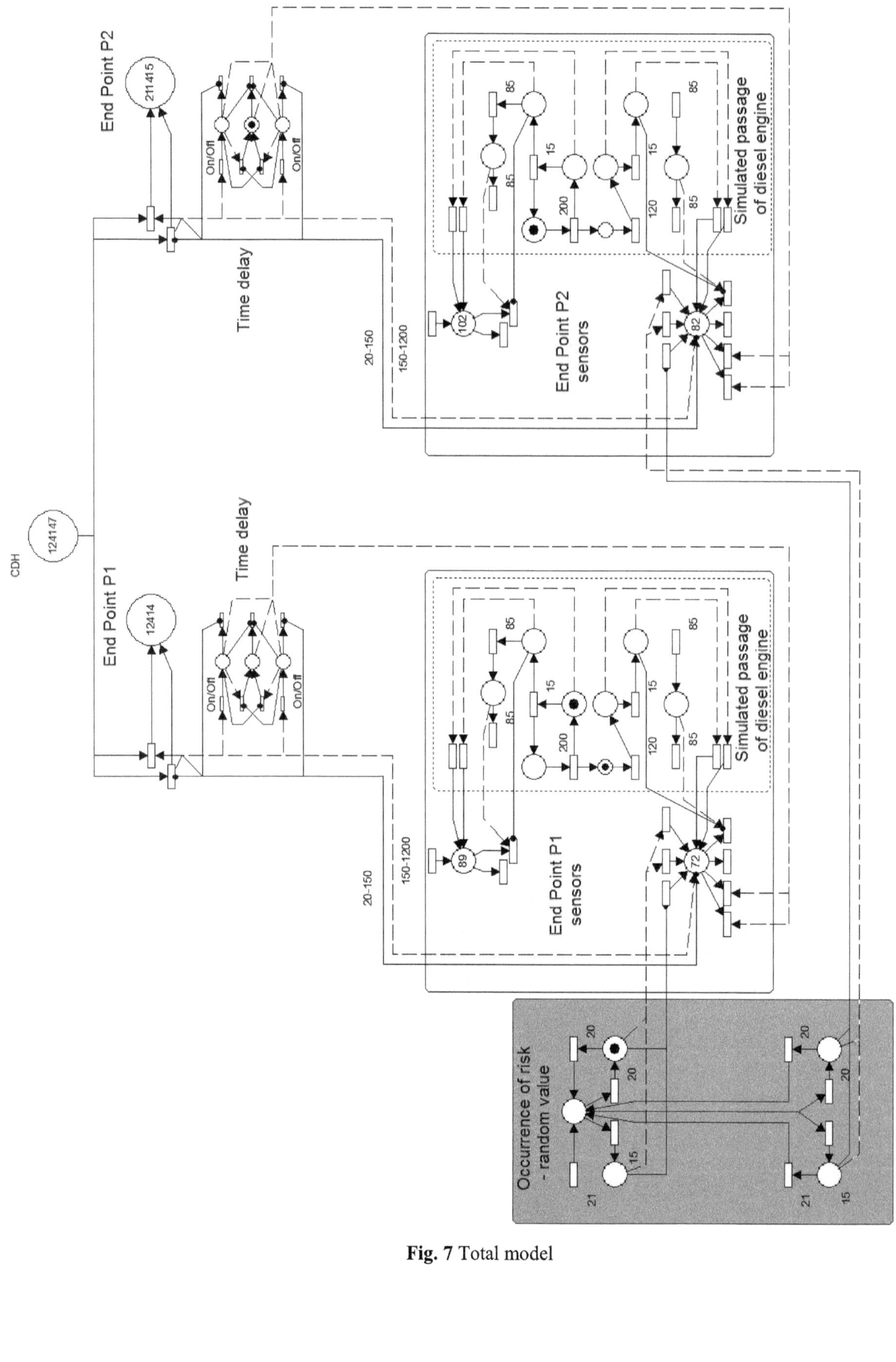

Fig. 7 Total model

27

AN ANALYSIS OF THE POSSIBILITY OF INTERSYSTEM TRANSFORMATIONS FOR PURPOSES OF GEO-REFERENCING OLD MINE WORKINGS

Hana STAŇKOVÁ [1], *Pavel ČERNOTA* [2], *Miroslav NOVOSAD* [3]

[1] *Ing. Ph.D., Institute of Geodesy and Mine Surveying, Faculty of Mining and Geology, VSB-Technical University of Ostrava, 17. listopadu 15, 708 33 Ostrava - Poruba, Czech Republic*
e-mail: hana.stankova@vsb.cz

[2] *Ing. Ph.D., Institute of Geodesy and Mine Surveying, Faculty of Mining and Geology, VSB-Technical University of Ostrava, 17. listopadu 15, 708 33 Ostrava - Poruba, Czech Republic*
e-mail: pavel.cernota@vsb.cz

[3] *Ing. Ph.D., Institute of Geodesy and Mine Surveying, Faculty of Mining and Geology, VSB-Technical University of Ostrava, 17. listopadu 15, 708 33 Ostrava - Poruba, Czech Republic*
e-mail: miroslav.novosad@vsb.cz

Abstract

The paper is dedicated to the issues of both planar and spatial transformations with respect to the territory affected by mining activities. The planar transformations have been applied among the planar Otto mine coordinate system, St. Stephen Datum of Cadastre Coordinates, and the Datum of Uniform Trigonometric Cadastral Network (hereinafter referred to as JTSK). Here basic transformation characteristics among the above systems are defined. The spatial transformation is performed through a point coordinate conversion between the JTSK and the European Terrestrial Reference System (hereinafter referred to as ETRS89) where conventional relations as well as other conversion possibilities were described. Part of the article is the execution of an experiment of a direct transformation between the St. Stephen Datum of Cadastre Coordinates and the ETRS98 system for the territory of Brno, which was not affected by mining activities. For the above transformation, 8 trigonometric points were used, for which the identity analysis had been done initially. Furthermore, the Interpolation and One Step Transformations were tested, which solve the conversion as divided tasks for both position and heights. The Interpolation Transformation showed better values for residues. The results of the above experiment will be applied for purposes of surveying old mine workings using the GNSS technology.

Abstract

Článek je věnován problematice rovinných a prostorových transformací s ohledem na území ovlivněném hornickou činností. Rovinné transformace byly aplikovány mezi rovinným Ottovým důlním souřadnicovým systémem, Katastrálním souřadnicovým systémem svatoštěpánským a systémem Jednotné trigonometrické sítě katastrální (JTSK). Zde jsou definovány základní charakteristiky převodu mezi těmito systémy. Prostorová transformace je realizována převodem souřadnic bodů mezi systémem JTSK a Evropským terestrickým referenčním systémem (ETRS89), kde jsou popsány konvenční vztahy a další možnosti převodu. Součástí článku je realizace experimentu přímé transformaci mezi katastrálním souřadnicovým systémem svatoštěpánským a systémem ETRS98 na území Brna, které není hornickou činností ovlivněno. Pro tuto transformaci bylo použito 8 trigonometrických bodů, u nichž byla původně provedena analýza identity. Testována byla Interpolační transformace a transformace One Step, které řeší převod jako rozdělené úlohy pro polohu a pro výšky. Lepší hodnoty reziduí vykazovala Interpolační transformace. Výsledky experimentu budou použity pro účely vytyčení starých důlních děl technologií GNSS.

Key words: geo-referencing, St. Stephen Datum of Cadastre Coordinates, ETRS89, coordinate transformation, Datum of Uniform Trigonometric Cadastral Network, abandoned working

1 INTRODUCTION

The issues of determining positions of old mine workings in the territory of cancelled mining areas in the western part of the Ostrava-Karvina Coalfield (OKC) is still a current problem. Abandoned old main mine workings from the 19th century, imperceptible in the landscape, became, after the backfilling of active pit shafts in connection with the termination of mining in the 90ies of the 20th century, a natural collector of methane exhalations from underground to the surface. So they become a security risk to human health and environment. Old mine workings, abandoned exploratory workings, abandoned mine workings and other objects are distinguished. An old mine working is understood to be an underground mine or quarry, whose original operator or its successor is unknown. Abandoned exploratory workings operated through state resources within geological survey, which were not passed for exploitation after the work completion. Abandoned mine workings are the workings out of operation, which have their owner or successor. Other objects are subsurface areas, which have been excluded for a purpose other than for mining and exploring mineral resources.

In the process of elimination of consequences of coal mining, it was necessary to find and cleanup old main mine workings and pits whose position had been originally determined in the Otto Coordinate System. The Otto District Coordinate System in the OKC was introduced in the mid-19th century by Ing. Francis Otto, the main mining surveyor of the Ferdinand's track. Positions of all significant points of in-those-days existing adits and pits, characteristic points of the terrain and other monumented points on the surface, were determined from a performed triangulation. The relation between the Otto Coordinate System and the St. Stephen Datum of Cadastre Coordinates is given by a linear transformation.

However, for the needs of state and local governments for planning and developing land affected by coal exploitation, it is necessary to convert (transform) the coordinates of positions of main mine workings and original pits from the 19th century to the coordinates of binding reference systems of the Czech Republic. The conversions between the systems are solved by transformations that can be performed in different ways with different results. The transformation type selection depends on required accuracy and the need to maintain the correct size of network.

An established way of converting point coordinates from the St. Stephen Datum of Cadastre Coordinates to the ETRS89 system is the transformation through S-JTSK. An experiment was carried out in Brno and its environs, consisted in a direct transformation between the St. Stephen Datum of Cadastre Coordinates and ETRS89 without a transition via the JTSK system. As identical points, the points of stable cadastre geodetic base were used, i.e. the points of cadastral triangulation (1822-1829) of order I, II and III.

2 TRANSFORMATIONS BETWEEN COORDINATE SYSTEMS

A transformation is generally understood to be a method of changing trigonometric coordinates or coordinates of other positional points – individually or collectively, for a given area, in fact without changing positions of points on the physical Earth's surface.

The main reasons for performing a transformation are according to [5] as follows:

1. Choice of a different reference ellipsoid when geographical point coordinates are being changed – to the same length of the trigonometric side different coordinate differences ($\Delta\varphi,\Delta\lambda$) correspond.
2. Shift, angular displacement or scale change in the entire network of permanently marked points when the scale specification, change in orientation or network approach to surrounding networks occur.
3. Change in the method for projecting an ellipsoid into a plane. The geographic coordinates of the point (φ,λ) remain the same, but can be transformed into a plane of e.g. conical or cylindrical projection. Therefore, it is a direct transformation $(x,y)\Rightarrow(X,Y)$ without an intermediate transition via the geographic coordinates, where x,y are the coordinates of the original (old) coordinate system and X,Y of the new one.
4. New surveying a part of the network. This includes a new triangulation (formerly), then new distance measurement, and currently measurements by means of GNSS (Global Navigation Satellite System).
5. The transition to new coordinate systems (geodetic system). This is a new measurement, new processing.

The first three cases above concerns a transformation of homogeneous coordinates. During this transformation, angle values remain constant and between the original (primary) and new (secondary) systems, precise mathematical relations apply. Cases 4 and 5 concern a transformation of non-homogeneous coordinates, when values of angles and lengths are randomly being changed, and there is no exact mathematical relation for converting coordinates from one system to another.

It is possible to transform coordinates not only of real, i.e. monumented points, but also of lost or notional points (e.g. corners of map sheets). The first step in a transformation is setting up a key of transformation, whose form depends on the applied transformation method and the selection of identical points. Identical points are such points, for which coordinates are known in both the original and new systems. If a necessary number of identical points are available, the calculation of other point coordinates is executed directly by means of a simple key of transformation.

The choice of identical points is crucial [13]. In order to maintain homogeneity and for possible shifts in certain point stabilisation, it is convenient to relate densification points (non-identical) to the given (identical) points so that they lie inside the polygon circumscribed by marginal identical points of the densified (transformed) area. Worse homogeneity may occur, for example, when relating the determined points only to one or two identical points. For assessing the quality of inserting new networks into original ones, or vice versa, in particular scale conversions, angular rotations and position deviations of identical points may be useful.

The number of identical points is usually higher than necessary, so to calculate adjusted coefficients of a key of transformation, a least squares adjustment method is used. Such transformed point coordinates of a new system, are burdened by residual deviations arising from differences between the given and transformed point coordinates. These deviations have the same value as coordinate corrections, but with an opposite sign. In order to avoid the lack of homogeneity (deformation) of a whole minor control towards identical points, it is vital to introduce corrections also in case of other detailed survey points. Among a number of methods permitting a classification of coordinate corrections there are:

- Graphic method (plotting a chart of corrections);
- Method of a general arithmetic mean of an identical point shift - Jung's transformation;
- Spline interpolation [23].

3 PLANAR TRANSFORMATION BETWEEN S-OTT, S-SK (ST. STEPHEN DATUM, GUSTERBERG DATUM) AND S-JTSK

The X-axis of Otto Coordinate System was parallel to the X-axis of the stable cadastre coordinate system passing through the main tower of St. Stephen's Cathedral in Vienna. This cadastral system was valid for the countries of the Austro-Hungarian Empire, Lower Austria, Moravia, Silesia and Dalmatia. The relation between the Otto Coordinate System and the St. Stephen Datum of Cadastre Coordinates is given by the following linear transformation:

$$y = y_0 + Y' \cos\psi + X' \sin\psi, \quad (1)$$

$$x = x_0 + X' \cos\psi - Y' \sin\psi, \quad (2)$$

where y_0 = -138443.332 m, x_0 = -183579.651 m are the coordinates of the Otto system origin in S-SK; Y', X' are the coordinates in St. Stephen Datum of Cadastre Coordinates; and $\psi = 180°$ is the angle of the meridian of the Otto Coordinate System against the meridian of the St. Stephen Datum of Cadastre Coordinates.

Connecting the Otto triangulation to the St. Stephen Datum of Cadastre Coordinates was solved by measuring the points of cadastral triangulation of order II in the “Ostrava Region”. It concerned churches (church towers) in municipalities Vratimov (Ratimow), Hošťákovice (Hostialkowitz) and Vrbice (Wirbitz). More information about the St. Stephen Datum of Cadastre Coordinates may be found in [22] and [23].

According to [1], the derivation of a genuine link between the systems S-SK (stable cadastre coordinate system) and S-JTSK encounters certain below mentioned drawbacks.

1. An error occurred in the determination of the Gusterberg datum direction as the X-axis does not identify with the Gusterberg point meridian, but forms a small angle (4′22.3′′) westwards. The original cadastral maps of Bohemia are thus oriented according to a different meridian [20].
2. In [2], it is stated that there is no established transition method from an ellipsoid to a reference surface or a reduction of bearings into a plane. A reduction of bearings into a plane is described

e.g. in [4], but it is not documented whether this reduction was introduced or not. Horizontal angles in triangles were corrected only by a spherical excess (in order I only) and the calculation was carried out as in planar nets.

3. The net was constructed according to the needs of detailed measuring, therefore the procedure from large to small was not observed.
4. The conversion coefficient value to convert from fathom to metric system is not identical in different literature sources. The coefficient value was identified in UAZK archive materials (Central Archives of Surveying, Mapping and Cadastre) and one value was always found out – 1°=1,896483843m. This value corresponds e.g. to the data in [19] and [3].
5. It is possible to meet different values of coordinate system origins, whether in the Gusterberg Datum or St. Stephen Datum.
6. The variability of scales of both projections **Q= M_{jtsk}:M_{sk}** causes that the direct point series of one system is transformed into the other system as a curve. This is most manifested in straight lines parallel to isometric lines $q = const.$ In perpendicular sides, straight line retention applies, however with a maximum failure of dividing ratio.

4 SPATIAL TRANSFORMATION BETWEEN S-JTSK AND ETRS89 SYSTEMS

Coordinates obtained from measuring using the GNSS technology in the ETRS89 system may be transformed into S-JTSK in the following ways [6], [7], [15].

1. Through transformations of rectangular coordinates of X, Y, Z in the ETRS89 to rectangular coordinates of the Bessel ellipsoid by means of: spatial identity transformation; spatial similarity transformation; general k-th degree polynomial transformation and a follow-up conversion of the transformed rectangular coordinates to geographic coordinates with a further conversion into a S-JTSK plane.
2. Converting rectangular coordinates of X, Y, Z in the ETRS89 system to geographic coordinates of the Bessel ellipsoid, followed by a cartographic conversion into a S-JTSK plane and through a planar similarity transformation of Y, X coordinates in a Krovak's projection with an additional height transformation.
3. Converting plane coordinates of Y, X and height h of identical points from the JTSK system to the Bessel ellipsoid, followed by an identity transformation to the centric Bessel ellipsoid, adjusting GNSS vectors on this ellipsoid with identical points as fixed ones and an inverse identity transformation of all the points back to the Bessel ellipsoid, finally classically converting into S-JTSK.
4. Through the reduction of the measured vectors using the GNSS technology to the Bessel ellipsoid, conversion to S-JTSK and classical adjustment as in terrestrial measuring.
5. Through an interpolation method without the knowledge of map projection or relevant reference ellipsoid dimensions. The ground of the solution is the identification of a projection relation between identical point coordinates.

4,1 Similarity transformation

The similarity (Helmert) transformation is used to convert geodetic networks, which approximately are of the same shape in both coordinate systems, but differ in size. The most commonly used spatial transformation is the one with the following parameters:

- $\Delta X, \Delta Y, \Delta Z$ are the coordinate differences reflecting a shift of the Bessel ellipsoid towards the GRS80 ellipsoid,
- α, β, γ are the angles expressing the angular displacement of the transformed system relative to the coordinate axes,
- Δm is the scale factor.

For that transformation two mathematical models are used as follows:

- Bursa – Wolf Model (3),

- Molodensky – Badekas Model (4).

The first (Bursa – Wolf) model solves the searched parameters in regard to the geocentre and is therefore suitable for transformations of larger networks. The latter (Molodensky – Badekas) model uses the coordinates of the centre of gravity of identical points instead of the geocentre and is suitable for smaller areas to be transformed.

Bursa – Wolf Model

$$\begin{pmatrix} X_i \\ Y_i \\ Z_i \end{pmatrix}_{Bess} = \begin{pmatrix} \Delta X \\ \Delta Y \\ \Delta Z \end{pmatrix}_{Bess} + (1+\Delta m)\cdot \begin{pmatrix} 1 & \gamma & -\beta \\ -\gamma & 1 & \alpha \\ \beta & -\alpha & 1 \end{pmatrix} \cdot \begin{pmatrix} X_i \\ Y_i \\ Z_i \end{pmatrix}_{ETRS89} \tag{3}$$

Molodensky – Badekas Model

$$\begin{pmatrix} X_i \\ Y_i \\ Z_i \end{pmatrix}_{Bess} = \begin{pmatrix} \overline{X}_t \\ \overline{Y}_t \\ \overline{Z}_t \end{pmatrix}_{Bess} + (1+\Delta m)\cdot \begin{pmatrix} 1 & \gamma & -\beta \\ -\gamma & 1 & \alpha \\ \beta & -\alpha & 1 \end{pmatrix} \cdot \begin{pmatrix} X_i - X_t \\ Y_i - Y_t \\ Z_i - Z_t \end{pmatrix}_{ETRS89} \tag{4}$$

$\overline{X}_t, \overline{Y}_t, \overline{Z}_t$ are the gravity centre coordinates of identical points in JTSK,

X_t, Y_t, Z_t are the gravity centre coordinates of identical points in ETRS89,

X_i, Y_i, Z_i are the coordinates of the points to be transformed.

In both models, the rotation matrix is simplified due to small values of angles of rotation. After performing the spatial transformation of ETRS89 coordinates to the Bessel ellipsoid and cartographic conversion into S-JTSK, the completion of transformation of coordinates Y, X is used [17].

The transformation schema according to [16] which represents a 7-element Helmert similarity transformation of spatial coordinates with a subsequent transformation completion that already takes place in a plane of the Krovak projection.

Transformation ($B,L,H_{el.}$) in ETRS-89 to (Y,X,h_N) in S-JTSK :

$(B,L,H_{el.})_{geoc.}$⇒①⇒ $(X,Y,Z)_{geoc.}$⇒②⇒ $(X,Y,Z)_{Bessel}$⇒③⇒ $(B,L,H_N)_{Bessel}$⇒④⇒ $(Y,X)_{Krovak}$
$(H_{el.})$ ⇒⑤⇒ (H_N)

① transfer of ellipsoidal coordinates to spatial rectangular ones (introduction of quasi-geoid height)
② seven-element Helmert Spatial Transformation,
③ inverse to ①,
④ equations of the Krovak projection with addition of corrections ($\Delta X, \Delta Y$),
⑤ simple adding the height of quasi-geoid. In ① and ④, $H_{el.}$ plays only a mediating role.

Transformation (Y,X,H_N) in S-JTSK to ($B,L,H_{el.}$) in ETRS-89:

$(Y,X)_{Krovak}$⇒④′⇒$(B,L,H_N)_{Bessel}$⇒③′⇒$(X,Y,Z)_{Bessel}$⇒②′⇒$(X,Y,Z)_{geoc.}$ ⇒①′⇒$(B,L,H_{el.})_{geoc.}$
(H_N) ⇒⑤′⇒$(H_{el.})$
Dashed transformations are inverse transformations to non-dashed ones.

4.2 Identity transformation

The identity transformation is used if a network shape has been changed in a new position and the transformed points should better comply with the given identical points. The objective of an identity transformation is to adjust the shape and dimensions of a pattern formed by points in one system to the shape and dimensions of a point pattern in the other system.

The transformation key includes the following parameters:

- *ΔX, ΔY, ΔZ* are coordinate differences which express a Bessel ellipsoid shift to a GRS80 ellipsoid;
- *α, β,γ* are angles expressing the transformed system rotation with respect to coordinate axes.

$$\begin{pmatrix} X_i \\ Y_i \\ Z_i \end{pmatrix}_{Bess} = \begin{pmatrix} \Delta X \\ \Delta Y \\ \Delta Z \end{pmatrix}_{Bess} + \begin{pmatrix} 1 & \gamma & -\beta \\ -\gamma & 1 & \alpha \\ \beta & -\alpha & 1 \end{pmatrix} \cdot \begin{pmatrix} X_i \\ Y_i \\ Z_i \end{pmatrix}_{ETRS89} \tag{5}$$

The transformation key does not include any coordinate scale change.

4.3 General polynomial transformation

This method of transformation may be used for the transformation of point position as well as for transformations of heights. Its principle is an interlay of a k-th degree planar polynomial function with identical points [18].

The transformation relation of B,L geographic coordinates is as follows:

$$B_{Bess} - B_{ETRS890} = \Delta B = \sum_{i=0}^{k} \sum_{j=0}^{i} a_{i-j,j} \cdot U^{i-j} \cdot V^{j} \tag{6}$$

$$L_{Bess} - L_{ETRS890} = \Delta L = \sum_{i=0}^{k} \sum_{j=0}^{i} b_{i-j,j} \cdot U^{i-j} \cdot V^{j}, \tag{7}$$

where: B, L, - parameters of geographic coordinates,
a, b, - searched transformation parameters,
U, V - reduced geographic coordinates in ETRS89,

$$U = p(B - B_0), V = p(L - L_0), p = c\pi / 180, \tag{8}$$

B_o, L_o - selected reduction values,
c - selected constant (reduced coordinates should be in the interval of <-1,1>),
k - index of general transformation degree, and
i, j - integer indices with a condition $i - j \cap j \leq 9$ limiting individual member powers

Increasing the number of polynomial function degrees, the planar polynomial more approaches all the identical points and thus it is possible to obtain lower values of transformation key accuracy characteristics. Using a sufficient number of quality identical points and their balanced distribution it is possible to give a true picture of a local network deformation via a polynomial function. This transformation between ETRS89 and S-42/83 was used for the whole Czech Republic.

4.4 Other methods of converting local planar systems into geocentric ones

Conformal transformation of higher orders

This transformation respects angles to avoid a new network shape disruption. Its application is exceptional, e.g., when inserting a more accurate network into an older, less accurate basis. The transformation is based on the condition stipulating that elementary patterns in both the coordinate systems are similar. Straight lines in one coordinate system are represented in the other as generally curved with a disrupted dividing ratio, but the angles between their images remain preserved. This method's disadvantage is ambiguous results on the borders of transformed areas and therefore, the method is not much suitable to convert points from a larger area [14]. This third-order transformation was applied to create the 1952 Coordinate System (S-52).

Transformation of system coordinates through a multiple regression analysis

According to [12], the Helmert transformation has a linear character. By means of its parameters, it is not possible to capture local inhomogeneity of a reference triangulation network well. The inhomogeneity of local triangulation networks can be reduced by means of multiple regression relations being used for the transformation.

The transformation can be described by means of the following multiple regression relations:

$$\Delta B = A_{0,0} + A_{1,0,}U + A_{0,1}V + A_{2,0}U^2 + A_{0,2}V^2 + A_{1,1}UV + ... = \sum_{p,q} A_{p,q}U^pV^q ,$$
$$\Delta L = B_{0,0} + B_{1,0,}U + B_{0,1}V + B_{2,0}U^2 + B_{0,2}V^2 + B_{1,1}UV + ... = \sum_{p,q} B_{p,q}U^pV^q , \quad (10)$$
$$\Delta H = C_{0,0} + C_{1,0,}U + C_{0,1}V + C_{2,0}U^2 + C_{0,2}V^2 + C_{1,1}UV + ... = \sum_{p,q} C_{p,q}U^pV^q ,$$

where $p,q = 0,1,2, ..., U=k(B-B_0)$ is normalized geodetic latitude. $V=k(L-L_0)$ is the normalized geodetic longitude, k is a scale factor and B_0, L_0 are geographical coordinates of a transformed area centre.

The author [12] states that the multiple regression transformation is more than four times more effective than the Helmert transformation into EUREF89 and as much as six times more effective in case of a transformation of S-42/83 to EUREF89.

Transformation of system coordinates by means of non-linearized rotation matrices and conformal projection

In [10], a method of transformation from WGS-84 geodetic datum coordinates into a local coordinate system is processed by means of non-linearized rotation matrices to ensure millimetre transformation accuracy resulting in high accuracy of point position determination using the NAVSTAR GPS technology. The method has several advantages. Through the transformation, high relative accuracy of point position determination using the GNSS technology is preserved, and next, the local system is implemented so that no complex relations are needed to reduce the measured parameters. Compared to a similarity transformation, instead of surveying in three identical points, one is sufficient.

The mathematical formulation of a coordinate transformation from a geocentric system to a local geocentric system by means of rotation matrices and a conformal projection [11] making use of P surface, osculating sphere and its subsequent transformation into the local system retains a high accuracy of determining coordinates using the GNSS technology. It does not distort angles or produce scale errors. In addition, it minimizes network point relative heights from the basic plane as a result of which distance reductions from relative height above sea level are mostly negligible. This also ensures sufficient compliance of the local planar system with S-JTSK by means of one identical point.

Non-linear 3D transformations

It is apparent from the results in [24] that the lowest positional deviation is reached by a general quadratic transformation and it is stated that the most suitable transformations between the geocentric coordinate system and S-JTSK are non-linear transformations. According to [9], the model of a local 3D non-linear transformation overcomes the problem of correlation between estimated parameters, which occurs during global non-linear transformations, and is suitable for small areas too.

Other possibilities of transformation, e.g., by means of robust methods are stated e.g. in [8], [17].

5 DIRECT TRANSFORMATION BETWEEN S-JTSK AND ETRS89 SYSTEMS

The direct transformation between the S-SK (St. Stephen) and ETRS89 systems has been implemented in Brno and its environs. A set condition was to start from the points of a geodetic base, and thus for the St. Stephen Datum, cadastral triangulation points (1822-1829) of order I-III within Brno were used, in case of which it was possible to document their origin and accuracy. The original point coordinates of the Stable Cadastre system, St. Stephen, in Brno, were obtained from archive materials of UAZK, in particular from protocols of triangulation calculations (1822-1827). After carrying out terrain reconnaissance and identity analysis [21], 8 points were identified, suitable for surveying using the GNSS technology.

As the reference station for point surveying using the GNSS technology, the south-west pillar on the roof of the Faculty of Civil Engineering (BUT) in Brno, was selected. This pillar was labelled under a working title of TUBO_JZ. The TUBO_JZ coordinates in ETRS89 were determined during earlier measurements (φ = 49°12′20,92837", λ = 16°35′34,25047", H =322,563 m). For measuring, 3 two-frequency apparatuses were used, and the mean time of measuring in individual points was about an hour. The following parameters were selected for calculations:

- Cut of angle: 10°,
- Ephemeris: Broadcast,
- Tropospheric model: Saastamoinen,
- Ionospheric model: Standard,
- Maximum baseline length : 20 km,
- Processing mode: all baseline.

Tab.1 Resulting geographical coordinates of points in the ETRS89 system

Point Id	Point Class	Epochs	Latitude	Longitude	Elip.Hgt.	S.d.Lat.	S.d.Lon.	S.d.H.
944210024	Measured	09/20/2003 06:48:50	49° 14' 06.86054" N	16° 36' 01.22831" E	344, 2271	0.0002	0.0001	0.0003
944230014	Measured	09/20/2003 09:50:20	49° 05' 18.45970" N	16° 42' 31.66072" E	261, 5801	0.0006	0.0004	0.0010
943070030	Measured	09/20/2003 10:06:50	49° 06' 58.22828" N	16° 25' 36.28757" E	495.4455	0.0015	0.0009	0.0023
944170014	Measured	09/20/2003 07:10:20	49° 10' 50.90132" N	16° 47' 01.74861" E	324, 4055	0.0016	0.0009	0.0025
944220029	Averaged	09/20/2003 08:41:00	49° 07' 54.44743" N	16° 39' 50.49924" E	271.7589	0.0005	0.0003	0.0006
943030010*	Averaged	09/20/2003 11:32:30	49° 05' 22.01223" N	16° 32' 02.93827" E	305, 9306	0.0007	0.0004	0.0006
944230028	Reference	09/20/2003 11:32:30	49° 02' 30.40543" N	16° 38' 20.76636" E	398, 9631	0.0044	0.0009	0.0011
943020011	Reference	09/20/2003 08:41:00	49° 10' 41.07452" N	16° 34' 54.54647" E	356, 3670	0.0006	0.0002	0.0021
TUBO_JZ	Control	09/20/2003 14:14:50	49° 12' 20.92837" N	16° 35' 34.25047" E	322, 1563	0.0000	0.0000	0.0000
S.d.Lat.	Standard Deviation Latitude							
S.d.Lon.	Standard Deviation Longitude							
S.d.H.	Standard Deviation Height							

The information on the course of calculation, or the computational record, including the resulting coordinates, variance-covariance matrices and other accuracy characteristics are available in [21].

The transformation was implemented using the software by Leica SKI-Pro, where an ellipsoid used for the St. Stephen Datum was defined, and the origin of the coordinate system (St. Stephen) in geographic coordinates was selected.

The applied software offers several types of transformations:

- *Classical 3D* is a classical spatial transformation according to the Bursa-Wolf method or Molodensky-Badekas method,
- *Interpolation* is a method that inserts the measured coordinates among the coordinates of a given system, i.e. their position and height independently from each other. Errors in heights are not projected into positions, and neither ellipsoid nor map projection must be known. There is no necessity to identify points for positional or height adjustments. The method is suitable for transformations within small areas (10-15 km^2).
- *OneStep* transforms positions and heights separately. Positions are transformed individually through a classical 2D transformation.
- *TwoStep* is a method of split transformations of position and height. Positions are determined by a 3D transformation; heights are calculated separately.

For direct transformations between the SK and ETRS89 systems, the *Interpolation* and *OneStep* transformations were chosen, due to the separate transformations of position and height. Then, transformation keys and residual values of dX and dY were calculated, for the *Interpolation* transformation see Tab.2, and for the *OneStep* transformation see Tab.3.

Tab. 2 Table of residual values of *Interpolation* transformation

Interpolation transformation				
A system	B system	dX [m]	dY [m]	Position [m]
944230028	944230028	-0.190	0.048	0.196
943010023	943010023	-0.044	0.042	0.061
943020011	943020011	-0.088	-0.082	0.120
943070030	943070030	0.058	-0.012	0.059
944170014	944170014	-0.093	-0.005	0.093
944210024	944210024	0.129	0.004	0.129
944220029	944220029	-0.150	0.085	0.172
944230014	944230014	0.377	-0.079	0.386

Tab. 3 Table of residual values of *OneStep* transformation

OneStep transformation				
A system	B system	dX [m]	dY [m]	Position [m]
944230028	944230028	-0.308	0.251	-0.398
943010023	943010023	0.044	-0.014	-0.046
943020011	943020011	-0.152	-0.121	-0.194
943070030	943070030	0.262	0.004	-0.262
944170014	944170014	-0.214	-0.131	-0.250
944210024	944210024	0.146	-0.049	-0.154
944220029	944220029	-0.226	0.089	-0.243
944230014	944230014	0.448	-0.029	-0.449

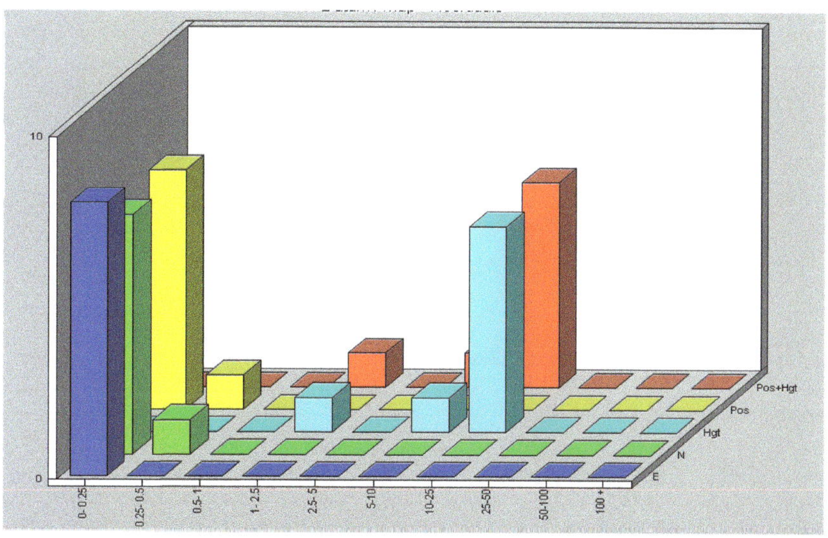

Graph 1: Residual values of Interpolation transformation

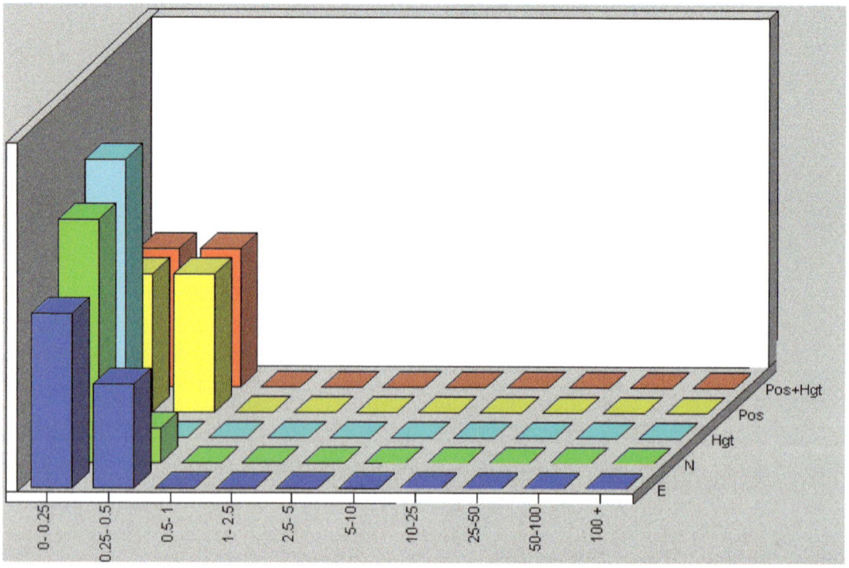

Graph 2: Residual values of Onestep transformation

Values of transformation parameters:

System A:	*GPS_0903*	*System B:*	*Brno_St. Stephen*
Ellipsoid:	*WGS 1984*		*Zach (S-SK)*
System:	*WGS84*		*St. Stephen*

Transformation parameters of *Interpolation transformation*:

Position:

0.00320050	-0.99994511
-0.99996042	0.00317242

Scale:

X: -49.7712 ppm
Y: -34.5489 ppm

Translation:

X: -18068.6773 m
Y: -126005.1615 m

Transformation accuracy: dX: 0.0554 m
dY: 0.1729 m
Mean scale between A and B: 0.99995507

Transformation parameters of *OneStep Transformation:*

2D_Helmert Transformation:
Number of common points: 8
Rotation origin: X_0: 4.7903 m
Y_0: 0.7171 m

No:	Parameter:	Value:	r.m.s.	Unit:
1	Shift dX	-104849.7163	0.0799	m
2	Shift dY	-18137.8178	0.0799	m
3	Rotation about Z	647341.81040	1.61955	′′
4	Scale	-45.8670	7.8521	ppm

6 CONCLUSIONS

The reason for geo-referencing old and abandoned main mine workings and pits is the threat of landscape and health of population due to possible gas leaks from underground to the surface. An equally serious problem is the real possibility of mechanical damage to buildings due to break and fall through of backfill materials in case of construction and other activities in these areas.

In the process of disposing old main mine workings, a final report is being prepared whose integral part must be, among others, the passportization of mine workings open to the surface or close to the surface. Part of the passportization is the registration list of a main mine working, which in addition to a pit (adit) name must contain coordinates x, y of the pit axis in a valid coordinate system in compliance with Decree of the Czech Mining Office (CBU) No. 435/1992 Coll., as amended. For purposes of the passportization, it is necessary to set out the mouth of all active and old main mine workings from the period of the oldest mining activities in the field. The position coordinates of the old main mine workings are determined via a graphical transformation of identical points of LC (Land Cadastre) and LR (Land Registry) maps or via a linear transformation to S-SK and then the Helmert Transformation to JTSK. After setting out and verifying the mouth position of a pit or adit in the terrain (by geophysical survey and drilling), the space on the surface is determined for safety reasons, in which the protruding mine atmosphere could endanger the life or health of persons, because the passportized main mine workings in the Ostrava part of OKC belongs among gassy coal mine workings with a possibility of methane exit to the surface.

During activities of liquidation and rehabilitation of old mine workings, the Otto Coordinate System, in which the positions of the main mine workings and pits had been identified, acquired its importance. Since the Otto Coordinate System is linearly linked to the Stephen Datum of Cadastre Coordinates, the transfer between these systems is simple. With the development and application of the GNSS technology (Global Navigation Satellite System) the possibility of setting out positions of old main mine workings and pits by methods of the technology, was analyzed. For the purpose of setting out the main mine workings and pits by the GNSS technology, an experiment was carried out, leading to creating a transformation key of a direct transformation between the St. Stephen Datum of Cadastre Coordinates and ETRS89.

A set of points of the St. Stephen Datum of Cadastre Coordinates, triangulation networks of the order I-III from the original adjustment was selected. From this set, 36 points were selected, which might be regarded as identical ones. After the reconnaissance of all 36 points, only 8 of them were selected, which satisfied the conditions for surveying using the GNSS technology. These points were then analysed for their identity with the St. Stephen Datum of Cadastre Coordinates. The points that were surveyed by the GNSS technology acquired their coordinates in the spatial European Terrestrial Reference System (ETRS89).

Two transformation methods were used for direct transformations between the SK and ETRS-89 systems,:

- OneStep transformation, and
- Interpolation transformation.

Both the transformations solve the key of transformation as separate tasks for positions and heights. The transformations were solved by means of the Leica SKI-Pro software, where an ellipsoid for the SK system had been defined and St. Stephen Datum origin coordinates had been inserted in geographic coordinates. For both the transformations, transformation keys and residual values of dX and dY were calculated. As expected, lower residual values were shown in case of the Interpolation transformation.

REFERENCES

[1] ČADA, V.: Obnova katastrálního operátu v lokalitách souřadnicových systémů stabilního katastru, *Geodetický a kartografický obzor,* roč.45/87, 1999, číslo 6, str.122, ISSN 0016-7096

[2] ČADA, V.: Robustní formy tvorby a vedení digitálních katastrálních map v lokalitách sáhových měřítek, Habilitační práce podaná na ČVUT Praha v oboru Geodézie a kartografie, Plzeň, 2003

[3] BÖHM, J.: Vyšší geodézie II, Souřadnicové soustavy, SNTL, Praha 1966

[4] BÖHM, J.: Matematická kartografie, díl II., Souřadnicové soustavy v geodézii a kartografii, Donátův fond, rektorát Vysoké školy technické Dr. E.Beneše v Brně, Brno 1951

[5] BÖHM, J., HORA, L., KOLENATÝ, E: Vyšší geodézie, díl I, II, Praha 1982, 1983

[6] BUČKO, E.: Princip a technológia merani GPS, Zborník prednášok Katedry geodézie Stavebnej fakulty STU Bratislava, 18.zošit, október 1996

[7] CIMBÁLNÍK, M., MERVART, L.: Vyšší geodézie 1, ČVUT Praha 1997

[8] GAŠINCOVÁ, S., GAŠINEC, J., TREMBECZKÁ, E.: Vyrovnanie 2D geodetických sietí pomocou robustných metod (Adjustment of 2D geodetic networks by robust estimations) - 2008. In: Geodézia, kartografia a geografické informačné systémy 2008. - Košice : GK a GIS, 2008 17 s. - ISBN 9788055300795

[9] HEFTY, J., FROHMAN, E.: Nelineárne trojrozmerné transformácie a ich využitie na prevod súradníc medzi S-JTSK a ETRS-89, *Geodetický a kartografický obzor*, roč. 44/86, 1998, č.6, pp.121-127, ISSN 0016-7096

[10] MELICHER, J., FLASSIK, T.: Transformácia súradníc zo Svetového geodetického systému do lokálneho súradnicového systému pomocou nelinearizovaných rotačných matíc, GaKo 44/86, č. 2, ISSN 0016-7096

[11] MELICHER, J., GALGONOVÁ, R.: Transformácia súradnic z geocentrického rovníkového systému do lokálneho rovinného systému pomocou rotačných matic a konformného zobrazenia, *Geodetický a kartografický obzor*, roč. 45/87, 1999, č. 2, pp. 25-29, ISSN 0016-7096

[12] MOJZEŠ, M.: Transformácia súradníc systémov multiregresnou analýzou, Kartografické listy, 1995, č. 5

[13] NEVOSÁD, Z.: K identitě bodů při spojování družicových a triangulačních sítí, sborník referátů Seminář s mezinárodní účastí Zkušenosti s využitím GPS pro bodová pole, VUT FAST Brno, 1998

[14] NEVOSÁD, Z.: Geodézie IV, Vyrovnání geodetických sítí, Vojenská akademie Brno 1994

[15] KOSTELECKÝ, J.: K převodu výsledků měření aparaturami GPS do souřadnicového systému S-JTSK, *Geodetický a kartografický obzor,* 39/1993 č.7, ISSN 0016-7096

[16] KOSTELECKÝ, J., CIMBÁLNÍK, M.: Převod souřadnic mezi S-JTSK a ETRS-89, *Geodetický a kartografický obzor*, ročník 42/84, č. 2,1996, PP.23-32, ISSN 0016-7096

[17] LABANT, S., WEISS, G., KUKUČKA, P.: Adjustment of a geodetic Network Measured by a Satellite Technology in the Dargovskych Hrdinov Suburb, Acta Montanistica Slovaca, Volume 16, Issue 3, 2011, pp. 229-237, ISSN 1335-1778

[18] OLŠOVSKÝ, V.: Analýza možnosti využití obecné transformace pro transformaci elipsoidických souřadnic a výpočtu relativních odlehlostí elipsoidů, kandidátská disertační práce

[19] PROVÁZEK, J.: Vývoj polohových základů na území ČR, Zeměměřický úřad, Praha, 2000

[20] RYŠAVÝ, J.: Geodesie, SNTL Praha 1955

[21] STAŇKOVÁ, H.: *Problematika identity trigonometrických bodů pro účely mezisystémových transformací v Brně a okolí*, Disertační práce, Brno: VUT FAST, 2006

[22] STAŇKOVÁ, H.. ČERNOTA, P.: A Principle of Forming and Developing Geodetic Bases in the Czech Republic, *Geodesy and Cartography*, Vilnius: Technica, 2010, Vol.36, No.3 p. 103-112, ISSN 1392-1541 print/ISSN 1648-3502 online

[23] ŠESTÁK, L.: Možnosti použití splajnů pro transformaci souřadnic, Písemná práce k rigorózní zkoušce, Vojenská akademie Brno 1998

[24] VAĽKO, J., DEMČÁKOVÁ, L.: Transformácia súradníc medzi geocentrickým súradnicovým systémom a S-JTSK, Kartografické listy, 1998, č. 6

RESUMÉ

Georeferencování, jinými slovy vyjádření prostorových referencí, je proces určení vztahu mezi polohou dat v přístrojovém souřadnicovém systému a geografickou, resp. mapovou polohou. Transformace souřadnic je převod souřadnic bodů z jednoho souřadnicového systému do druhého. Proces převodu souřadnic je řešen pomocí exaktně vyjádřených transformačních rovnic a následným výpočtem transformačního klíče. Transformací jsou hledány optimální vztahy mezi dvěmi souřadnicovými systémy. Starým důlním dílem se rozumí opuštěné důlní dílo v podzemí nebo opuštěný lom po těžbě vyhrazených nerostů. V obou případech není znám původní provozovatel ani jeho nástupce. Registr starých důlních děl je Ministerstvem životního prostředí pověřena vést Česká geologická služba – Geofond. Proces transformace mezi dvěmi rovinnými souřadnicovými systémy a následně i prostorovým systémem a určením optimální vztahů mezi nimi je jedním z výsledků předloženého článku. Součástí je rovněž experiment o přímou transformaci mezi rovinným katastrálním souřadnicovým systémem svatoštěpánským a prostorovým systémem ETRS89, který mohl být proveden až po definování základních a charakteristických vlastností jednotlivých zobrazení, tzn. použitého elipsoidu a typu zobrazení.

Permissions

All chapters in this book were first published in GE, by De Gruyter; hereby published with permission under the Creative Commons Attribution License or equivalent. Every chapter published in this book has been scrutinized by our experts. Their significance has been extensively debated. The topics covered herein carry significant findings which will fuel the growth of the discipline. They may even be implemented as practical applications or may be referred to as a beginning point for another development.

The contributors of this book come from diverse backgrounds, making this book a truly international effort. This book will bring forth new frontiers with its revolutionizing research information and detailed analysis of the nascent developments around the world.

We would like to thank all the contributing authors for lending their expertise to make the book truly unique. They have played a crucial role in the development of this book. Without their invaluable contributions this book wouldn't have been possible. They have made vital efforts to compile up to date information on the varied aspects of this subject to make this book a valuable addition to the collection of many professionals and students.

This book was conceptualized with the vision of imparting up-to-date information and advanced data in this field. To ensure the same, a matchless editorial board was set up. Every individual on the board went through rigorous rounds of assessment to prove their worth. After which they invested a large part of their time researching and compiling the most relevant data for our readers.

The editorial board has been involved in producing this book since its inception. They have spent rigorous hours researching and exploring the diverse topics which have resulted in the successful publishing of this book. They have passed on their knowledge of decades through this book. To expedite this challenging task, the publisher supported the team at every step. A small team of assistant editors was also appointed to further simplify the editing procedure and attain best results for the readers.

Apart from the editorial board, the designing team has also invested a significant amount of their time in understanding the subject and creating the most relevant covers. They scrutinized every image to scout for the most suitable representation of the subject and create an appropriate cover for the book.

The publishing team has been an ardent support to the editorial, designing and production team. Their endless efforts to recruit the best for this project, has resulted in the accomplishment of this book. They are a veteran in the field of academics and their pool of knowledge is as vast as their experience in printing. Their expertise and guidance has proved useful at every step. Their uncompromising quality standards have made this book an exceptional effort. Their encouragement from time to time has been an inspiration for everyone.

The publisher and the editorial board hope that this book will prove to be a valuable piece of knowledge for researchers, students, practitioners and scholars across the globe.

List of Contributors

Jana DADOVÁ
Department of Environmental Sciences, Faculty of Natural Sciences, Matej Bel University,
Tajovského 52, Banská Bystrica

Jiří KUPKA
Institute of Environmental Engineering , Faculty of Mining and Geology, VŠB -Technical
University of Ostrava, 17. listopadu 15, Ostrava

Peter BARTO
Department of Geography, Faculty of Natural Sciences, Matej Bel University Tajovského 52, Banská Bystrica

Erika REMEŠICOVÁ
Institute of Environmental Engineering , Faculty of Mining and Geology, VŠB -Technical University of Ostrava, 17. listopadu 15, Ostrava,

Tomáš ŠTRBA
Department of Environmental Sciences, Faculty of Natural Sciences, Matej Bel University
Tajovského 52, Banská Bystrica

Xiyang FENG
Institute of Safety Technology Management, Faculty of Prospecting and Mining Engineering, Yunnan
Land and Resources Vocational College, 2. Jingniu Road, Yunnan, China,

Alois ADAMUS
Institute of Mining Engineering and Safety, Faculty of Mining and Geology, VŠB - Technical University of Ostrava, 17. listopadu 15, 708 33 Ostrava - Poruba, Czech Republic,

Martin RELI
Department of Physical Chemistry and Theory of Technological Processes, Faculty of Metallurgy and Materials Engineering, VSB-Technical University of Ostrava 17. listopadu 15, Ostrava

Kamila KOČÍ
Department of Physical Chemistry and Theory of Technological Processes, Faculty of Metallurgy and Materials Engineering, VSB-Technical University of Ostrava 17. listopadu 15, Ostrava

Vlastimil MATĚJKA
Nanotechnology center, VSB-Technical University of Ostrava 17. listopadu 15, Ostrava

Pavel KOVÁŘ
The Czech Technological Centre for Inorganic Pigments, Nábřeží Dr. E. Beneše 24, Přerov

Lucie OB1ALOVÁ
Department of Physical Chemistry and Theory of Technological Processes, Faculty of Metallurgy and Materials Engineering, VSB-Technical University of Ostrava 17. listopadu 15, Ostrava

Jana MAGNUSKOVÁ
Institute of Economics and Control Systems, Faculty of Mining and Geology, VŠB - Technical University of Ostrava, 17. listopadu 15/2172, Ostrava

Zdeněk PAVELEK
OKD, HBZS, a.s. Ostrava

Lucie KRČMARSKÁ
Institute of Economics and Control Systems, Faculty of Mining and Geology, VŠB - Technical University of Ostrava 17. listopadu 15/2172, Ostrava

Vlastimil KAJZAR
Institute of Geonics AS CR, v. v. i.Stud, entska 1768 Ostrava-Poruba

Hana DOLEŽALOVÁ
Institute of Geonics AS CR, v. v. i. Drobneho 28, Brno

Jiří KUPKA
Institute of Environmental Engineering , Faculty of Mining and Geology, VŠB -Technical University of Ostrava, 17. listopadu 15, Ostrava

Hana ŠVEHLÁKOVÁ
Institute of Environmental Engineering, Faculty of Mining and Geology, VŠB -Technical
University of Ostrava, 17. listopadu 15, Ostrava

Kamila KAŠOVSKÁ
Dean´s Office, Faculty of Mechanical Engineering, VŠB -Technical University of Ostrava, 17.
listopadu 15, Ostrava

Lukáš SNOPEK
Institute of Mining Engineering and Safety, Faculty of Mining and Geology, VŠB - Technical University of Ostrava, 17. listopadu 15, 708 33 Ostrava - Poruba, Czech Republic,

Alois ADAMUS
Institute of Mining Engineering and Safety, Faculty of Mining and Geology, VŠB - Technical University of Ostrava, 17. listopadu 15, 708 33 Ostrava - Poruba, Czech Republic,

Radim PAVLÍK
OKD, plc., Karviná Mine, Lazy Plant, Karviná doly, ZIP 73 506

Vlastimil ŘEPKA
Institute of Mining Engineering and Safety, Faculty of Mining and Geology , VŠB - Technical University of Ostrava, 17. listopadu 15, Ostrava

Jitka HAJDUKOVÁ
Institute of Clean Technologies for Mining and Utilization of Raw Materials for Energy Use, Faculty of Mining and Geology, VŠB - Technical University of Ostrava, 17. listopadu 15, Ostrava - Poruba

Jan THOMAS
Institute of Clean Technologies for Mining and Utilization of Raw Materials for Energy Use, Faculty of Mining and Geology, VŠB - Technical University of Ostrava, 17. listopadu 15, Ostrava - Poruba

Petra MALÍKOVÁ
Institute of Clean Technologies for Mining and Utilization of Raw Materials for Energy Use, Faculty of Mining and Geology, VŠB - Technical University of Ostrava, 17. listopadu 15, Ostrava - Poruba

Veronika MATÚŠKOVÁ
Institute of Clean Technologies for Mining and Utilization of Raw Materials for Energy Use, Faculty of Mining and Geology, VŠB - Technical University of Ostrava, 17. listopadu 15, Ostrava - Poruba

Alena ORLÍKOVÁ
Institute of Environmental Engineering, Faculty of Mining and Geology, VSB - Technical University of Ostrava, tř. 17. listopadu 15, 708 33 Ostrava-Poruba

Ilona V. KOCHURA
Production Management Department, Mine Faculty, Donetsk National Technical University, Artema str. 58, Donetsk 83000, UKRAINE

Václav DOMBEK
Institute of Clean Technologies for Mining and Utilization of Raw Materials for Energy Use,
Faculty of Mining and Geology, VŠB - Technical University of Ostrava, 17.listopadu 15/2172, Ostrava-Poruba, 708 33

Remini BOUALEM
Département des sciences de l'eau et de l'environnement, Faculté de Technologie, Université Blida 1
BP 270, Blida, Algéri

Bensafia DJILLALI
Département des sciences de l'eau et de l'environnement, Faculté de Technologie, Université Blida 1
BP 270, Blida, Algéri

Missoum MOHAMED
Département des sciences de l'eau et de l'environnement, Faculté de Technologie, Université Blida 1
BP 270, Blida, Algéri

MATÚŠ GALÍK
Faculty of Civil Engineering, Slovak University of Technology, Radlinskeho 11, 813 68 Bratislava, Slovak Republic

JOZEF KRIŠ
Faculty of Civil Engineering, Slovak University of Technology, Radlinskeho 11, 813 68 Bratislava, Slovak Republic

JÁN ILAVSKÝ
Faculty of Civil Engineering, Slovak University of Technology, Radlinskeho 11, 813 68 Bratislava, Slovak Republic

Danka BARLOKOVÁ
Department of Sanitary and Environmental Engineering, Faculty of Civil Engineering, Slovak University of Technology Radlinského 11, 81368 Bratislava

Ján ILAVSKÝ
Department of Sanitary and Environmental Engineering, Faculty of Civil Engineering, Slovak University of Technology Radlinského 11, 81368 Bratislava

Marek SOKÁČ
Department of Sanitary and Environmental Engineering, Faculty of Civil Engineering, Slovak University of Technology Radlinského 11, 81368 Bratislava

Marcela GERGEĽOVÁ
Institute of Geodesy, Cartography and Geographic Information Systems, Faculty of Mining, Ecology, Process Control and Geotechnologies, Technical University of Košice, Park Komenského 19, 043 84 Košice, Slovak Republic

Žofia KUZEVIČOVÁ
Institute of Geodesy, Cartography and Geographic Information Systems, Faculty of Mining, Ecology, Process Control and Geotechnologies, Technical University of Košice, Park Komenského 19, 043 84 Košice, Slovak Republic

Štefan KUZEVIČ
Institute of Business and Management, Technical University of Košice, Faculty of Mining, Ecology, Process Control and Geotechnologies, Park Komenského 19 042 00 Košice, Slovak Republic

Vancho ADJISKI
Faculty of Natural and Technical Sciences, Mining engineering, "Goce Delchev" University, P.O. Box 201, 2000 Shtip, Republic of Macedonia

Marianna HUNDÁKOVÁ
Nanotechnology Centre, IT4 Innovations Centre of Excellence, VŠB-Technical University of Ostrava 17. listopadu 15, Ostrava

Marta VALÁŠKOVÁ
Nanotechnology Centre, IT4 Innovations Centre of Excellence, VŠB-Technical University of Ostrava 17. listopadu 15, Ostrava

Magda SAMLÍKOVÁ
Nanotechnology Centre, IT4 Innovations Centre of Excellence, VŠB-Technical University of Ostrava 17. listopadu 15, Ostrava

Erich PAZDZIORA
Institute of Public Health in Ostrava Partyzánské náměstí 7, Ostrava

Jaroslavav DVOŘÁČEK
Institute of Economics and Control Systems, Faculty of Mining and Geology, VŠB-Technical University of Ostrava 17. listopadu 15, Ostrava-Poruba

Elena V. MARTYAKOVÁ
Production Management Department, Mine Faculty, Donetsk National Technical University Artema str. 58, Donetsk 83000, Ukraina

Ilona V. KOCHURA
Production Management Department, Mine Faculty, Donetsk National Technical University Artema str. 58, Donetsk 83000, Ukraina

Radmila SOUSEDÍKOVÁ
Institute of Economics and Control Systems, Faculty of Mining and Geology, VŠB-Technical University of Ostrava 17. listopadu 15, Ostrava-Poruba

Igor SAVIČ
Cybex Industrial LTD, C/O Columbus Trading-Partners GMBH Riedinger str. 18, 95448 Bayreuth, German

Remini Boualem
Department of Water Sciences and Environment, Faculty of Engineering, Blida University, Blida 9000, Algeria

Achour Bachir
Department of Hydraulic, Larhyss laboratory, Biskra University, Biskra, 7000, Algeria

Kechad Rabah
Department of Economics, Faculty of Engineering, Blida University, Blida, 9000, Algeria

Anna HANUSOVÁ
Ing, Department of Sanitary and Environmental Engineering, Faculty of Civil Engineering, Slovak University of Technology Radlinského 11, 813 68 Bratislava

Michal PORZER
Institute of Geological Engineering, Faculty of Mining and Geology, VŠB - Technical University of Ostrava 17. listopadu 15, Ostrava-Poruba

Petr BUJOK
Institute of Geological Engineering, Faculty of Mining and Geology, VŠB - Technical University of Ostrava 17. listopadu 15, Ostrava-Poruba

Martin KLEMPA
Institute of Geological Engineering, Faculty of Mining and Geology, VŠB - Technical University of Ostrava 17. listopadu 15, Ostrava-Poruba

Petr PÁNEK
Department of Chemistry, Faculty of Metallurgy and Materials Engineering, VŠB - Technical University of Ostrava 17. listopadu 15, Ostrava-Poruba

Jana MAGNUSKOVÁ
Institute of Economics and Control Systems, Faculty of Mining and Geology, VŠB - Technical University of Ostrava 17. listopadu 15/2172, Ostrava

Jana BARTOŇOVÁ
Institute of Economics and Control Systems, Faculty of Mining and Geology, VŠB - Technical University of Ostrava 17. listopadu 15/2172, Ostrava

Lucie KRČMARSKÁ
Institute of Economics and Control Systems, Faculty of Mining and Geology, VŠB - Technical University of Ostrava 17. listopadu 15/2172, Ostrava

Dagmar LÉTAVKOVÁ
Institute of Economics and Control Systems, Faculty of Mining and Geology, VŠB - Technical University of Ostrava 17. listopadu 15/2172, Ostrava

Josef CHOVANEC
Institute of Mining Engineering and Safety, Faculty of Mining and Geology, VSB-TU Ostrava, 17. listopadu 15/2172, 708 33 Ostrava-Poruba, Czech Republic

Jaroslav ZÁVADA
Department of Environmental Engineering, Faculty of Mining and Geology, VSB-TU Ostrava, 17. listopadu 15/2172, 708 33 Ostrava-Poruba, Czech Republic

Tomáš BOUCHAL
Department of Environmental Engineering, Faculty of Mining and Geology, VSB-TU Ostrava, 17. listopadu 15/2172, 708 33 Ostrava-Poruba, Czech Republic

ladislav BLAŽEK
Department of Environmental Engineering, Faculty of Mining and Geology, VSB-TU Ostrava, 17. listopadu 15/2172, 708 33 Ostrava-Poruba, Czech Republic

Kamil OŽANA
Institute of Mining Engineering and Safety, Faculty of Mining and Geology, VSB-TU Ostrava, 17. listopadu 15/2172, 708 33 Ostrava-Poruba, Czech Republic

Silvie Heviánková
Environmental engineering department, Faculty of Mining and Geology VSB Technical University of Ostrava, 17. listopadu 15, 708 00, Ostrava- Poruba, Czech Rebublic

Miroslav Kyncl
Environmental engineering department, Faculty of Mining and Geology VSB- Technical University of Ostrava, 17. listopadu 15, 708 00, Ostrava- Poruba, Czech Rebublic

Jana Kodymová
Environmental engineering department, Faculty of Mining and Geology VSB- Technical University of Ostrava,17. listopadu 15, 708 00, Ostrava- Poruba, Czech Rebublic

Lukas Chabal
Department of Sanitary and Environmental Engineering, Faculty of Civil Engineering, Slovak University of Technology in Bratislava Radlinského street 11, Bratislava

Stefan Stanko
Department of Sanitary and Environmental Engineering, Faculty of Civil Engineering, Slovak University of Technology in Bratislava Radlinského street 11, Bratislava

Lukáš OTTE
Institute of Economics and Control Systems, Faculty of Mining and Geology , VŠB - Technical University of Ostrava, 17. listopadu 15, Ostrava

Vladislav VANČURA
Institute of Economics and Control Systems, Faculty of Mining and Geology , VŠB - Technical University of Ostrava, 17. listopadu 15, Ostrava

Roman DANEL
Institute of Economics and Control Systems, Faculty of Mining and Geology , VŠB - Technical University of Ostrava, 17. listopadu 15, Ostrava

Michal ŘEPKA
Institute of Economics and Control Systems, Faculty of Mining and Geology , VŠB - Technical University of Ostrava, 17. listopadu 15, Ostrava

Hana STAŇKOVÁ
Institute of Geodesy and Mine Surveying, Faculty of Mining and Geology, VSB-Technical University of Ostrava, 17. listopadu 15, 708 33 Ostrava - Poruba, Czech Republic

Pavel ČERNOTA
Institute of Geodesy and Mine Surveying, Faculty of Mining and Geology, VSB-Technical University of Ostrava, 17. listopadu 15, 708 33 Ostrava - Poruba, Czech Republic

Miroslav NOVOSAD
Institute of Geodesy and Mine Surveying, Faculty of Mining and Geology, VSB-Technical University of Ostrava, 17. listopadu 15, 708 33 Ostrava - Poruba, Czech Republic